What is the place of Europe in the origin of humankind? Whilst our earliest human ancestors may have come out of Africa, many of our more recent relatives, and those of other primates, left their fossil remains in Europe and the Near East. Hominoid primates including *Dryopithecus* in Spain and Hungary, *Oreopithecus* in Italy, *Ankarapithecus* in Turkey and *Ouranopithecus* in Greece flourished in the Miocene, between about 12 and 8 million years ago. This volume examines these and other hominoid fossils found in Eurasia and discusses what we can learn from them using the biostratigraphic and ecological frameworks established in the first volume of this set. In addition, new methods of analysing and visualising fossil hominoids are explored, including computer tomography-based and computer-assisted virtual reconstruction of fossils to allow three-dimensional images of external and internal morphology of even fragmentary or distorted fossils. This volume will therefore be invaluable for practising palaeoanthropologists and palaeontologists whatever their specialism.

LOUIS DE BONIS is Professor of Vertebrate Palaeontology and Palaeoanthropology in the Laboratoire de Géobiologie, Biochronologie & Paléontologie Humaine at the University of Poitiers, France. He works on fossil primates and is involved in understanding the origins of hominids and problems of palaeoenvironments. He is also interested in the evolution of carnivores.

GEORGE D. KOUFOS is Professor of Palaeontology and Stratigraphy in the Department of Geology in the Aristotle University of Thessaloniki, Greece. He works on fossil primates, equids and carnivores, plus Neogene/Quaternary biochronology, biostratigraphy and palaeoenvironments.

PETER ANDREWS is a research scientist at the Natural History Museum in London, where he works on fossil primates, taphonomic and palaeoecological issues relating to the early stages of human evolution.

HOMINOID EVOLUTION AND CLIMATIC CHANGE IN EUROPE

VOLUME 2

Phylogeny of the Neogene Hominoid Primates of Eurasia

HOMINOID EVOLUTION AND CLIMATIC CHANGE IN EUROPE

VOLUME 2

Phylogeny of the Neogene Hominoid Primates of Eurasia

Edited by

LOUIS DE BONIS

GEORGE D. KOUFOS

and

PETER ANDREWS

CAMBRIDGE
UNIVERSITY PRESS

PUBLISHED BY THE PRESS SYNDICATE OF THE UNIVERSITY OF CAMBRIDGE
The Pitt Building, Trumpington Street, Cambridge, United Kingdom

CAMBRIDGE UNIVERSITY PRESS
The Edinburgh Building, Cambridge CB2 2RU, UK
40 West 20th Street, New York, NY 10011-4211, USA
10 Stamford Road, Oakleigh, VIC 3166, Australia
Ruiz de Alarcón 13, 28014 Madrid, Spain
Dock House, The Waterfront, Cape Town 8001, South Africa

http://www.cambridge.org

First published 2001

Printed in the United Kingdom at the University Press, Cambridge

Typeset in Utopia 9/13pt [VN]

A catalogue record for this book is available from the British Library

Library of Congress Cataloguing in Publication data

Hominoid evolution and climate change in Europe: phylogeny for the Neogene
hominoid primates of Eurasia/edited by Louis de Bonis, George D. Koufos, and Peter Andrews.
p. cm.
Includes bibliographical references and index.
1. Primates, Fossil – Eurasia. 2. Paleontology – Neogene. 3. Climatic changes – Eurasia.
4. Hominids – Evolution. I. Bonis, Louis de. II. Koufos, George D. III. Andrews, Peter.

QE882.P7 H66 2001
599.93'8 – dc21 00-060802

ISBN 0 521 66075 0 hardback

Contents

Contributors

Jorge Agustí
Institut de Paleontologia "M. Crusafont", c/Escola Industrial 23, E-08201 Sabadell, Spain

D. M. Alba
Institut de Paleontologia "M. Crusafont", c/Escola Industrial 23, E-08201 Sabadell, Barcelona, Spain

Berna Alpagut
Ankara Universitesi, Sihhiye/Ankara, Turkey

Peter Andrews
Department of Palaeontology, Natural History Museum, Cromwell Road, London SW7 5BD, UK

D. R. Begun
Department of Anthropology, University of Toronto, Toronto, ON M5S 3G3, Canada

Luca Bondioli
Museo Nazionale Preistorico Etnografico "L. Pigorini", Sezione di Antropologia, P. le G. Marconi 14, 00144 Roma, Italy

José Braga
Laboratoire d'Anthropologie, Université Bordeaux 1, U.M.R. 5809 du CNRS, Avenue des Facultés, 33405 Talence Cedex, France

L. Cabrera
Grup de Geodinàmica i Anàlisi de Conques, Departament de Estratigrafia i Palaeontologia, Universitat de Barcelona, Campus Pedralbes, E-08028 Barcelona, Spain

Mark Collard
Department of Anthropology, University College London, Gower Street, London WC1E 6BT, UK

Louis de Bonis
Laboratoire de Geobiologie, Biochronologie et Palaeontologie humaine, Université de Poitiers, 40 Avenue du Recteur Pineau, F-86022 Poitiers, France

Mikael Fortelius
Department of Geology, Box 11, University of Helsinki, FIN-00014, Finland

Ekaterine Gabashvili
Georgian State Museum, 3 Purtseladze, 380007 Tbilisi, Republic of Georgia

Leo Gabunia
Institute of Paleobiology, Georgian Academy of Sciences, 4A Niagvris, 380004 Tblisi, Republic of Georgia

M. Garcés
Laboratory of Paleomagnetism, University of Barcelona, Campus de Pedralbes, E-08028 Barcelona, Spain

Arja Hokkanen
Department of Geology, Box 11, University of Helsinki, FIN-00014, Finland

Jay Kelley
Department of Oral Biology, College of Dentistry, University of Illinois at Chicago, 801 S. Paulina, Chicago, IL 60612, USA

Tania King
Department of Palaeontology, The Natural History Museum, Cromwell Road, London SW7 5BD, UK

Meike Köhler
Institut de Palaeontologia "M. Crusafont", c/ Escuola Industrial 23, E-08201 Sabadell, Barcelona, Spain

George D. Koufos
Aristotle University of Thessaloniki, School of Geology, Department of Geology & Physical Geography, 54006 Thessaloniki, Greece

David Lordkipanidze
Georgian State Museum, 3 Purtseladze, 380007 Tbilisi, Republic of Georgia

Roberto Macchiarelli
Museo Nazionale Preistorico Etnografico "L. Pigorini", Sezione di Antropologia, P. le G. Marconi 14, 00144 Roma, Italy

Salvador Moyà-Solà
Institut de Paleontologia "M. Crusafont", c/ Escuola Industrial 23, E-08201 Sabadell, Spain

David R. Pilbeam
Department of Anthropology, Harvard University, Cambridge, MA 02138, USA

Marcia S. Ponce de León
Anthropological Institute and Institute of Computer Science, MultiMedia Laboratory, University of Zurich-Irchel, CH-8057 Zurich, Switzerland

Brian G. Richmond
Department of Anthropology, George Washington University, 2110 G Street, NW, Washington, DC 20052, USA

Lorenzo Rook
Dipartimento di Scienze della Terra e Museo di Storia Naturale, Sezione di Geologia e Paleontologia, Università di Firenze, via G. la Pira 4, 50121 Firenze, Italy

Abesalom Vekua
Institute of Paleobiology, Georgian Academy of Sciences, 4A Niagvris, 380004 Tbilisi, Republic of Georgia

Michael Whalen
Sociology and Anthropology Department, Front Range Community College, 3645 W 112th Ave, Westminster, CO 80030, USA

Bernard Wood
Department of Anthropology, George Washington University, 2110 G Street NW, Washington, DC 20052, USA and Human Origins Program, National Museum of natural History, Smithsonian Institution, Washington DC 20560, USA

Nathan M. Young
Department of Anthropology, Harvard University, Cambridge, MA 02138, USA

Christoph P. E. Zollikofer
Anthropoligical Institute and Institute of Computer Science, MultiMedia Laboratory, University of Zurich-Irchel, CH-8057 Zurich, Switzerland

Acknowledgements: The European Science Foundation

The European Science Foundation (ESF) acts as a catalyst for the development of science by bringing together leading scientists for funding agencies to debate, plan and implement pan-European scientific and science policy initiatives.

ESF is the European association of more than 60 major national funding agencies devoted to basic scientific research in over 20 countries. It represents all scientific disciplines: physical and engineering sciences, life and environmental sciences, medical sciences, humanities and social sciences. The Foundation assists its Member Organisations in two main ways: by bringing scientists together in its scientific programmes, networks, exploratory workshops and European research conferences, to work on topics of common concern; and through the joint study of issues of strategic importance in European science policy.

It maintains close relations with other scientific institutions within and outside Europe. By its activities, the ESF adds value by cooperation and coordination across national frontiers and endeavours, offers expert scientific advice on strategic issues, and provides the European forum for fundamental science.

PART I

Chronology and environment

1 Chronology and zoogeography of the Miocene hominoid record in Europe

J. Agustí, L. Cabrera and M. Garcés

Since their first appearance in the Middle Miocene, Western Eurasian Miocene hominoids were characterised by increasing levels of diversity and abundance, ranging from Spain to Pakistan, through France, Germany, Austria, Slovakia, Greece, Turkey and Georgia. As a consequence, during the Vallesian mammal stage, this region had a variety of fossil apes, including quadrupedal semiterrestrial forms (*Griphopithecus*), slender suspensory forest-dwellers (*Dryopithecus*), robust gorilla-like hominoids (*Ouranopithecus* or *Graecopithecus*) and arboreal pongids (*Sivapithecus*). Therefore, the Middle Miocene Eurasian hominoid radiation parallels the one observed in Africa in Pliocene times, which finally gave rise to bipedal hominids. However, this radiation was abruptly interrupted in early Late Miocene times, in the frame of the late Vallesian Crisis (9.6 to 9.3 Ma), an extinction event which affected most of the forest-adapted European faunas (Agustí & Moyà-Solà, 1990; Agustí *et al.*, 1997). Testing the link between this set of events and the environmental changes that affected Europe in the Miocene is usually hampered by the absence of accurate datings of most of the localities bearing hominoid remains. However, recent magnetobiostratigraphic progress in a number of Miocene basins and sections enables one to offer a more robust chronological background for the hominoid record in Europe and to test its correlation with the main palaeogeographic and palaeoenvironmental events which affected this continent.

The chronological framework

The Miocene hominoid record in Eurasia includes several species, based on sets of remains varying in quality from isolated dental remains to rather well documented cranial and post-cranial skeletal remains (Bonis *et al.*, 1990; Begun, 1992; Moyà-Solà & Kohler, 1993, 1996; Alpagut *et al.*, 1996; Andrews *et al.*, 1996). These species are grouped into five main genera: *Griphopithecus, Dryopithecus, Sivapithecus*, and the closely related *Ankarapithecus* and *Ouranopithecus* (or *Graecopithecus*). The currently available biochronological, biostratigraphic and magnetostratigraphic data that provide information to establish the chronological framework of these genera are discussed in this section and are summarised in Figure 1.1. The highly specialized genus *Oreopithecus* will not be discussed here as it is an example of insular evolution during early–middle Turolian times, which is

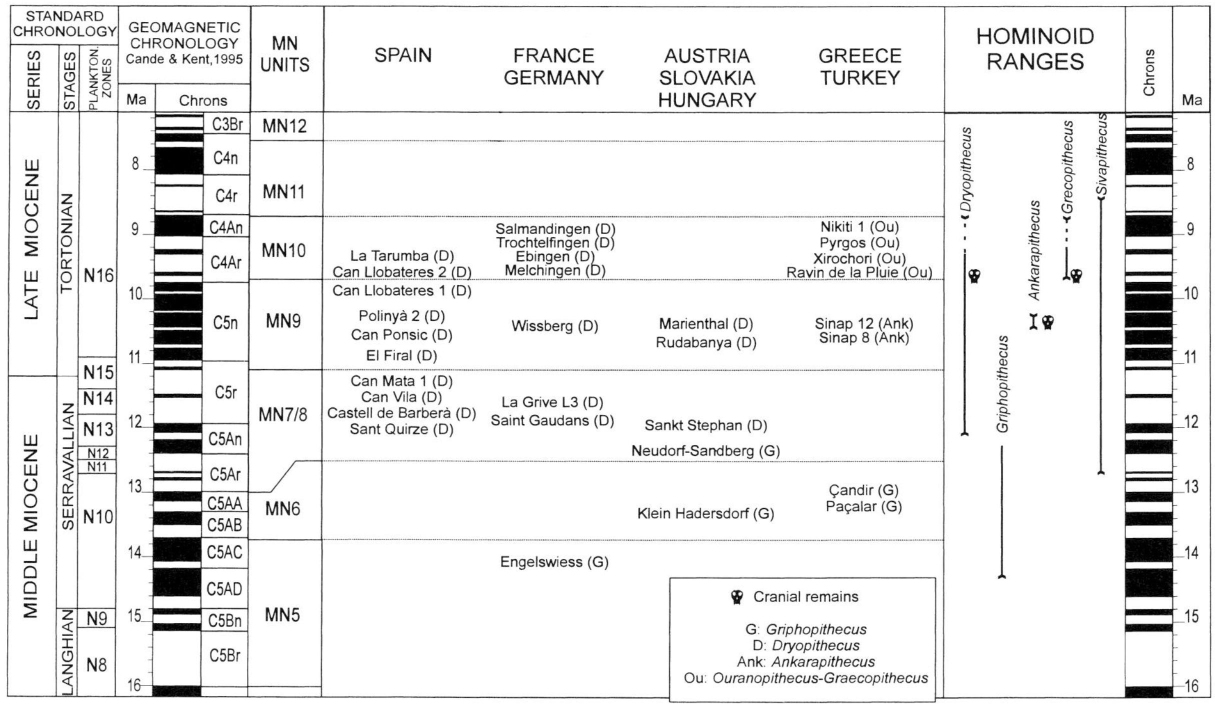

[Figure 1.1]
Biochronologic, magnetostratigraphic and chronostratigraphic background of the Miocene hominoid record in Europe.

not in the context of the Middle–Late Miocene continental evolution (Köhler & Moyà-Solà, 1997).

Griphopithecus

The first hominoid remains reported in Europe were referred to the genus *Griphopithecus* (species *G. darwini*). This taxon was first defined in the locality of Neudorf-Sandberg (Slovakia; Abel, 1902) on the basis of dental remains, which have been reassigned to the genus *Kenyapithecus* on the basis of the thick enamelled dentition and other morpohological features (Begun, 1992; Andrews *et al.*, 1996). The age of the first appearence of this hominoid in Eurasia has been a matter of discussion, mainly because of the controversial chronological position of Neudorf-Sandberg. Until recently, the circulating faunal lists of this locality included taxa characteristic of different MN mammal units (for instance, *Bunolistriodon lockharti* – MN 4 or MN 5 – and *Protragocerus* – MN 7 to MN 9). As a consequence, an 'average' early middle Miocene age (MN 5 unit) was proposed for this site. However, it has been shown that the material originally assigned to *Bunolistriodon lockharti* had to be referred to the late Aragonian species *Listriodon splendens* (Made, 1996). This taxon is found in localities included in mammal units MN 6 to MN 9. The co-occurrence of *Protragocerus* (MN 7 to MN 9) and *Thetytragus* (MN 5 to MN 7) strongly supports the final assigment of this locality to the MN 7 unit. Furthermore, the presence of *Parachleuastochoerus steinheimensis* in this locality also indicates an age close to Steinheim (type-locality of MN 7).

In fact, most of the other localities with *Griphopithecus* are late Aragonian in age and have been usually referred to the MN 6 unit. This is the case, for instance, for Klein-Hadersdorf in Austria, and Çandir and Paşalar in Turkey. All these localities have been referred to MN 6 on the basis of common taxa shared with the reference locality of this unit, Sansan (France): *Listriodon splendens, Anchitherium aurelianense, Chalicotherium grande, Acerorhinus tetradactylum* and others (Bernor & Tobien, 1990). However, in all of the above mentioned localities *Thetytragus* is present. This is a bovid of eastern affinities which is absent from Sansan and appears in western Europe at around 13.5 Ma (in the locality of Manchones, equivalent to the level of Las Planas 5K in the Calatayud-Daroca Basin, chron C5Abn; Krijgsman *et al.*, 1996). In the case of Klein-Hadersdorf, this post-Sansan age is also confirmed by the presence of the cervid *Stehlinoceros elegantulus*. Therefore, the presence of *Thetytragus* suggests a post-Sansan age for all the *Griphopithecus* sites in eastern Europe. The only exception would be half a tooth coming from the MN 5 German locality of Engelswiess which undoubtedly indicates the presence of a thick-enamelled hominoid in this site (Begun, Chapter 10).

The MN-6 type locality of Sansan has been recently proposed to be correlated to the top of chron C5B, close to 15 Ma (Sen, 1998). However, it must be emphasised that this chronological attribution is based on a very short stratigraphic succession and that it does not fit with the MN 6 magnetostratigraphic calibrations proposed by Krijgsman *et al.* (1996) in the Calatayud-Daroca basin, which are based on thick, continuous and well contrasted biostratigraphic successions. In this basin the MN-6 mammal unit is located between the top of the chron C5ACn and the middle part of the chron C5Ar, i.e. in a time span ranging from 13.75 ± 0.03 to 12.75 ± 0.25 Ma. Therefore, this range marks a lower limit for the entry of *Griphopithecus* in Europe. However, if the assignment of Engelswiess to MN 5 is confirmed (no detailed faunal analysis has yet been published), the FAD (first appearance datum) of *Griphopithecus* could be older than 15 Ma, back to the age of the lower boundary of the MN 5 unit, which has been established at 16 Ma (base of chron C5Br) by Krijgsman *et al.* (1996) and at 17 Ma (within chron C5Cr) by Steininger *et al.* (1996). A FAD of 16 Ma for *Griphopithecus* in Europe would be more consistent with the age of the earliest finds of *Kenyapithecus* in East Africa (Pickford, 1986).

With regards to the upper chronological limit of *Griphopithecus*, this one could be extended to the lower part of the chron C5Ar (i.e. 12.5 Ma). However, if we take into account the foraminifera present in the marine nearshore sands of Neudorf-Sandberg, which indicates a middle Badenian age (N9-N13; Cicha *et al.*, 1972; Steininger *et al.*, 1996), this locality would be older than the Badenian–Sarmatian boundary, placed at around 13 Ma (Steininger *et al.* , 1996).

Dryopithecus

Dryopithecus is well represented in a number of early Vallesian basins and sites from Europe: Vallès-Penedès Basin in Spain (several localities from the *Cricetulodon* Zone – Can Ponsic, Can Llobateres 1, Polinyà; also El Firal, in the Seu d'Urgell Basin), Wissberg (Germany); Mariathal (Austria); and the several findings in the Rudabanya section (Hungary). However, its presence in well-dated pre-Vallesian localities use to be rare. The type-locality of the species *Dryopithecus fontani*, Saint Gaudens (France), is certainly late Aragonian in age, but the few mammal elements associated with this find do not permit more accurate dating. In the Vallès-Penedès Basin, the oldest *Dryopithecus* find is one tooth from the Sant Quirze site (= Trinxera del Ferrocarril = San Quirico; Golpe, 1993), assigned to MN 7. Although no direct palaeomagnetic calibration has been established for this locality, its age is older than the lower boundary of the Vallesian (established in the Vallès-Penedès Basin at 11.1 Ma; Agustí *et al.*, 1997; Garcés *et al.*, 1997) and younger than 12.4 Ma. This older age corresponds to the Font del Ferro 3C locality, which is placed at the base of chron C5An and presents a MN 7 rodent association older than Sant Quirze. Moreover, a number of *Dryopithecus* localities in the Vallès-Penedès Basin younger than Sant Quirze have to be placed between this site and those recording the FAD of *Hippotherium* at 11.1 Ma – Castell de Barberà, Can Vila and Can Mata 1 (lower levels of the Hostalets de Pierola section). Interpolation between Font del Ferro and the first levels with *Hippotherium* of the intermediate faunal associations of Sant Quirze, Castell de Barberà and Lower Hostalets would result in an approximate age of 11.7 ± 0.5 for the first *Dryopithecus* record in the Vallès-Penedès.

Outside of Spain, pre-*Hippotherium* localities with *Dryopithecus* are La Grive (France) and St. Stephan (Austria). At La Grive, *Dryopithecus* is represented by an upper M3 and by an upper incisor, which is preserved in the Basel Museum (Mein, 1986). The two teeth present the characteristic pattern of the fissure La Grive L3 (Mein, 1986). There are few data reported from this locality, although they include the cricetid species *Hispanomys bijugatus* (Mein & Freudenthal, 1971), which indicates an age younger than the classical La Grive infillings and, probably, younger than the remainder of the infillings of the La Grive complex (La Grive L7 and La Grive M). The evolutionary grade of this species probably indicates an age immediatelly anterior to the FAD of *Hippotherium*, as in the case of other pre-Vallesian localities of the Vallès-Penedès Basin such as Castell de Barberà and the lower Hostalets levels. As in La Grive L3, all these localities have *Hispanomys* present, a genus which is still absent from Sant Quirze. These data suggest that La Grive L3 is in fact younger than Sant Quirze.

The locality of St. Stephan (Austria) yielded a mandible assigned to the species *Dryopithecus carinthiacus* (Mottl, 1958). This site has been indirectly correlated with the early Sarmatian and is, therefore, of approximate late Serravallian age, close to the inferred age of Sant Quirze. In this way, it is difficult to decide wether St. Stephan is older than Sant Quirze or vice versa and which locality presents the oldest record of *Dryopithecus*. However, there are some indications that would favour an older age for the Austrian locality. The first one is the persistence of the cervid *Lagomeryx pumilio*, absent from Sant Quirze. A second argument has been provided by Made (1996), on the basis of the smaller size of the St. Stephan's *Listriodon splendens*, a species which shows a trend towards increasing size through time.

In conclusion, the oldest record of *Dryopithecus* in Europe (Sant Quirze, St. Stephan) can be dated as 11.7 ± 0.5 Ma, and would be close to the latest record of *Griphopithecus* (Klein-Hadersdorf).

Ankarapithecus – Ouranopithecus – Graecopithecus

The generic assignment of a number of middle and early late Miocene findings in Turkey remained an open question until the discovery in 1995 of a well-preserved skull in level 12 of the Sinap Formation (Alpagut *et al.*, 1996). Previous findings in this Formation, including a fragmentary face from the same level, were assigned to the genus *Sivapithecus*, because of the zygomatic shape and the thick enamelled dentition (Andrews & Cronin, 1982). A mandible described in 1957 was erected as the holotype of the genus *Ankarapithecus* (Ozansoy, 1955, 1965), which was later synonimised with *Sivapithecus*. The new findings in 1996 revealed that the hominoid material from the Sinap Formation did not belong to the genus *Sivapithecus*, which led to the formal re-establishment of the taxon *Ankarapithecus* (Alpagut *et al.*, 1996). *Ankarapithecus* comes from two different levels in the Sinap Formation – Sinap L8 and Sinap 12. Recent palaeomagnetic analysis in the area enabled the placing of these levels towards the middle upper part of chron C5n.2n, between 9.92 and 10.95 Ma (Kappelman *et al.*, 1996; Pekka Lunkka *et al.*, 1999). This chronological position correlates well with the early Vallesian levels in the Vallès-Penedès Basin which has the highest number of *Dryopithecus* finds (localities of the *Cricetulodon* Zone, such as Can Ponsic and Can Llobateres 1). The difference with these levels, despite their early Vallesian age, is that the murids are already abundant in the Sinap L8 and L12 levels. It must be pointed out that, despite several findings of large mammals above L8 and L12, *Ankarapithecus* is absent from those levels of the Sinap Formation, which are equivalent to those of early late Vallesian age from western Europe (where hominoids are rare but still present).

In Greece, a number of cranial remains have been assigned to a robust, thick enamelled hominoid, known either as *Graecopithecus* (Martin & Andrews, 1984; Andrews *et al.*, 1996) or *Ouranopithecus* (Bonis *et al.*, 1990) from Ravin de la Pluie, Xirochori 1, Nikiti 1, Pyrgos. On the basis of their large mammal association, most of these localities have been referred to the late Vallesian, except in the case of Pyrgos which could be referred either to the Vallesian or the Turolian (De Bonis & Koufos, 1999). A late Vallesian age is also supported in the case of Ravin de la Pluie by the presence of *Progonomys cathalai* (Bonis & Melentis, 1975). On the other hand, new magnetostratigraphic data are now available from a number of these sections (Ravin de la Pluie, Xirochori 1, Prochoma and others) and a correlation to the GTPS has been proposed by Sen *et al.* (1998). Although these results have to be taken with some caution, given the shortness of the sections, the overall results are consistent with what is observed in the Vallès-Penedès Basin (Garcés *et al.*, 1996; Agustí *et al.*, 1997). Therefore, the base of the section of Xirochori is correlated to chron C4Ar.2n (9.58–9.64 Ma), while the section of Ravin de la Pluie 1 is correlated to chron C4Ar.1n (9.23–9.31 Ma; Sen *et al.*, 1998). Should these correlations be confirmed, the type-species fragment skull of *Ouranopithecus macedoniensis* from Xirochori would be almost coeval with the CL-18000 skeleton of *Dryopithecus laietanus* from Can Llobateres 2 (the Greek locality being placed at the base of the late Vallesian, exactly between Can Llobateres 1 and Can Llobateres 2). Ravin de la Pluie could have the same age as Xirochori or be somewhat younger. In any case, and despite their similarity, all the findings of *Ouranopithecus* are younger than the levels with *Ankarapithecus* of the Sinap Formation in Turkey.

Sivapithecus

The *Sivapithecus* chronological data, as reported by several authors (e.g. Kappelman *et al.*, 1991, for extended references), are discussed here just for comparison of this western Eurasian taxon with the former European genera. The oldest well-calibrated occurence of *Sivapithecus* in the Chinji Formation (Siwaliks) is at the locality Y750, placed in chron C5Ar-1, between 12.71 Ma and 12.68 Ma (Kappelman *et al.*, 1991; although these authors do not discount an older age for some other badly dated finds). On the other hand, the youngest record of *Sivapithecus* in the Siwaliks is reported at the chron C4r, between 8.70 Ma and 8.07 Ma. The age of the famous U level of Siwaliks, which records the better preserved craniofacial remains of *Sivapithecus,* is placed at the base of chron C4Ar, which correlates with the base of the late Vallesian (Kappelman *et al.*, 1991; Garcés *et al.*, 1996). This is surprising, since it almost coincides with the age of the fragment skull of *Ouranopithecus macedoniensis* from Xirochori and the CL-18000 skeleton of *Dryopithecus laietanus* from Can Llobateres 2. Therefore, *Sivapithecus* craniofacial derived features were already present shortly after the first record of *Ankarapithecus,* at the time of the first *Ouranopithecus* (or *Graecopithecus*) remains.

Miocene hominoid dispersals and vicarious evolution

As mentioned above, the oldest record of *Dryopithecus* in Europe (Sant Quirze, St. Stephan) is close to the younger reported occurrence of *Griphopithecus* (Klein-Hadersdorf). In this context, the simplest and mostly parsimonious hypothesis to account for the origin of the later Miocene European hominoids is to assume that, after the entry of *Griphopithecus* sometime in the middle Miocene between 16 and 14 Ma, a proccess of vicarious evolution took place in the late middle Miocene times, leading to the thin enamelled *Dryopithecus* in Western Europe and to the thick enamelled *Ankarapithecus* in Eastern Europe. It is well known that late Aragonian faunas in Western Europe (Spain) are characterized by a return to the forest conditions existing in this area in the early Miocene, after the middle Miocene aridity interval (Agustí *et al.*, 1984; Daams & Meulen, 1984). In this case, some populations of *Griphopithecus* could have re-adapted to a more frugivorous diet, leading to the thin-enamelled *Dryopithecus*. Thick enamel in *Ankarapithecus–Ouranopithecus* would be a primitive character shared with the original *Griphopithecus*, while the thin enamel would be a derived feature developed independently by *Dryopithecus*.

Against this hypothesis is, however, the very different anatomy and locomotory adaptations of *Dryopithecus* (Moyà-Solà & Köhler, 1993, 1996) and *Kenyapithecus* (McCrossin & Benefit, 1997), the supossed sister-genus of *Griphopithecus*. Moreover, as we have seen, the last *Griphopithecus* are found in MN 6 localities, the upper limit of which is placed at 13–12.5 Ma. On the other hand, the oldest record of the thick-enamelled robust forms like *Ankarapithecus*, is at the L8 and L12 early Vallesian levels in the Sinap Formation, towards the middle upper part of chron C5n.2n and close to 10.6 Ma (the base of this chron is dated at 11 Ma, a date which represents the maximum age for these findings). Therefore, the gap between the last *Griphopithecus* and the first *Ankarapithecus* in eastern Europe is much more significant than is the case for *Dryopithecus* (almost 2 million years).

An alternative vicarious model proposed by Agustí *et al.* (1996) tries to account for the diversification of Miocene hominoids in western Eurasia. According to this model, the late Aragonian–Vallesian radiation of thick-enamelled hominids in western Eurasia has its roots in a thin-enamelled form close to *Dryopithecus* and not to *Griphopithecus*. Thus, at about 12 Ma, an orthograde African ape similar to *Dryopithecus* would have settled on Eurasia from Africa, after the Arabian–European collision. In fact, there is strong palaeogeographic evidence supporting the definitive re-establishment of the Arabian–Anatolian bridge in the early Serravallian, after the ephemeral disconnection during the Langhian transgression (Steininger *et al.*, 1985; Rögl, 1998). Faunal evidence of a late Aragonian Africa–Eurasia interchange is the entry of *Listriodon* and the hyaenid *Protictitherium* in Africa (early Vallesian of Bled Douarah, Tunisia) and that of *Tetralophodon* and, perhaps, *Albanohyus* in Europe (*Albanohyus* is a suid of possible African affinities, according to Fortelius *et al.*, 1996a). Other elements of possible African affinities in Europe are the slender palaeotragines found at a number of pre-Vallesian localities in the Vallès-Penedès Basin, such as Castell de Barberà, Can Missert and the lower Hostalets level of Can Mata-2. However, all these localities are younger than that of Sant Quirze, where *Dryopithecus* already occurs. Thus, the dispersal of these first *'Paleotragus'* might not to be associated with *Dryopithecus*.

After its entry in Europe, a first isolation process led to the differentiation of the *Dryopithecus–Pongo* clade to the north of the Mediterranean Sea (according to the cladogram proposed by Moyà-Solà & Köhler, 1993; ① in Figure 1.2). A posterior splitting process led to the appearance of derived thick-enamelled forms, such as *Ankarapithecus* and *Graecopithecus* in the Southern Paratethys, in relation to the spread of esclerophyllous vegetation in this region (② in Figure 1.2). In fact, a trend towards more arid conditions is clearly visible in the late Serravallian terrestrial record of western Eurasia compared with conditions in western and central Europe (Bernor *et al.*, 1984). The 'open' character of this woodland biotope is best exemplified by the faunal associations of the subparatethyan province, from Greece to Iran (Bernor *et al.*, 1984, Bonis *et al.*, 1992; Fortelius *et al.*, 1996b). This open woodland seems to have been also present in eastern and, probably, northern Africa, from 14 Ma onwards (as in the case of Fort Ternan; Andrews & Evans, 1978; Dugas & Retallack, 1993; Harris, 1993).

Finally, the geographic isolation caused by activation of the Hymalayan uplift could have provoked the differentiation of the *Sivapithecus–Pongo* clade in the East of the Zagros Chain. Since derived craniofacial *Sivapithecus* remains are reported in the Chinji Formation as old as 12.7 Ma, we have to assume that this is also the minimum age for the common ancestor of the *Ankarapithecus–Graecopithecus* clade and the *Sivapithecus–Pongo* clade (node C). This means that a primitive thick-enamelled form close to *Ankarapithecus* should have been present in the Subparathetyan Province at around 12.7 Ma.

This alternative vicarious model (Agustí *et al.*, 1996) assumes that the *Dryopithecus* ancestor evolved somewhere in the low latitudes of Africa, and poses some important questions about the timing of its migration into Eurasia. In fact, there are strong ecological arguments against the idea of a second dispersal of African hominoids during the Serravallian. Thus, postcranial anatomy of *Dryopithecus* revealed significant locomotory

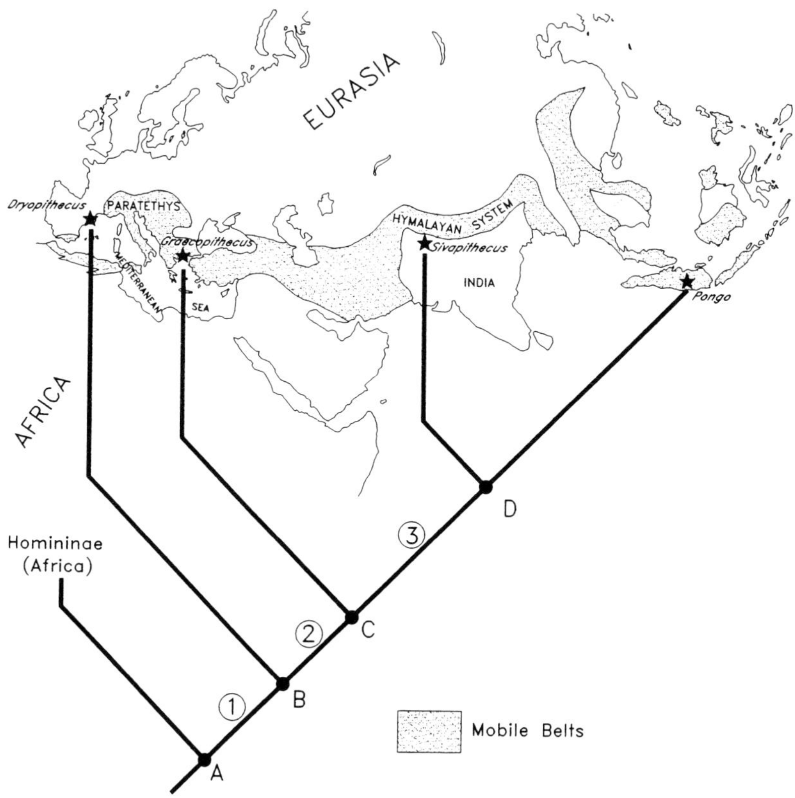

[Figure 1.2]
Pattern of vicarious speciation explaining the independent evolution of the Neogene Eurasian hominoids. After the settlement on this continent by *Dryopithecus*, isolating processes caused by the rising of the Alpine belt led to the diversification of more specialized taxa (*Graecopithecus*, *Sivapithecus*). The Vallesian extinction event affected all these forms, although some descendants of the *Sivapithecus–Pongo* clade could persist in refugial areas. Key derived characters for each node (Moyà-Solà & Kohler, 1993): A: derived pongine zygomatic morphology (robustness of the bone, zygomaxillary foramina located high on its frontal proccess); B: derived pongine dentognathic morphology (thick enamel, great size discrepancy between upper incisors, alveolar prognatism and associated characters); and C: derived pongine facial morphology (narrow interorbital pillar, airorhynchy, lack of glabellar thickening, lack of fronto-etmoidal sinus, high and narrow orbits).

adaptations for climbing in the context of a forest environment (Moyà-Solà & Köhler, 1996). This is confirmed by the fact that most of the localities with *Dryopithecus* also contain a variety of arboreal rodents (such as flying-squirrels, dormice and eomyids; Agustí *et al.*, 1984). The tropical forest biotope which could be the origin of the dietary and locomotory adaptations of *Dryopithecus* must have been at that time far away in central Africa, with the protosavanna biotope being much more extended in northern and eastern Africa than during early Miocene times (Harris, 1993). The hominoid remains of similar age in Africa are closer to quadrupedal semiterrestrial forms of the genus *Kenyapitheus*. Certainly, forest adapted hominoids with the suspensory abilities of *Dryopithecus* evolved somewhere in the low latitudes of Africa, but a late Aragonian dispersal of these primates throughout the middle latitudes of Africa into western Europe is in any case a long journey accros an unfavourable biotope.

There is, however, a way to account for the *Dryopithecus* dispersal, by admiting an earlier, late Burdigalian exit of this ape from Africa, in the context of the still mild conditions of the early–middle Miocene. According to this scenario, *Dryopithecus* could have settled on Eurasia together with *Griphopithecus* in early Aragonian times, *c.* 14 Ma, taking part in the same dispersal wave. However, in this case the absence of any *Dryopithecus* finds before 11.7 ± 0.5 Ma remains to be explained. One explanation arises from the possibility of both taphonomical and palaeoenvironmental factors which would have favoured the preservation of *Griphopithecus* remains instead of *Dryopithecus*. Several late Aragonian and Vallesian western European *Dryopithecus* localities occur in alluvial dominated successions which developed at high altitude ranges (i.e. St. Gaudens in the northern Pyrenean foreland basin, Can Llobateres and several other localities in the Vallès-Penedès rift-related basin) or in intramontainous fault-bounded basins (i.e. El Firal, located in the Seu d'Urgell Basin). The existence of high altitude ranges would result in forced orographic precipitation that would have favoured the development of evergreen highland forests, which probably retained most of their humidity throughout the year, as in today's broad-leaved laurophyllous forest of the Canary Islands (Bessedik, 1985). Evergreen forests would have only developed in the lowlands either during specially humid climatic episodes or as gallery forests related with zones of alluvial fan and fluvial water table discharge in low-lying zones. On the contrary esclerophyllous forests, where seasonallity was probably more accentuated, would be more dominant in the lowland throughout the middle Miocene.

Folivorous middle to late Miocene primates (i.e. *Pliopithecus, Griphopithecus*) probably remained as basic dwellers of the esclerophyllous forests which were mainly widespread and areally extensive in the low-lying zones where alluvial-dominated environments existed. On the contrary, *Dryopithecus* could have populated the restricted evergreen forests, which developed in watertable fed, low lying zones, while the more extensive evergreen forests developed in the high altitude zones. In such a case, the probabilities of fossilisation would have been much higher in the case of the primates living close to the lowland depocentres (*Pliopithecus, Griphopithecus*) than in the case of *Dryopithecus*. This scenario accounts also for the exclusion observed between *Dryopithecus* and *Pliopithecus* in the Middle to Late Miocene localities in the Vallès-Penedès basin. Since they tended to occupy different biotopes, the abundance, occurrence or absence of these primates in a site would depend on the spreading to lower or higher altitudes of the boundary bewteen the lowland esclerophyllous and the highland laurophyllous forests. It can be hypothesised that the abundance of thin-enamelled hominoids in the Vallès-Penedès localities such as, Can Ponsic or Can Llobateres 1 would have resulted from either generalized or more areally restricted extension to lower altitudes of the evergreen forest during early Vallesian times, rather than be the effect of purely taphonomic factors. This could have been also the case for Rudabanya, in the Carpathian Basin, associated to the higlands of the Carpathian Range, and that of the early Turolian thin-enamelled hominoids from the Caucasus (*Udabnopithecus* from Udabno, Georgia; Burtschak-Abramovitsch and Gabachvili, 1945; Gabunia *et al.*, Chapter 14).

These factors would suggest that the spread and evolution of Miocene hominoid throughout western Eurasia could have been largely driven by regional climatic, physiographic and palaeoenvironmental changes related to the Alpine tectonic evolution in the Western Tethys.

Hominoid extinctions

As in the case of several early Vallesian fruit-eaters, the high diversity of *Dryopithecus* suddenly fell after the late Vallesian Crisis (Agustí & Moyà-Solà, 1990; Agustí *et al.*, 1997). However, *Dryopithecus* persisted in some late Vallesian (MN 10) localities such as Can Llobateres 2, La Tarumba (Viladecavalls section, Vallès-Penedès Basin) and Salmendingen (Germany). Can Llobateres 2 is famous since this locality delivered the so far most complete cranial and postcranial remains of the species *Dryopithecus laietanus* (Moyà-Solà & Köhler, 1993, 1996). The small mammalian fauna associated with these finds include *Progonomys* sp. as one of the most common rodents, toghether with *Cricetulodon sabadellensis* from the previous biozone. This association indicates an early late Vallesian age (*Cricetulodon–Progonomys* Zone, chron C4Ar.3r to C4Ar.1n; Agustí *et al.*, 1997). This is also the position of the locality of La Tarumba, type-locality of the species *Dryopithecus laietanus. Dryopithecus* is already absent in the levels of the Vallès-Penedès Basin which overlay litho- and biostragraphically those of La

Tarumba (Can Tarumbot in Viladecavalls and TSA, TNA and TF in the Terrassa section, assigned to the *Rotundomys* Zone; the pliopithecids are still present in the upper part of the Terrassa section). Therefore, the level of La Tarumba, referred to the base of chron C4Ar.2r (between 9.6 and 9.3 Ma), is the youngest record of *Dryopithecus* in the Vallès-Penedès Basin.

The fissure infilling of Salmendingen, in the Suebian Alb (Germany) is the type-locality of the species *Dryopithecus brancoi* (based on one lower molar) and has been quoted as the youngest record of *Dryopithecus* in western Europe. Another fissure infilling in the Suebian Alb, Melchingen, also delivered two upper and two lower molars of *Dryopithecus brancoi*. An early Turolian age was suggested for Salmendingen, because of the presence of the castorid *Dipoides problematicus*, a species which is common in the European Turolian fauna (Mein, 1986). However, these localities have proven to be late Vallesian (MN 10) in age (Franzen & Storch, 1999), although they are probably younger than the last late Vallesian localities with *Dryopithecus* in the Vallès-Penedès.

The last record of *Ouranopithecus* is placed at 9.23 Ma, an age somewhat younger than that of the last hominoid localities in the Vallès-Penedès Basin (La Tarumba, placed in chron C4Ar.2r), but again late Vallesian in age (with the possible exception of Pyrgos, which could prove to be Turolian). In any case, there is a remarkable coincidence between the last hominoid records in western and eastern Europe at around 9.5 Ma.

In the case of *Sivapithecus*, as we have seen, the youngest record in the Siwaliks is reported from chron C4r, between 8.70 Ma and 8.07 Ma (Kappelman *et al.*, 1991). From these data it arises that, as in the case of *Dryopithecus*, *Sivapithecus* records one of the more extended hominoid time spans, between chron C5Ar-1 (about 12.7 Ma) and chron C4r (about 8.7 Ma). Another significant fact is that the chronological range of these two genera almost coincide, *Sivapithecus* becoming extinct in the early Turolian, about 1 Ma later than the last *Dryopithecus*.This date coincides fairly well with the age of the main C3–C4 vegetational change, dated between 8 and 6 Ma and marking the moment in which grasses became dominant and replaced the Miocene subtropical forest biotope in Eurasia (Cerling *et al.*, 1993). The delay existing between the last records of the western Asian apes, on the one hand, and the European apes, on the other hand, reflects in fact a difference in the main factors acting in each extinction event – increase of seasonality and thermal gradient, in the case of the Late Vallesian Crisis, and spread of grasses associated with moonson dynamics, in the case of southwestern Asia.

Acknowledgements

This paper has been supported by the projects DGICYT-PB 97–0157 PB 97–0882–C03–01 and SGR 97–73.

References

Abel, O. (1902). Zwei neue Menschenaffen aus den Leithakalkbildingen des wienner Bekkens. *S. Ber. Akad. Wiss. Wien, math.-nat.*, **1**: 1171–202.

Agustí, J., Cabrera, L., Garcés, M. & Parés, J. M. (1997). The Vallesian mammal succession in the Vallés-Penedés basin (northeast Spain): paleomagnetic calibration and correlation with global events. *Palaeogeogr. Palaeoclim. Palaeoecol.* **133**.

Agustí, J., Kohler, M., Moyà, S., Cabrera, L., Garcés, M. & Parés, J.M. (1996). Can Llobateres: the pattern and timing of the Vallesian hominoid radiation reconsidered. *J. Hum. Evol.* **31**: 143–55.

Agustí, J. & Moyà-Solà, S. (1990). Mammal extinctions in the Vallesian (Upper Miocene). *Lect. Notes Earth Sci.* **30**: 425–32.

Agustí, J., Moyà-Solà, S. & Pons-Moyà, J. (1984). Mammal distribution dynamics in the eastern margin of the Iberian Peninsula during the Miocene. *Paléobiol. continent.* **14**: 33–46.

Alpagut, B., Andrews, P., Fortelius, M., Kappelman, J., Temizsoy, I., Celebi, H. & Lindsay, E. (1996). A new specimen of *Ankarapithecus meteai* from the Sinap Formation of central Anatolia. *Nature*, **382**, 349–51.

Andrews, P. & Evans, E. M. N. (1978). The environment of *Ramapithecus* in Africa. *J. H. Evol.* **5**: 22–30.

Andrews, P. & Cronin, J. (1982). The relationships of *Sivapithecus* and *Ramapithecus* and the evolution of the orangutan. *Nature*, **297**: 541–6.

Andrews, P., Harrison, T., Delson, E. Bernor, R. L. & Martin, R. (1996). Distribution and biochronology of European and Southwest Asian Miocene catharrines. In Bernor, R., Fahlbusch, V. & Mittman, W. (eds.), *The Evolution of Western Eurasian Neogene Mammal Faunas*, pp. 168–207. New York: Columbia University Press.

Begun, D. (1992). Miocene fossil hominids and the chimp-human clade. *Science*, **247**: 1929–33.

Bernor, R. L. (1984). A zoogeographic theater and biochronologic play: the time/biofacies phenomena of Eurasian and African Miocene mammal provinces. *Paleobiologie continentale* **14**: 121–42.

Bernor, R. & Tobien, H. (1990). The mammalian geochronology and biogeography of Paçalar (Middle Miocene, Turkey). *Journal of Human Evolution*, **19**: 551–68.

Bessedik, M. (1985). *Reconstitution des environments Miocenes des regions Nord-Ouest Mediterraneennes a partir de la palynologie.* Thèse USTELA, Montpellier, 162 pp.

Bonis, L. de, Bouvrain, G., Geraards, D. & Koufos, G. (1990). New hominid skull material from late Miocene of Macedonia in Northern Greece. *Nature*, **345**: 712–14.

Bonis, L. de, Brunet, M., Heintz, E. & Sen, S. (1992). La province greco-irano-afghane et la repartition des faunes mammaliennes au Miocène supérieur. *Paleont. i Evol.* **24–25**: 103–12.

Bonis, L. de & Koufos, G. (1999). The Miocene large mammal succession in Greece. In Agustí, J., Rook, L. & Andrews, P. (eds.), *Evolution of Neogene Terrestrial Ecosystems in Europe*, pp. 205–37. Cambridge: Cambridge University Press.

Bonis, L. & Melentis, J. (1975). Première decouverte de Muridés (Mammalia, Rodentia) dans el Miocène de la région de Thessalonique. Precisions sur l'âge géologique des Dryopithecinés de Macédonie. *C. R. Acad. Sc. Paris*, **280**: 1233–6.

Burtschak-Abramovitsch, N. O. and Gabachvili, E. G. (1945). Anthropoid ape from upper Tertiary deposits of Eastern Georgia, *Comm. Acad. Nat. Georgia*, **6**.

Cerling T. E., Wang. Y. & Quade, J. (1993). Expansion of C 4 ecosystems as an indicator of global ecological change in the Late Miocene. *Nature* **361**: 344–5.

Cicha, I., Fahlbusch, V. & Fejfar, O. (1972). Biostratigraphic correlation of some late Tertiary vertebrate faunas in Central Europe. *N. Jb. Paläont. Abh.* **140**: 129–45.

Daams, R. & Meulen, A. van der (1984). Paleoenvironmental and paleoclimatic interpretation of micromammal faunal successions in the Upper Oligocene and Miocene of North Central Spain. *Paléobiol. continent.* **14**: 241–57.

Dugas, D. P. & Retallack, G. J. 1993. Middle Miocene fossil grsses from Fort Ternan. *J Paleont.* **67**: 113–28.

Fortelius, M., van der Made, J. & Bernor, R. L. (1996a). Middle and Late Miocene Suoidea of Central Europe and the Eastern Mediterranean: Evolution, Biogeography and Paleoecology. In Bernor, R., Fahlbusch, V. & Mittman, W. (eds.), *The Evolution of Western Eurasian Neogene Mammal Faunas*, pp. 348–77. New York: Columbia University Press.

Fortelius, M., Wenderlin, L., Andrews, P., Bernor, R. L., Gentry, A., Humphrey, L., Mittmann, H. W. & Viranta, S. (1996b). Provinciality, diversity, turnover, and paleoecology in land mammal faunas of the late Miocene of western Eurasia. In Bernor, R., Fahlbusch, V. & Mittman, W. (eds.), *The Evolution of Western Eurasian Neogene Mammal Faunas*, pp. 348–77. New York: Columbia University Press.

Franzen, J. L. & Storch, G. (1999). The Late Miocene Mammals from Central Europe. In Agustí, J., Rook, L. & Andrews, P. (eds.), *Evolution of Neogene Terrestrial Ecosystems in Europe*, pp. 165–90. Cambridge: Cambridge University Press.

Garcés, M., Agustí, J., Cabrera, L. & Parés, J. M. (1996). Magnetostratigraphy of the Vallesian (late Miocene) in the Vallès-Penedès Basin (northeast Spain). *Earth Planet. Sci. Lett.* **142**: 381–96.

Garcés, M., Cabrera, L., Agustí, J. & Parés, J. M. (1997). Old World first appearence datum of '*Hipparion*' horses: late Miocene large mammal dispersal and global events. *Geology*, **25**: 19–22.

Golpe. J. M. (1993). Los hispanpitecos (Primates, Pongidae) de los yacimientos del Vallès-Penedès (Cataluña, España). II: Descripción del material existente en el Instituto de Paleontología de Sabadell. *Paleont. i Evol.* **26–27**: 151–224.

Harris, J. (1993). Ecosystem structure and growth of the African savanna. *Glob. Planet. Change* **8**: 231–48.

Kappelman, J., Kelley, J., Pilbeam, D., Khalid, A.S., Ward, S., Anwar, M., Barry, J. C., Brown, B., Hake, P., Johnson, N. M., Raza, S. M. & Shah, S. M. I. (1991). The earliest occurrence of *Sivapithecus* from the middle Miocene Chinji Formation of Pakistan. *Journal of Human Evolution*, **21**: 61–73.

Kappelman, J. (1996).

Köhler, M. & Moyà-Solà, S. (1997): Ape-like or hominid-like? The positional behavior of *Oreopithecus bambolii* reconsidered. *Proc. Nat. Acad. Sci. USA*, **94**: 11747–50.

Koufos, G. , Bonis, L. de & Sen, S. (1995). *Lophocyon paraskevaidisi*, a new viverrid (Carnivora, Mammalia) from the middle Miocene of Chios Island, Greece. *Geobios*, **28**: 511–23.

Krigjsman, M., Garcés, M., Langeries, C. G., Daams, R., Dam, J. van, Meulen, A. van der, Augustí, J. & Cabrera, L. (1996). A new chronology for the middle to late Miocene continental record in Spain. *Earth Planet. Sci. Lett.* **142**: 367–80.

Made, J. van der (1996). Listriodontinae (Suidae, Mammalia), their evolution systematics and distribution in time and space. *Contr. Tert. Quat. Geol.* **33**: 3–160.

Martin, R. & Andrews, P. (1984). The phyletic position of *Graecopithecus freybergi* v. Koenigswald. *Courier Forsch.-Ins. Senckenberg*, **69**: 25–40.

McCrossing, M. L. & Benefit, B. R. (1997). On the relationships and adaptations of *Kenyapithecus*, a large bodied hominoid from the middle Miocene of Eastern Africa. In Begun, D., Ward, C. & Rose, M. (eds.), *Function, Phylogeny and Fossils: Miocene Hominoid Evolution and Adaptation*, pp. 241–67. New York: Plenum Press.

Mein, P. (1986): Chronological succession of hominoids in the European Neogene. In J. G. Else & P. C. Lee (eds.): *Primate Evolution*, vol. 1, Cambridge: Cambridge University Press.

Mein, P. & Freudenthal, M. (1971). Une nouvelle classification des Cricetidae (Mammalia, Rodentia) du Tertiare de l'Europe. *Scr. Geol.* **2**: 1–37.

Mottl, M. (1958). Weitere Säugetierereste aus der Sarmat von St.-Staphan im Lavanttal, Kärntern. *Carinthia II, Mitt. Naturwiss. Ver. Karnten* **68**: 46–8.

Moyà, S. & Kohler, M. (1993). Recent discoveries of *Dryopithecus* shed new light on evolution of great apes. *Nature*, **365**: 543–5.

Moyà, S. & Köhler, M. (1996). A *Dryopithecus* skeleton and the origin of great-ape locomotion. *Nature*, **379**: 156–9.

Ozansoy, F. (1955). Faunes des mamifères du Tertiare de Turquie et leur revision stratigraphique. *Bull. Min. Res. Expl., Ankara* **49**: 29–48.

Ozansoy, F. 1965. Etude des gisements continentaux et des mamifères du Cénozoïque de Turquie. *Mém. Soc. Géol. France*, **44**: 1–92.

Pekka Lunkka, J., Fortelius, M., Kappelman, J. & Sen, S. (1999). Chronology and mammal faunas of the Miocene Sinap Formation, Turkey. In Agustí, J., Rook, L. & Andrews, P. (eds.), *Evolution of Neogene Terrestrial Ecosystems in Europe*, pp. 238–64. Cambridge: Cambridge University Press.

Pickford, M. (1986). Geochronology of the hominoidea: a summary. In Else, J. G. & Lee, P. C. (eds.), *Primate Evolution*, vol. 1, pp. 123–7. Cambridge: Cambridge University Press.

Sen, S., Koufos, G., Kondopoulo, D. & Bonis, L. de (1998). Magnetostratigraphy of late Miocene continental deposits of the Lower Axios Valley, macedonia, Greece. *Proc. RCMNS Inter-Coll. Mediterranean Neogene cyclostratigraphyin marine-continental deposits:* 56. Patras, Greece.

Steininger, F. F., Rabeder, G. & Rögl, F. (1985). Land mammal distribution in the Mediterranean Neogene A consequence of geokinematic and climatic events. In Stanley, D. J. & Wezel, F. C. (eds.), *Geological Evolution of the Mediterranean basin*, pp. 559–71. Berlin: Springer-Verlag.

Steininger, F. F., Berggren, W. B., Kent, D. V., Bernor, R. L., Sen, S. & Agustí, J. (1996). Circum-Mediterranean Neogene (Miocene and Pliocene) Marine-Continental Chronologic Correlations of European Mammal Units. In Bernor, R., Fahlbusch, V. & Mittman, W. (eds.), *The Evolution of Western Eurasian Neogene Mammal Faunas*, pp. 7–46. New York: Columbia University Press.

2 The trophic context of hominoid occurrence in the later Miocene of western Eurasia: a primate-free view

Mikael Fortelius and Arja Hokkanen

Introduction

The term 'hominoid locality' is common enough in the informal jargon of students of fossil mammals, and is frequently used in a sense that somehow implies more than just 'a locality from which hominoid material has been recovered'. It is a natural and intuitively attractive expectation that hominoid primates would occur in characteristic habitats and taxonomic settings. One might, for example, expect to find hominoids at localities that have produced large numbers (individuals or species) of other frugivorous, omnivorous or arboreal mammals. The impulse for the present study arose out of a wish to investigate whether, in fact, the pattern of hominoid occurrence differs from that of non-hominoid mammals in some easily demonstrable way. The trivial answer to this trivial question appears to be 'no', but as it turns out this 'no' hides a remarkably strong pattern that can be related both to the occurrence and the disappearance of hominoid primates from western Eurasia.

The opportunity to carry out such a study, without an inordinate data compilation effort, has recently arisen through the creation of the NOW database of Neogene Old World fossil mammals (Fortelius *et al.*, 1996b; see also under 'Material'). One of the main objectives of the European Science Foundation Network on Hominoid Evolution and Environmental Change in the Neogene of Europe (HOMINET) was the revision and updating of the NOW database, especially with regard to 'hominoid localities'. To some extent, this chapter may be regarded as a pilot study and summary presentation of the state of the database at the close of the HOMINET in the autumn of 1998.

In order to study the context of occurrence of a taxon it is obvious that the effect of the taxon itself must be excluded from the analysis. In order to create a simple and robust procedure we decided to eliminate the primate order entirely from all studies investigating context – hence the 'primate-free view' of the title. The study was carried out as a simple comparison of two sets of localities defined by whether they did or did not include hominoid primates (family Hominidae in the wide sense of Andrews *et al.*, 1996). For a more comprehensive view, sets of localities with pliopithecid and cercopithecid primates were also defined and investigated. In preliminary analyses we charted taxonomic, trophic and locomotory aspects of the occurrence of hominoids and non-hominoids. Based on these preliminary studies we selected trophic structure, distribution of molar crown height among herbivores and omnivores, and body size distributions as the main targets of investigation

There is no problem-free way of extracting ecologically interpretable information from the fossil record. In our view, methods that rely on species richness are particularly vulnerable to sampling biases as well as difficult to interpret, if only because the ecology of species richness in the living world is far from clear (e.g. Rosenzweig, 1995). We have therefore ignored species richness here, and instead approached the problem in terms of community structure and other 'taxon-free' methods (Damuth, 1992). Such studies of fossil mammals have usually relied either on faunal lists from individual localities or on various kinds of composite lists and on comparison with similar lists from modern habitats (Andrews *et al.*, 1979; Bonis *et al.*, 1992a,b, 1999; Andrews 1996; Fortelius *et al.*, 1996b). Except for small mammal studies (e.g. van Dam, 1997), abundance data have not usually been used (or available) for analyses that concern more than single localities. Since abundance plays a central role in community ecology (for example through estimates of biomass or equability), abundance data would be highly useful for broader studies of extinct communities, and any reasonable proxy for abundance would be worth exploring.

We propose that the potential for developing such proxies is offered, at least in principle, by the advent of databases that record taxon-occurrence matrices such as the one analysed here. There are several possibilities, but the simplest one would seem to be to recognise each single occurrence of a species within a defined area and interval, instead of collapsing the occurrences into a list of species. In this way a data matrix is generated where the species are weighted in proportion to their number of occurrences. All else being equal, the probability of recording a taxon from a locality should be proportional to its abundance when alive.

All else is very far from equal in this case, of course, but it still seems reasonable to assume a strong positive correlation between abundance and number of species-locality occurrences (splocs), even if exceptions are to be expected. As long as comparisons are relative within the dataset studied the issue of geographic and temporal extent of the data does not constitute a sampling problem (and ways of standardising for those effects can certainly be developed if different datasets are to be compared). It is a matter for future testing whether 'sploc analysis' could, with more complete sampling of the record, be used to bypass the lack of abundance data for large fossil mammals. We predict that such will be the case at least for large datasets from areas that are not much larger than the average geographic distribution of the species studied.

In the present analysis we would not argue that we are sampling abundance, only that we are introducing a partial weighting that reflects abundance more than any other factor. Other factors that will influence the number of occurrences are: how well the group has been studied; and how well its species-level taxonomy is resolved. The most serious problem we can think of at this stage is that common but poorly resolved taxa (such as '*Hipparion* indet.') would be given too little weight, but in the present case this problem cannot be avoided anyway, since all 'indet.' and 'cf.' identifications were excluded to begin with (see under 'Material').

Sploc analysis offers two additional advantages. First, it genuinely increases sample size, which is otherwise often rather small for statistical significance. Unless it can be shown that the occurrences represent mostly noise, which appears most unlikely, comparing the number of occurrences makes at least as good ecological sense as comparing the presence and absence of species. Secondly, the best known and least ambiguous species are emphasised in the analysis. It therefore appears that sploc analysis has much to say for it, in addition to the fact that it offers a way of bringing abundance data into the study of large mammal palaeoecology. This study is a first attempt at evaluating and applying this method in order to resolve the main features of the occurrence context and history of hominoid primates in western Eurasia in relation to the overall large mammal pattern. It is not by any means intended to investigate hominoid occurrence or palaeoenvironments as such, for which we refer readers to the recent detailed reviews of Andrews (1996) and Andrews *et al.* (1996), as well as to other contributions in this volume.

Material and methods

Material

Data were derived from the NOW database (Fortelius *et al.*, 1996b). The dataset used for these analyses was downloaded on October 31, 1998 and is archived with the NOW database (gem_dig@gaia.pc.helsinki.fi). The dataset was limited geographically to the Eurasian continent and islands west of 60° E. The temporal range was set from the base of the middle Miocene to the end of the late Miocene, 15.2–5.3 Ma according to the currently used Time Unit Table of the NOW database (essentially, the correlations and boundaries of Steininger *et al.*, 1996, with pre-Volhynian Eastern Paratethys stages according to Alexey Tesakov, pers. comm., 1998). Localities were

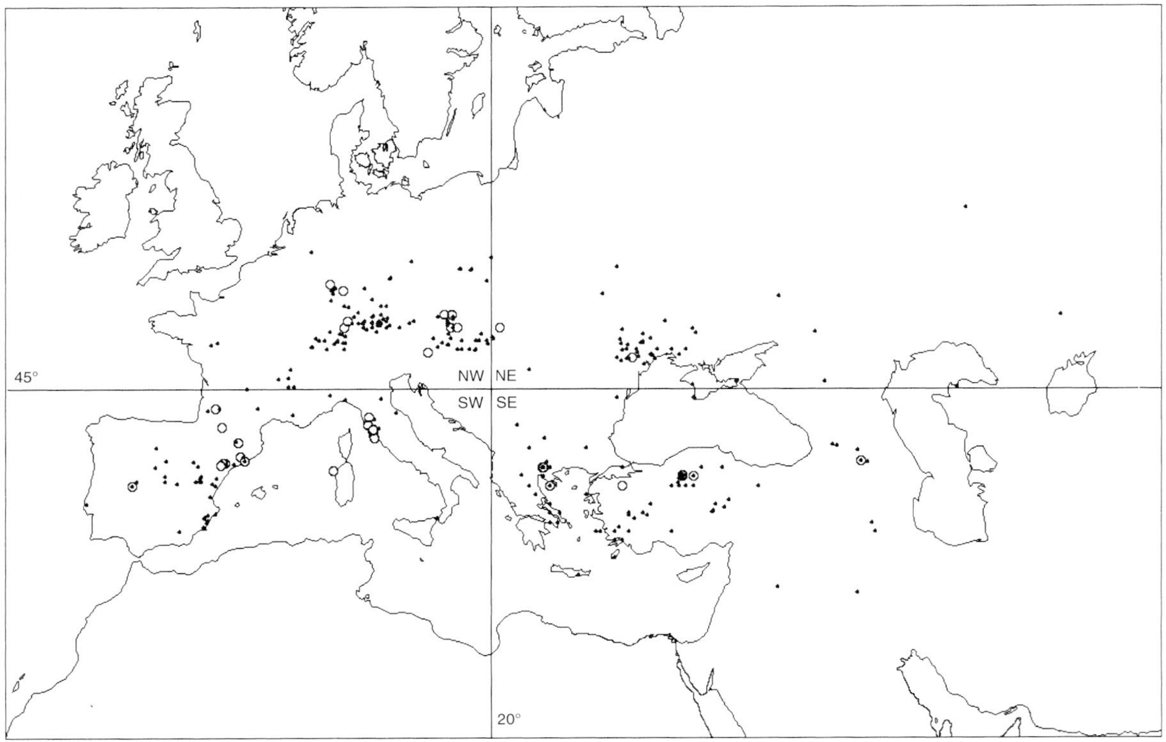

[Figure 2.1]
Map of localities included. The lines separating geographical halves (North, South, West, East) and quadrants (NW, SW, NE, SE) are indicated. Open circles are hominoid, small dots non-hominoid localities.

Table 2.1. *Breakdown of the dataset studied. Splocs: species-locality occurrences*

Subset	Localities	Species	Splocs
All	378	579	2150
Non-hominoid	341	518	1796
Hominoid	37	181	354
Pliopithecid	19	134	218
Cercopithecid	25	125	285
Griphopithecus	3	28	30
Dryopithecus	14	102	187
Graecopithecus	3	17	20
Oreopithecus	6	11	16

included when their age span (maximum to minimum age estimate) included these limits, which means that a few possibly early Miocene and early Pliocene localities were included among the 378 localities selected (Figure 2.1). The localities selected include a few cases of partial duplication, when a 'general' and one or more single-quarry localities are included (cf. Damuth, 1997), but the effect of this is minimal, since the number of taxa recorded from the time and place in question is not inflated by the procedure, and the species-occurrence pattern is affected only minimally.

All taxa of Primates, Carnivora, Artiodactyla, Perissodactyla, Hyracoidea, Proboscidea and Tubulidentata were included, except for taxa lacking genus or species level identifications (e.g. 'Bovidae indet.' was excluded). All 'cf.' identifications at the species level were also excluded (e.g., *Gazella* cf. *ancyrensis*), but such identifications at the genus level, which reflect taxonomic rather than identification uncertainty were included (e.g. cf. *Oioceros*).

In order to explore the trophic context of hominoid occurrence, the data were coded into four main subsets (Table 2.1): localities with hominoids (coded as 'homi'); localities without hominoids ('noho'); localities with pliopithecids ('plio'); and localities with cercopithecids ('cerco'). In this study, these subsets are also referred to as hominoid localities, pliopithecid localities, etc. Four smaller subsets similarly derived for localities with selected individual hominoid genera (*Griphopithecus*, *Dryopithecus*, *Graecopithecus* and *Oreopithecus*) were also compared (Table 2.1). Primate species (a total of 16) were excluded from all analyses except that of temporal and spatial occurrence of the primates themselves.

The NOW data follow the ETE database standards (Damuth, 1997) and are thus coded for inferred diet at three levels. For the statistical analyses we used only 'diet_1', with three values: animal-eater; omnivore; or plant-eater. For the graphic comparisons we used an idiosyncratic combination of 'diet_1', 'diet_2' and 'diet_3', which enabled us to show separately the animal-eater category 'invertebrate-eater' and the plant-eater category 'grazer'. Qualitative molar crown height (ETE field 'tht') is given as brachydont, mesodont or hypsodont. We analysed the distribution of molar crown heights in species entered as plant-eaters or omnivores. Body mass estimates in NOW (ETE field 'body_mass') are based on regression equations from living species (Damuth & MacFadden, 1990), or on other direct comparisons with living analogues. The origin of the dietary, molar crown height and body mass information in NOW is as reported in Fortelius *et al.* (1996b), with further updates by the same team (especially, Peter Andrews, Raymond L. Bernor, Alan Gentry, Mikael Fortelius, Elmar P. J. Heizmann, Gertrud Roessner, Suvi Viranta and Lars Werdelin). It should be emphasised that the dietary category 'grazer' was not based on molar crown height but on dental wear, using principles detailed in Fortelius & Solounias (2000).

The localities were coded for temporal and geographical position. Two temporal blocks ('Pre' and 'Post') were defined relative to the 'Vallesian Crisis' (Agustí & Moyà-Solà, 1990; Moyà-Solà & Agustí, 1990; Fortelius *et al.*, 1996b) such that 'Pre' localities had minimum age estimates of more than 9.5 Ma and 'Post' localities had maximum age estimates of less than 9.5 Ma. The localities that did not fall into either group were treated as 'crisis time' localities. Localities were also coded for 'MN-unit equivalents', defined using the MN unit boundaries and correlations of the Time Unit Table of the NOW database. This coding was strict, accepting only localities dated with the precision of a single MN unit or better, and not straddling the boundary, and thus unlike the 'inclusive' formula used by Fortelius *et al.* (1996b), where a locality spanning two MN units was counted for both. This strict coding resulted in the loss of a significant number of localities (and thus 'known occurrences') in the analyses of temporal sequences. All other comparisons, however, included less precisely dated localities.

'North' and 'South' halves were divided such that 'North' includes all localities north of 45° N, 'South' the rest. Similarly, an 'East' half was defined to include localities east of 20° E and a 'West' half for the rest (note that these 'East' and 'West' differ slightly from the politically defined blocks of Fortelius *et al.*, 1996b). Finally, the same boundaries were used to define four quadrants ('NW', 'SW', 'NE' and 'SE'). It goes without saying that such divisions are highly arbitrary and fail to account for palaeogeographic features and changes, but nothing else is as yet readily available, especially for coarse-scale comparisons involving long intervals of time during which geographic changes took place.

Alroy (1996, unpublished data) has compared different ways of correcting

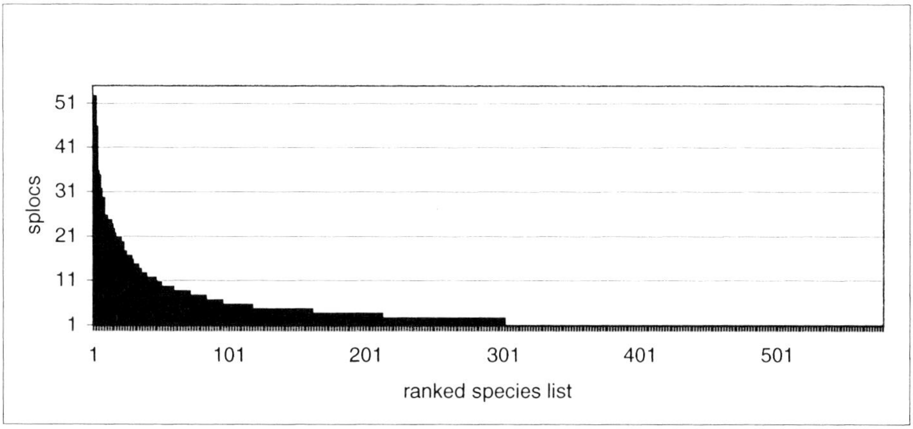

[Figure 2.2]
Ranked distribution of locality occurrences of the species in the dataset. Splocs: species-locality occurrence counts.

for sampling bias and recommends rarefaction for cases when faunal lists are used, or else methods based on counting Lazarus taxa. Using an earlier version of the dataset analysed here, Fortelius *et al.* (1996b) used the latter approach to gage the level of sampling (which appeared mostly adequate) but did not use it to correct the diversity curve. Instead, they concluded that the geographic area sampled was a major determinant of species diversity observed within individual MN units of the (politically defined) West and East blocks that were contrasted in that study, and corrected the diversity for each area. This correction modified the pattern somewhat but did not change the main relationships. The present data matrix presents a significantly better state of sampling than that of Fortelius *et al.* (1996b), and since we are not here concerned with diversity but only with relative proportions of eco(morpho)logically defined categories within data sets we feel that further investigation of sampling is uncalled for at this stage.

The distribution of splocs is highly skewed to the right, with only a few species having tens of occurrences and the vast majority less than three (Figure 2.2). The mean number is only 3.71 for all species and 9.63 for those that have more than three occurrences, so the occurrence matrix is obviously far from saturation.

Methods

The comparative analyses were based on composite species lists as well as on sploc lists – i.e. lists including all the individual locality occurrences of all the species. As discussed in the Introduction, the latter were emphasised in this study, but it should be understood that species and splocs give highly similar patterns in all the comparisons reported here, and indeed in all comparisons we have so far undertaken. These data subsets described above were compared with each other following the basic logic that a finer geographical and temporal subdivision was undertaken each time a significant difference was discovered, but not when a broad comparison revealed no significant differences. For these finer divisions sploc analysis was chosen as the more conservative alternative (more likely to still show a significant difference, which was the null hypothesis). The statistical methods employed were chi squared analysis and Kruskal–Wallis analysis, using the default settings on the statistical package 'Systat for Windows 7.0'.

Results

Taxonomic contrasts

Contrasting all localities from which hominoid primates have been collected with all localities that have not delivered hominoids shows a clear difference for both species counts and sploc counts – hominoid localities having proportionally more animal-eaters and omnivores (Figure 2.3, Table 2.2). Pliopithecid localities are like hominoid localities, but even more strongly dominated by animal-eaters, while cercopithecid localities represent the other extreme, with few omnivores and strong plant-eater dominance. The non-hominoid localities show a trophic pattern closer to that of cercopithecid than hominoid localities, and the pattern shown by all localities shows only a slightly higher proportion of animal-eaters. The same pattern is repeated in the proportion of brachydont, mesodont and hypsodont molars of plant-eaters and omnivores, with pliopithecid and hominoid localities having the least and cercopithecid localities the most hypsodont and mesodont species and species-locality occurrences. The temporal sequence reveals a shift from a 'hominoid-like' middle Miocene pattern to a 'cercopithecid-like' late Miocene pattern. The statistical significance of these differences is mostly high as shown in Table 2.2.

The difference between sploc-analysis and conventional analysis of composite species lists is shown by the two pairs of graphs in Figure 2.3. For trophic structure, the differences between the homi, plio, cerco, noho, and all subsets are very similar in both graphs (Figures 2.3A and B). For the temporal sequence the species-based graph (Figure 2.3B) shows a stronger but also more irregular pattern, with extreme animal-eater and omnivore dominance in the 11.8 Ma (MN 7 + 8) interval and the extreme plant-eater dominance in the 8.9 Ma (MN 11) interval. There is a clear difference

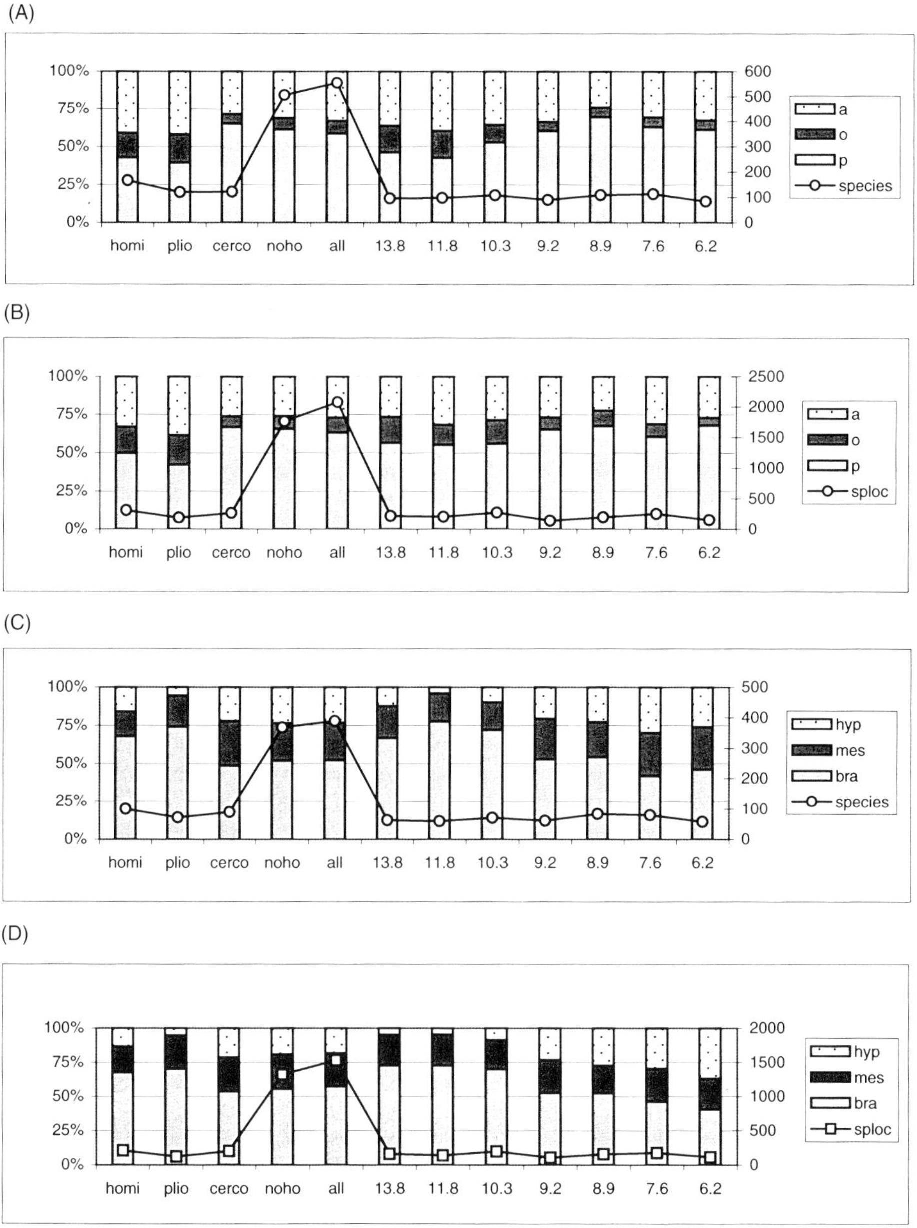

[Figure 2.3]

Distribution of species and species-locality occurrences by trophic affinity (A, B) and molar crown height (C, D) in taxonomically and temporally defined subsets of localities (all: all localities; cerco: localities with cercopithecoids; homi: localities with hominoids; noho: localities without hominoids; plio: pliothecids. a: animal eaters; o: omnivores; p: plant eaters; bra: brachydont; mes: mesodont; hyp: hypsodont; species: total number of species in each subset. (A) and (C) are based on species counts, (B) and (D) on species-locality occurrence counts (sploc). The values are the midpoints (in Ma) of time intervals corresponding to MN units 6 to 13 as calibrated by Steininger *et al*. (1996).

Table 2.2. *Chi squared tests of differences in trophic structure and omnivore/plant-eater molar crown height distribution between taxonomic, geographic and temporal subsets of the data*

		Trophic structure			Hypsodonty (omnivore, plant-eater)		
Comparison	Method	Total *N*	χ^2	Probability	Total *N*	χ^2	Probability
North / south	Species count	231 / 405	0.170	0.919	129 / 218	8.499	0.014
East / west	Species count	306 / 272	48.614	**0.000**	187 / 126	10.269	**0.006**
Pre / post	Species count	212 / 308	11.950	**0.003**	115 / 181	25.742	**0.000**
North / south	Sploc count	683 / 1326	3.975	0.137	479 / 871	57.893	**0.000**
East / west	Sploc count	1026 / 983	67.681	**0.000**	713 / 637	63.298	**0.000**
Pre / post	Sploc count	754 / 1105	31.238	**0.000**	505 / 752	111.822	**0.000**
Homi / noho	Species count	161 / 458	19.949	**0.000**	81 / 258	6.372	0.041
Homi / plio	Species count	161 / 114	0.406	0.816	81 / 54	3.526	0.172
Homi / cerco	Species count	161 / 112	14.521	**0.001**	81 / 72	6.129	0.047
Homi / noho	Sploc count	308 / 1701	34.252	**0.000**	186 / 1164	9.714	**0.008**
Homi / plio	Sploc count	308 / 187	2.813	0.050	186 / 94	4.862	0.088
Homi / cerco	Sploc count	308 / 256	20.028	**0.000**	186 / 199	7.838	0.020

df: 2 in all cases.

Abbreviations (see also 'Material'): cerco: cercopithecid localities; homi: hominoid localities; plio: pliopithecid localities; north: localities north of 45° N; south: localities south of 45° N; east: localities east of 20° E; west: localities west of 20° E; pre: localities older than the Vallesian Crisis; post: localities younger than the Vallesian Crisis; sploc: species-locality occurrence.

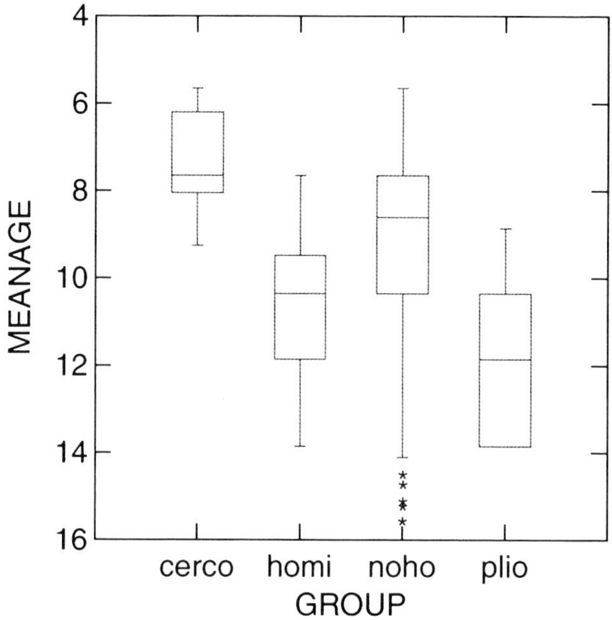

[Figure 2.4]
Box-and-whiskers plot of the mean age of the species-locality occurrences in the four taxonomic subsets of data compared. MEANAGE: mean age in Ma, GROUP: taxonomically defined subset of data (abbreviations as in Figure 2.3).

between the middle Miocene and late Miocene intervals, but the first interval of the late Miocene (10.3 Ma; MN 9) is intermediate, suggesting a gradual transition rather than an abrupt crisis.

For molar crown height, we see the same trend – the species-based analysis shows a stronger and less regular pattern, with the extremes at 11.8 and 8.9 Ma (Figure 2.3C), instead of at the endpoints as in the sploc-based graph (Figure 2.3D). The strong increase in percentage of hypsodont species during the late Miocene that one would expect from experience with the fossil material is only seen in the sploc-based graph, in which the Late Miocene intervals from 8.9 Ma (MN 11) onwards show more occurrences of high-crowned forms than is seen in the cercopithecid subset. For both trophic structure and molar crown height the sploc-based analysis (Figures 2.3B and D) shows a more distinct shift at the time of the Vallesian Crisis, which divides the graphs neatly into 'pre-crisis' and 'post-crisis' portions (between the 10.3 and 9.2 Ma intervals, at the MN 9/10 boundary).

Since the temporal range of apes is primarily middle and early late Miocene, while that of monkeys is exclusively late Miocene (Figure 2.4), the correspondence observed between temporal and taxonomic groupings suggests that the difference between taxonomic groupings might be primarily or entirely an age effect. This indeed appears to be the case. The difference between hominoid and non-hominoid localities is significant only for pre-crisis time, and when the halves and quadrants are analysed individually for this temporal subset, no area shows a significant hominoid/non-hominoid contrast for either trophic structure or hypsodonty (Table 2.3). The tests for individual quadrants are evidently dubious owing to reduced sample sizes, even for sploc analysis, but this problem affects only East (i.e. of North, South, West and East, only East is affected). The age difference is significant for hominoids versus non-hominoids, pliopithecids and cercopithecids for species ages and sploc ages, but the locality age difference is not significant for the hominoid/non-hominoid contrast, although the direction of the difference is the same (Table 2.4).

Geographic and temporal contrasts

The temporal change in trophic structure is shown separately for each geographical quadrant in Figure 2.5. Two environmentally indicative dietary subcategories have been added to the basic scheme to help the interpretation – invertebrate-eater and grazer (grass eater).

Strong temporal and geographical signals are immediately revealed. The temporal trend towards increasing dominance of plant-eaters is seen to be primarily a phenomenon of the eastern quadrants, especially the well-sampled SE that dominates the late Miocene in the whole sample. The trend, however, is reversed in the NW quadrant, which is very poorly sampled for the last intervals. The decrease in the proportion of omnivorous species appears to occur in all quadrants, but the level is higher throughout in pre-crisis time and in the western half of the area. Invertebrate-eaters were also more numerous in the western and southern halves of the area, but their relative occurrence does not appear to have been affected by the crisis. Grazers were more numerous in the southern and eastern halves and during post-crisis time and increased strongly in the last interval in all quadrants except the poorly sampled NW.

The evolution of crown height in plant-eaters and omnivores is shown by quadrant in Figure 2.6. It is immediately obvious that the main pattern seen for trophic structure is repeated – hypsodont forms dominate in East, South and post-crisis, and a gradual decrease of brachydont forms is seen in all quadrants except perhaps NW and except for a reversal in the last time slice (6.2 Ma; MN 13) in SE.

As already noted the contrast between East and West is highly significant for both trophic structure and crown height distribution (Table 2.2). This

Table 2.3. *Chi squared tests of differences in trophic structure and omnivore/plant-eater molar crown height distribution between pre- and post-Vallesian Crisis and between geographically defined subsets of the pre-crisis dataset*

		Trophic structure			Hypsodonty (omnivore, plant-eater)		
Comparison	Subset	Total N	χ^2	Probability	Total N	χ^2	Probability
Homi / noho	Pre crisis	225 / 259	14.244	**0.001**	135 / 370	1.664	0.435
Homi / noho	Post crisis	44 / 1061	2.765	0.251	33 / 719	5.286	0.071
Homi / noho	Pre + north	73 / 308	4.169	0.124	53 / 240	5.809	0.055
Homi / noho	Pre + south	152 / 221	3.650	0.161	82 / 130	0.702	0.704
Homi / noho	Pre + west	164 / 410	8.917	0.012	102 / 290	1.801	0.406
Homi / noho	Pre + east	61 / 119	7.254	0.027	33 / 80 (s)	0.283	0.868
Homi / noho	Pre + NW	48 / 268	2.140	0.343	40 / 211	7.237	0.027
Homi / noho	Pre + SW	116 / 142	0.507	0.776	62 / 79 (s)	0.464	0.793
Homi / noho	Pre + NE	23 / 40 (s)	5.740	0.057	13 / 29 (s)	0.223	0.895
Homi / noho	Pre + SE	36 / 23 (s)	2.883	0.239	20 / 51 (s)	0.986	0.611

s: more than 1/5 of fitted cells are sparse ($N < 5$); df: 2 in all cases.

Abbreviations as in Table 2.2 (see also 'Materials') and NW, SW, NE and SE are quadrants defined by 45 degrees northern latitude and 20 degrees eastern longitude.

Table 2.4. *Kruskal–Wallis tests of differences in the age distribution of localities, species-locality occurrences and species of taxonomically defined subsets of the data*

Comparison	Mean age (Ma)	Counts	Mann-Whitney U	Probability
Taxon ages				
Homi / noho	10.87 / 9.29	234 / 1057	483456	**0.000**
Homi / plio	10.87 / 11.99	234 / 163	37.586	**0.000**
Homi / cerco	10.87 / 7.45	234 / 206	1359.5	**0.000**
Taxon occurrence ages				
Homi / noho	10.81 / 9.31	313 / 1796	406566.5	**0.000**
Homi / plio	10.81 / 12.13	313 / 218	20590.0	**0.000**
Homi / cerco	10.81 / 7.45	313 / 285	3806.0	**0.000**
Locality ages				
Homi / noho	10.21 / 9.92	37 / 341	7000	0.272
Homi / plio	10.21 / 13.8	37 / 19	170.5	**0.001**
Homi / cerco	10.21 / 6.99	37 / 25	67.50	**0.000**

Abbreviations as in Table 2.2 (see also 'Materials').

was also the case for the pre- and post-crisis contrast, while for the North–South contrast only crown height distribution was highly significantly different. The proportional occurrence of omnivores and invertebrate-eaters was significantly higher and the proportion of grazers significantly lower in West than in East (Table 2.5). The proportion of omnivores, invertebrate eaters and grazers was significantly lower in North. Omnivores decreased significantly relative to other groups after the crisis, and grazers increased. Only two comparisons of subgroup contrasts returned non-significant differences – the proportion of omnivores appears to have been the same in North and South, and the crisis apparently did not affect the proportion of invertebrate-eaters in the faunas.

Body mass patterns

Mean body mass (of the 310 species, 54% of all species included for which an estimate was available) was higher in East than in West, and higher after than before the crisis ($P = 0.000$ in both cases), but no significant difference between North and South was detected. A breakdown of the temporal contrasts by the major families is shown in Figure 2.7 and Table 2.6. This reveals that most families do show the general trend of increasing body mass, strongest in the hyaenids, felids, ursids and suids, but bovids and equids do not, nor do hominoids. Using median rather than mean values amplifies the contrasts somewhat and affects only minor details, mostly

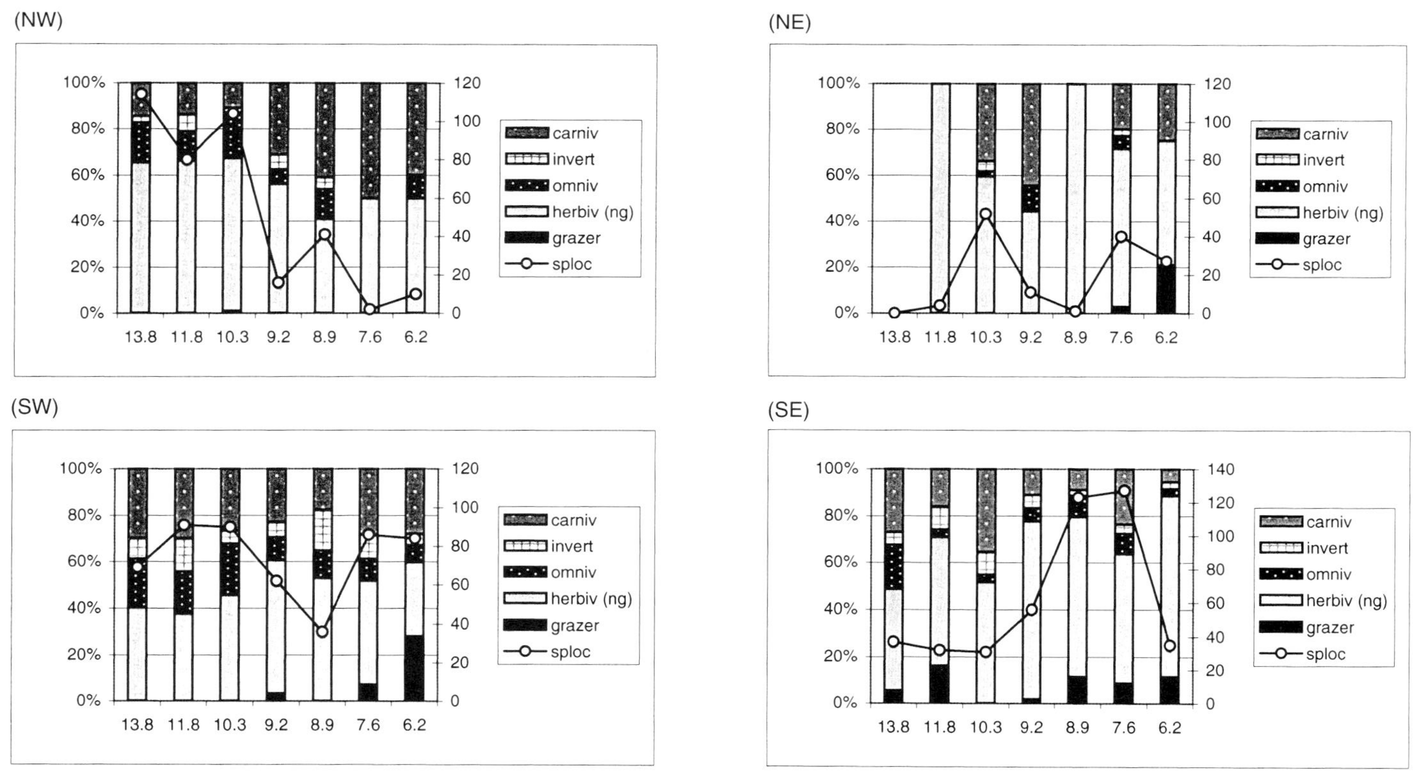

[Figure 2.5]
Distribution of species-locality occurrences by trophic affinity and interval in the four geographic quadrants. The time intervals are MN-unit equivalents labelled by the midpoint of their age range. Abbreviations as in Figure 2.3, and carniv: carnivore; invert: invertebrate-eater; omniv: omnivore; herbiv (ng): herbivore (non-grazer); sploc: species locality occurrence counts.

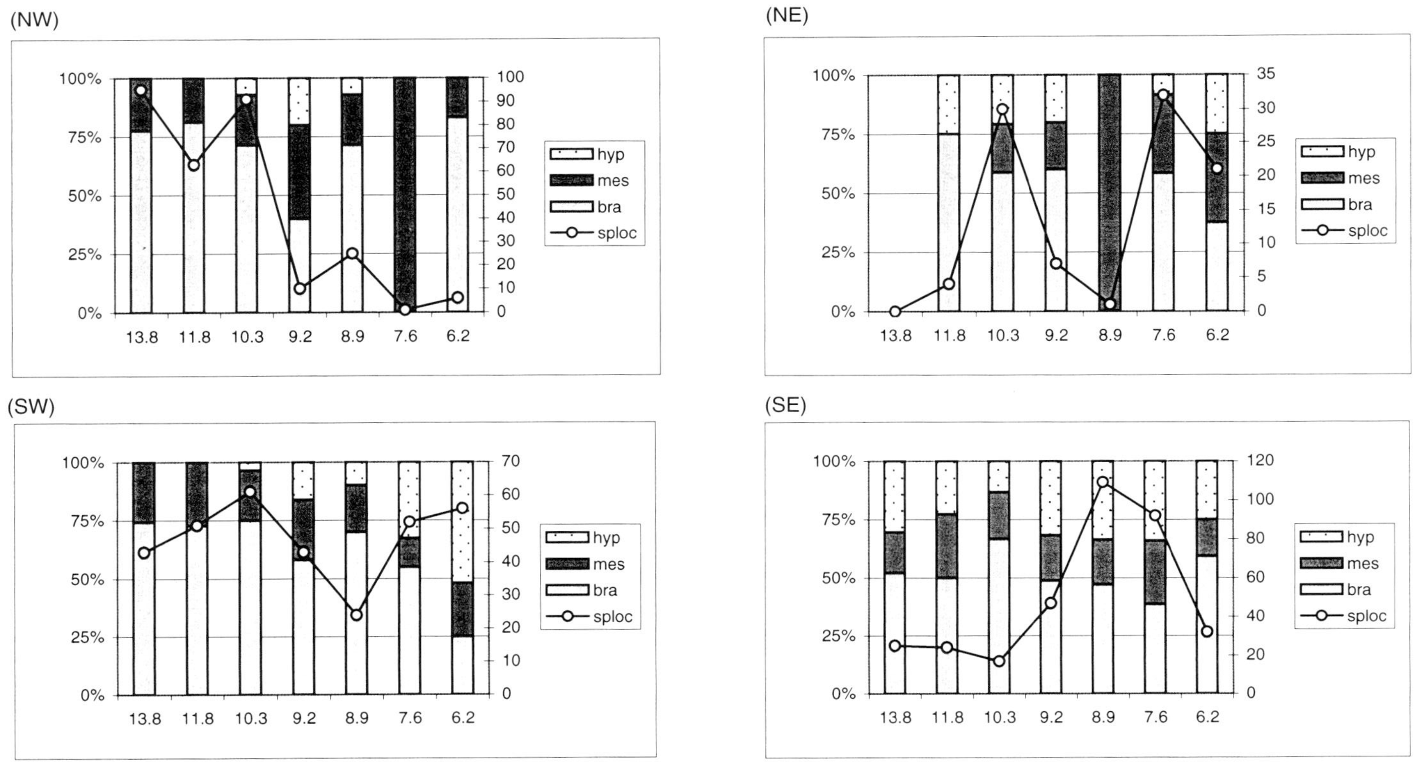

[Figure 2.6]
Distribution of species-locality occurrences by herbivore molar crown height and interval in the four geographic quadrants. The time intervals are MN-unit equivalents labelled by the midpoint of their age range. Abbreviations as in Figures 2.3 and 2.5.

Table 2.5. *Chi squared tests of differences in the species-locality occurrences of grazers, invertebrate-eaters and omnivores in geographically and temporally defined subsets of the data*

Comparison	Subset	Total *N*	χ^2	Probability
North / south	Omnivore / other	195 / 1814	0.042	0.837
North / south	Invertebrate-eater / other	104 / 1893	8.069	**0.005**
North / south	Grazer / other	96 / 1720	18.867	**0.000**
East / west	Omnivore / other	195 / 1814	49.328	**0.000**
East / west	Invertebrate-eater / other	104 / 1893	10.343	**0.001**
East / west	Grazer / other	96 / 1720	6.959	**0.008**
Pre / post	Omnivore / other	113 / 1676	28.681	**0.000**
Pre / post	Invertebrate-eater / other	97 / 1752	2.031	0.154
Pre / post	Grazer / other	93 / 1601	40.372	**0.000**

df: 1 in all cases.
Abbreviations as in Table 2.2 (see also 'Materials').

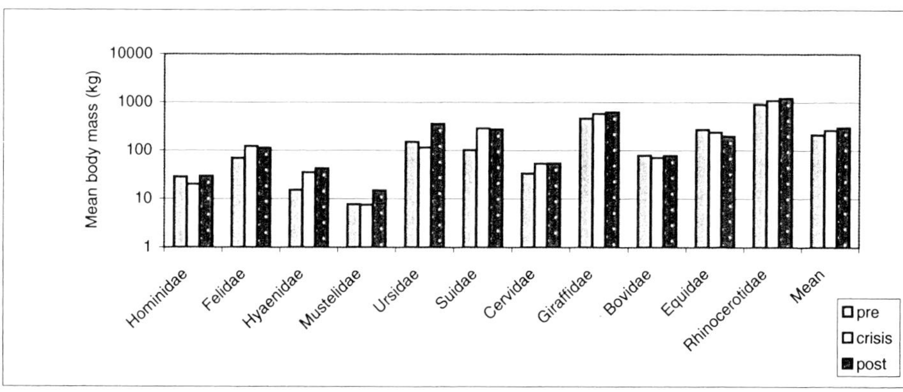

[Figure 2.7]
Mean body mass of selected large mammal families before, during and after the Vallesian Crisis (see under 'Material').

involving the small samples of crisis-time occurrences (Table 2.6). The increase in mean size of carnivores relative to the main ungulate families is in accordance with the gradual decrease in the proportion of animal-eaters to plant-eaters observed in the trophic comparisons, but demonstrating a link would require at least a species-by-species calculation of occurrence, body mass and generation length.

The coefficient of variance (cv) is highest for the body mass estimates of occurrences of mustelids and felids, both of which contain small species that are lost at about crisis time as confirmed by the decreased positive

Table 2.6. *Body size statistics for selected families before, during and after the Vallesian Crisis. The 'crisis' localities are ones that fall between the pre- and post-crisis localities*[a]

Family	Time	*N*	Median	Mean	cv	g1
Hominidae	pre	9	35	28.3	0.28	−0.27
	crisis	1	20	20	1.0	—
	post	10	30	29.5	0.12	−1.91
Felidae	pre	40	30	68.8	1.15	1.11
	crisis	4	129	122	0.93	−0.05
	post	67	85	112	0.67	0.37
Hyaenidae	pre	39	6.5	15.3	0.81	0.88
	crisis	12	34	35.2	0.68	0.40
	post	118	38	42.2	0.61	−0.10
Mustelidae	pre	49	4.5	7.79	1.38	3.01
	crisis	8	15	7.6	2.26	2.82
	post	34	5.0	14.9	1.24	1.18
Ursidae	pre	25	120	149	0.51	1.83
	crisis	3	90	115	0.45	—
	post	16	350	350	0.45	−0.14
Suidae	pre	113	91	.02	0.81	4.45
	crisis	6	330	285	0.55	−0.01
	post	61	270	272	0.25	−1.25
Cervidae	pre	80	35	33.6	0.75	0.001
	crisis	7	65	53.4	0.42	−2.35
	post	33	60	54.0	0.46	−0.87
Giraffidae	pre	9	400	464	0.37	0.88
	crisis	9	600	584	0.49	0.34
	post	78	600	627	0.44	0.02
Bovidae	pre	43	80	78.5	0.43	0.91
	crisis	22	41	70.9	0.86	1.05
	post	212	65	77.4	0.69	0.94
Equidae	pre	8	274	269	0.05	−2.83
	crisis	1	236	236	1.0	—
	post	64	187	195	0.27	−0.012
Rhinocerotidae	pre	96	585	904	0.47	0.23
	crisis	12	1139	1091	0.25	−1.06
	post	88	1286	1219	0.16	−1.57

[a] See under 'Materials' for definitions.
Abbreviations as in Table 2.2 (see also 'Materials'), and cv: coefficient of variation; g1: skeweness of distribution.

skeweness of the distribution (Table 2.6). The skeweness values for suids similarly reflect their transition from dominance by small to large forms. Equids, rhinocerotids and hominoids are among the families with the least body mass variance, and they are also among the groups showing the most negative skeweness values, indicating dominance of relatively large species.

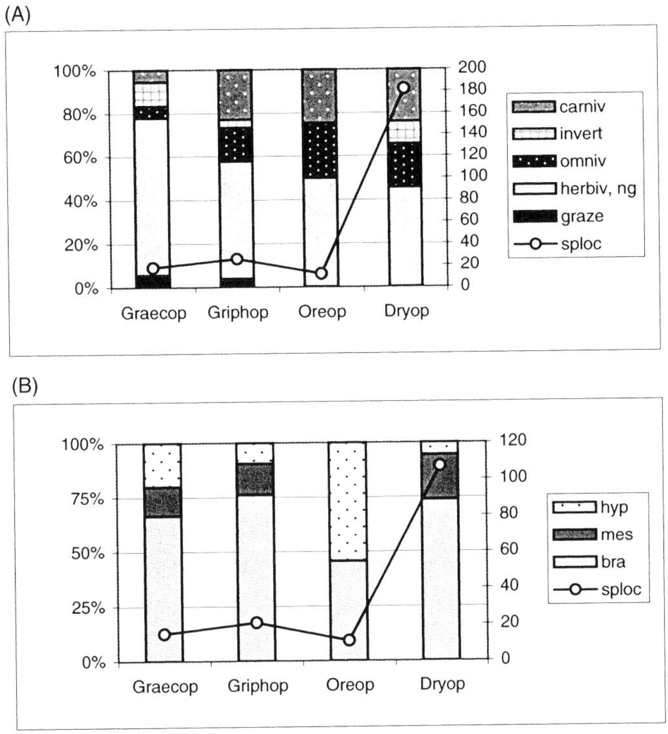

[Figure 2.8]
Distribution of species-locality occurrences by trophic affinity (A) and herbivore and omnivore molar crown height (B) distribution in sets of localities defined by presence of selected genera of hominoid primates. Abbreviations as in Figures 2.3 and 2.5, and Graecop: *Graecopithecus*; Griphop: *Griphopithecus*; Oreop: *Oreopithecus*; Dryop: *Dryopithecus*.

Trophic and crown height context of hominoid genera

Trophic structure and molar crown height distribution of occurrences associated with selected hominoid genera are shown in Figure 2.8. It is obvious that the pattern repeats the results described above in that each pattern is the one expected from the temporal and geographical distribution of the genus. The strongly western *Dryopithecus* shows the high proportion of animal-eaters and omnivores characteristic of pre-crisis West, while *Graecopithecus* shows the 'noho-like' south-eastern and post-crisis pattern that reflects its distribution (compare Figures. 2.3, 2.4 and 2.8). The predominantly south-eastern *Griphopithecus*, the oldest of the genera included, shows more plant-eater dominated pattern than does the younger but strongly western *Dryopithecus*. The enigmatic insular form *Oreopithecus* shows a *Dryopithecus*-like trophic context. The proportion of crown height classes (Figure 2.8B) repeats the pattern, with *Dryopithecus* having the lowest proportion of hypsodont forms. The other genera show the increase in mesodont and hypsodont forms expected from their general occurrence contexts, except that *Oreopithecus* is seen to be associated with a very high proportion of hypsodont forms (the peculiar bovids known from these insular localities).

Occurrence pattern of primate families

A breakdown of primate occurrence by quadrant and MN-unit equivalent is shown in Figure 2.9. It is immediately clear that the pattern is patchy and uneven, which is due partly to the incomplete nature of the database (especially for the important French and Spanish datasets), and in part to the fact that records dated with less precision than a single MN unit are not included. It is also clear, however, that even this incomplete and distorted view still shows the main features of the western Eurasian primate record. The pattern of primate occurrence clearly is not a reflection of sampling, at least not in the sense that more localities would mean more primates. If anything, the proportion of primates seems to have an inverse relationship to the number of localities, probably because species-poor localities have tended to become included in the database if one of the few species is a primate. A feature emphasised by this figure is the earlier disappearance of hominoid primates from the northern quadrants.

Discussion

The main features of the results described above are similar to those described by Fortelius *et al.* (1996b) and, to the extent that they can be compared, to those described from a much smaller dataset by Fortelius *et al.* (1996a). They are also in general agreement with the conclusions of the multivariate studies of taxonomic composition reported by Bonis *et al.* (1992a,b), especially regarding the existence of two markedly different sets of mammal assemblages corresponding roughly to the Vallesian and Turolian mammal ages, and which Bonis *et al.* (1999) refer to as hominoid and cercopithecoid faunas, respectively. Such general congruence of results obtained from datasets of strongly varying completeness and using different analysis techniques has been observed repeatedly in palaeontology (see Sepkoski 1993, 1997 for discussion), showing that the results are robust at least in the sense that the same pattern is readily reproducible from the data.

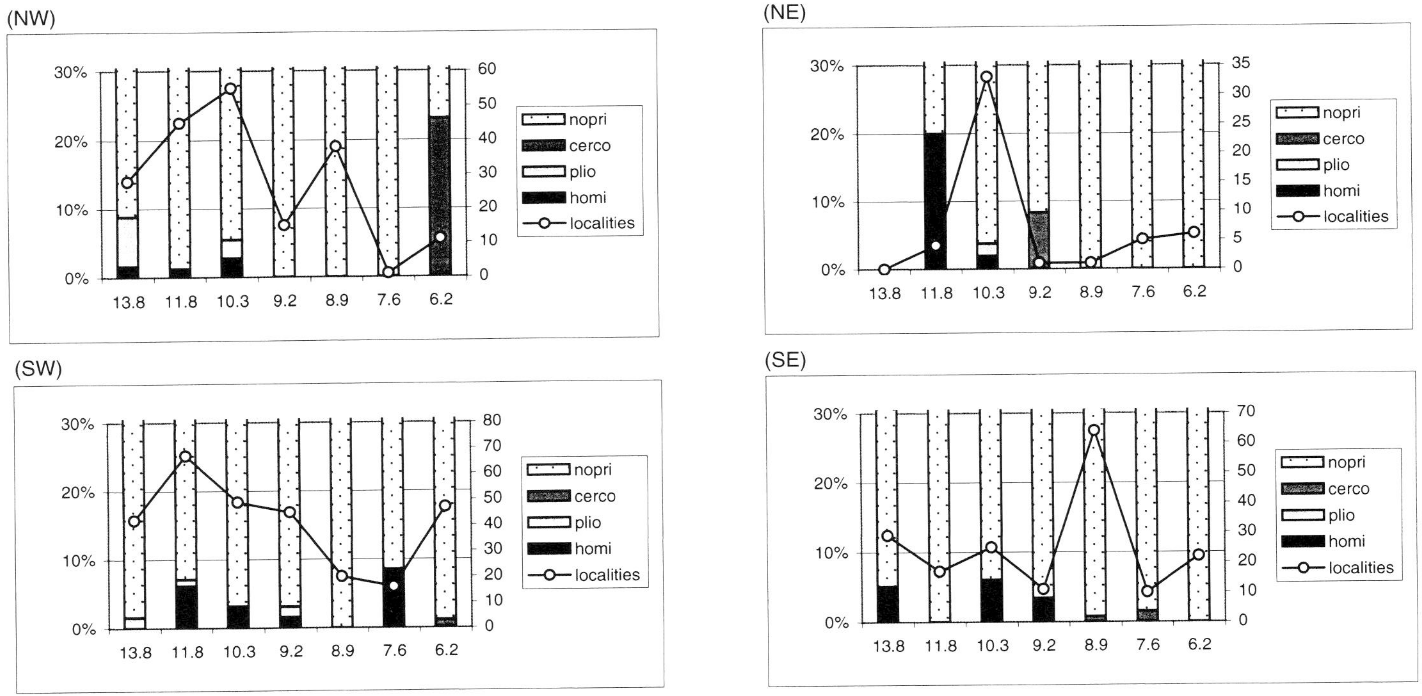

[Figure 2.9]
Pattern of occurrence of hominoid species in the four geographic quadrants. Abbreviations as in Figure 2.3.

This robusticity may accommodate substantial biases, but it means that with care relative statements at least may be made with some confidence. Fortunately, relative statements about fossil assemblages often go a long way towards explaining temporal and regional patterns of environmental and evolutionary change.

How does sploc analysis perform?

Before discussing the substance of the results we would like to comment briefly on the differences seen between conventional species counting and sploc analysis. Basically, the differences are minor and do not in any way affect the conclusions, except to the extent that sploc analysis gives higher statistical significance. We repeat the argument from the 'Methods' section that this improved significance is not spurious, as the test is of genuine, separate occurrences. It might even be argued that testing splocs rather than species is a more ecological way of looking at the data, but we do not wish to press the issue here, given the low level of sampling.

The main differences between analysis based on species and sploc counts can be seen in Figure 2.3. For trophic structure, the pattern appears somewhat subdued using splocs, but the differences are statistically stronger (Table 2.2). There seems to be less fluctuation between adjacent time units when using splocs, and the difference between pre- and post-crisis units is therefore clearer. For crown height the pattern appears stronger using splocs, and the difference between blocks appears more marked, with a clearer trend of gradually increasing hypsodonty during the interval studied. Since sampling error is more likely to cause random fluctuations than spuriously stable patterns it seems likely that sploc analysis is at least not more prone to sampling error than species-based analysis.

It is difficult to say whether sploc analysis is, in fact, capturing a significant portion of the abundance signal. If it were true that abundance data genuinely have such a small effect, palaeontological community structure analysis would obviously turn out to be less problematical than has been thought. This can be tested when real abundance data are compiled, but until then we must simply accept that the difference between sploc analysis and conventional species counting appears to be small, and that the difference at least in this study does not affect the conclusions in any way.

Are hominoid localities different?

The original impetus from this study was the question whether a distinct mammalian occurrence context can be defined for the hominoid primates of the Miocene of western Eurasia. This question has been firmly answered in the negative. The apparent difference seen when the hominoid and non-hominoid datasets are contrasted can easily be shown to be an artefact of temporal (and to some extent geographical) distribution of the localities. That is, hominoid localities show the characteristic signal of pre-crash West simply because most hominoid localities are from that subset. When the dataset is subdivided within pre-crash time, no significant differences between hominoid and non-hominoid localities remain (Table 2.3). Furthermore, the individual hominoid genera also show the trophic structure and crown height distribution of their temporal and geographic context. If hominoids occur in a special subset of circumstances, this is not picked up by general comparisons of trophic structure and ecomorphology.

In one sense, hominoid localities do appear to differ persistently from non-hominoid localities – they appear to include more rare taxa (analysis not included in this study). We assume that this is because they themselves are usually rare, and so tend to be found at well-sampled localities that also record other rare taxa. Another reason for this relationship could be that hominoid localities have attracted more than average attention, and therefore have produced more rare taxa subsequent to the discovery of hominoid material. This would certainly be the case for a locality like Paşalar (Turkey), where hominoids are among the most common mammals (Andrews, 1990).

In tracking down the causes of the apparent differences between hominoid and non-hominoid localities we have generated a more detailed picture of the trophic evolution of western Eurasian large land mammal faunas than has previously been available. The addition of dietary categories to the trophic analysis and breakdown of the temporal sequence into MN-unit equivalents and individual quadrants adds considerable detail to the coarse picture derived from earlier analyses, although declining sample size inevitably erodes some of the significance of minor details. In the following we summarise our main interpretations of the results described above, drawing especially on the analysis of individual quadrants. Finally, we turn to the question of hominoid disappearance.

Static geographic gradients

There is by now good evidence for the presence of a static geographical gradient that persists in the face of roughly synchronous and unidirectional change. A static east–west contrast was noticed by Fortelius *et al.* (1996b), who showed that the mean body size of omnivores and generally of 'large mammals of less than 30 kg body mass' was higher throughout the later Miocene in East than in West. The impression of a standing gradient is strongly apparent from comparison of the trophic and crown height patterns of the individual quadrants shown in Figures 2.4 and 2.5. It is obvious that what is detected at this scale must in reality be the sum of several interacting gradients, but as the phenomenon is only seen through its effect on mammal faunas, and since this effect is what we wish to study, this problem does not influence the conclusions.

The interpretation seems relatively straightforward as far as it goes, especially for molar crown height. The proportion of mesodont and hypsodont occurrences reflects the average wear regime of available plant foods, a somewhat elusive quality that however must reflect primarily a complex of interrelated factors such as degree of openness of the vegetation and seasonality of moisture (relative importance of a dry season). The smaller the proportion of soft plant parts available year round, the smaller is the expected proportion of brachydont forms in the fauna. A glance at Figure 2.6 shows that the proportion of brachydont forms declined through the interval in all quadrants (with a possible reversal in the last unit in the NW and SE quadrants), and that the proportion was lower throughout (except possibly for the last unit) in the two eastern quadrants. In other words, a static 'gradient of harshness' is seen to underlie the overall trend towards harsher conditions. The NW quadrant seems almost detached from the overall change, but is firmly part of the static difference.

The trophic structure is somewhat more difficult to interpret on its own, but the occurrence patterns of individual categories are revealing. True grazers (grass eaters) occur first in SE and hardly at all in NW, and except for SE they become common only in the last unit (6.2 Ma, MN 13). Invertebrate-eaters (which depend on year-round availability of invertebrates) decline in all quadrants but especially the eastern ones, and are more common throughout in the southern quadrants. Omnivores show a similar pattern, except that the decline is stronger and the main contrast is between East and West rather than between North and South. The proportion of plant-eaters to animal-eaters is also generally higher in the eastern and, to a lesser degree, southern quadrants, but the significance of this fact alone is unclear and we shall return to it later in the context of body mass.

The Vallesian Crisis

Overlain on the background of a static pattern is a dynamic one, including the main breakpoint known as the Vallesian Crisis (the MN 9/10 transition, between the time units labelled 10.3 and 9.2 Ma) (Agustí & Moyà-Solà, 1990; Moyà-Solà & Agustí, 1990). Apart from obvious artefacts of poor sampling in the two northern quadrants, the Vallesian Crisis constitutes the main step in the change of crown height proportions in all quadrants (Figure 2.6) and very clearly in the summary diagram (Figures 2.3C and D). This step does not reflect the entry of hipparionine horses, which happened at the preceding boundary, but no doubt does reflect their adaptive radiation into Eurasian communities. Whether the radiation was driven by the entry itself or by unrelated climatic change is not known, but bovids radiated almost as strongly at the same time, as seen in species richness plots (Fortelius *et al.*, 1996b, fig. 31.42), so at least the effect is not due only to hipparions. The effect is more marked for splocs than for species counts (Figure 2.3C versus 2.3D), and the sploc analysis more clearly suggests a subsequent gradual increase in mean crown height.

Trophic structure changes abruptly at the Vallesian Crisis only in the SE quadrant (Figure 2.5), a surprising result in view of the analysis of Fortelius *et al.* (1996b) which showed that the crisis was predominantly a western phenomenon. This is not the place to explore the reasons for this discrepancy in detail, but it should be said that the previous result was primarily based on raw species richness, and, despite some indications to the contrary, may well be in part a sampling artefact. The present result is comparably robust and indicates that the Vallesian Crisis had a major ecological impact in East as well as West – possibly even that the result was more pronounced in East. In the overall pattern, the main change in trophic structure is seen to occur at the Vallesian Crisis (Figure 2.3A and B), with the sploc analysis (Figure 2.3B) showing a remarkably simple pattern suggesting two intervals of stable proportions separated by the crisis.

Increasing body mass

The nearly universal increase of mean body mass observed in this material is in concordance with Cope's Rule generally and specifically with numerous studies showing this to be a general trend among large land mammals of the Neogene and, indeed, the entire Cenozoic (see Alroy, 1998). A family-level breakdown (Table 2.6, Figure 2.7), however, reveals interesting exceptions to the general trend. In particular, the most diverse and abundant ungulate families, bovids and equids, fail to show an increase in size, and equids even show a decrease. This result takes on particular interest when seen in relation to the fact that all main predator families show a strong increase in mean size, while at the same time the relative diversity and occurrence of animal-eaters goes down (Figures 2.3A and B). These results cannot be conclusively interpreted without calculations of biomass, which in turn demand better estimates of abundance than we now have available (as well as estimates of species-specific generation lengths, which would not be difficult to generate in principle; see Damuth, 1982). As far as they go, however, they do suggest the reasonable relationship that the decrease in predator diversity (and, hopefully, abundance) was accompanied by a corresponding increase in size.

Hominoid disappearance

It is strikingly obvious that the pattern of hominoid disappearance from western Eurasia does not match the main pattern of change, the progressively more seasonal conditions advancing from the continental interior towards the western ocean. Instead, the disappearance seems to follow a north–south gradient, with hominoids lingering on the Mediterranean long after their disappearance from the northern quadrants. It appears that hominoids were able at least initially to respond adaptively to the increasing seasonality expressed primarily in the east–west gradient, as suggested by their faunal occurrence context and independently by dental microwear analysis of their diets (King, Chapter 5). Without trespassing into the realm of hominoid palaeoecology we would like to suggest that this disappearance pattern argues for temperature being the main limiting factor. Theoretically this might be temperature as such (a cooling trend reaching some critical threshold value) or a combined seasonality-temperature effect (the seasonality effect being amplified in some critical way in a colder regime). The fact that the seasonality-related effects on the mammal communities appears to have been greater in the southern than northern quadrants could indicate that the primary reason for the disappearance of hominoids from Europe north of the Mediterranean was the declining temperature itself.

Conclusions

The geographic and temporal patterns revealed in this study can be interpreted as reflecting partly a geographically based, static climate gradient of increasing seasonality from west–east and north–south, and partly the global Neogene trend of climatic change towards increasing seasonality and lower temperatures, progressively driving conditions over the entire area towards 'harsher' conditions. Within this frame, hominoid primates occur most commonly in association with a trophic structure showing relatively high numbers of animal-eaters and omnivores and relatively low numbers of hypsodont plant eaters – i.e. at the low end of the inferred seasonality spectrum. Individual hominoid genera show the full spectrum of trophic structure and crown height distribution patterns, however, in each case closely matching those of the area and interval of occurrence. Pliopithecids have an even more strongly 'non-harsh' occurrence profile, while cercopithecids show the opposite profile, as expected from their later occurrence. The fact that hominoids appear to have survived longer in the south than in the north of Europe suggests that a north–south temperature gradient may ultimately have been more decisive than the east–west seasonality gradient in restricting hominoid distributions. There is no evidence at this scale of analysis that primate localities differ from non-primate localities of the same interval and geographic area.

Acknowledgements

We thank Louis de Bonis for kindly insisting that we should produce a synthesis of the NOW data pertaining to hominoid occurrence. MF thanks all organisers of the Nikiti workshop for an excellent meeting. We would like to express particular gratitude to the following individuals for recent efforts that have improved the dataset used here: Jorge Agustí, Peter Andrews, Raymond L. Bernor, Louis de Bonis, Jens Lorenz Franzen, Alan Gentry, Elmar P. J. Heizmann, Laszlo Kordos, George Koufos, Diana Pushkina, Lorenzo Rook, Gertrud Rössner, Alexey Tesakov, Suvi Viranta, Inessa Vislobokova and Lars Werdelin. Special thanks to John Damuth for discussion of community structure and the implications of sploc analysis. We also thank the HOMINET coordination committee as a body, the NOW advisory board, and all contributors to the NOW database. We would like to express our gratitude to the European Science Foundation for supporting the NOW database through the Network on Hominoid Evolution and Environmental Change in the Neogene of Europe, and to the Academy of Finland for support through grant 34080.

References

Agustí, J. & Moyà-Solà, S. (1990). Mammal extinctions in the Vallesian (Upper Miocene). In Kauffman, E. G. & Walliser, O. H. (eds.), *Extinctions, Events in Earth History. Lecture Notes in Earth Sciences*, **30**, pp. 425–32. Berlin: Springer-Verlag.

Alroy, J. (1996). Four methods of correcting diversity curves for sampling effects: which is best? *Geological Society of America Abstracts with Programs* **28**: A107.

Alroy, J. (1998). Cope's Rule. *Science* **282**: 50–1.

Andrews, P. (Ed.) (1990). The Miocene hominoid site at Pasalar, Turkey. (Special issue.) *J. Hum. Evol.* **19**: 335–588.

Andrews, P. (1996). Palaeoecology and hominoid palaeoenvironments. *Biol. Rev. (Cambridge)* **71**: 257–300.

Andrews, P., Lord, J. & Evans, E. M. N. (1979). Patterns of ecological diversity in fossil and recent mammal faunas. *Biol. J. Linn. Soc.* **11**: 177–205.

Andrews, P., Harrison, T., Delson, E., Bernor, R. L. & Martin, L. (1996). Distribution and biochronology of European and Southwest Asian Miocene catarrhines. In Bernor, R. L., Fahlbusch, V. & Mittmann, H.-V. (eds.) *The Evolution of Western Eurasian Neogene Mammal Faunas*, pp. 168–207. New York: Columbia University Press.

Bonis, L. de., Bouvrain, G., Geraads, D. & Koufos, G. (1992a). Diversity and paleoecology of Greek late Miocene mammalian faunas. *Palaeogeogr. Palaeoclimatol. Palaeoecol.* **91**: 99–121.

Bonis, L. de., Bouvrain, G., Geraads, D. & Koufos, G. (1992b). Multivariate study of late Cenozoic mammalian faunal compositions and paleoecology. *Paleontol. Evol.* **24/25**: 93–101.

Bonis, L. de, Bouvrain, G. & Koufos, G. (1999). Palaeoenvironment of the late Miocene primate localities in Macedonia, Greece. In Agustí, J., Rook, L. & Andrews, P. (eds.), *Evolution of Neogene Terrestrial Ecosystems in Europe*, pp. 413–435, Cambridge: Cambridge University Press.

Dam, J. van (1997). The small mammals from the Upper Miocene of the Teruel-Alfambra region (Spain). Paleobiology and paleoclimatic reconstructions. *Geol. Ultraiect.* **156**: 1–204.

Damuth, J. (1982). Analysis of the preservation of community structure in assemblages of fossil mammals. *Paleobiology* **8**: 434–46.

Damuth, J. (1992). Taxon-free characterization of animal communities. In Behrensmeyer, A. K., Damuth, J. D., DiMichele, W. A., Potts, R., Sues, H. & Wing, S. L. (eds.), *Terrestrial Ecosystems Through Time*, pp. 183–203. Chicago: University of Chicago Press.

Damuth, J. (1997). *ETE Database Manual.* Washington, DC: Evolution of Terrestrial Ecosystems Consortium.

Damuth, J. & MacFadden, B. J. (1990). *Body Size in Mammalian Paleobiology: Estimation and Biological Implications.* Cambridge: Cambridge University Press.

Fortelius, M., Andrews, P., Bernor, R. L., Viranta, S. & Werdelin, L. (1996a). Preliminary analysis of taxonomic diversity, turnover and provinciality in a subsample of large land mammals from the later Miocene of western Eurasia. In Nadachowski, A. & Werdelin, L. (eds.), *Neogene and Quaternary Mammals of the Palaearctic. Acta Zool. Cracov.* **39**: 167–78.

Fortelius, M., & Solounias, N. (2000). Functional characterization of ungulate molars using the abrasion–attrition wear gradient: a new method for reconstructing paleodiets. *Am. Mus. Novit.* **3301**: 1–36.

Fortelius, M., Werdelin, L., Andrews, P., Bernor, R. L., Gentry, A., Humphrey, L., Mittmann, H.-W. & Viranta, S. (1996b). Provinciality, diversity, turnover and paleoecology in land mammal faunas of the later Miocene of Western Eurasia. In Bernor, R. L., Fahlbusch, V. & Mittmann, H.-V. (eds.) *The Evolution of Western Eurasian Neogene Mammal Faunas*, pp. 414–48. New York: Columbia University Press.

Moyà-Solà, S. & Agustí, J. (1990). Bioevents and mammal successions in the Spanish Miocene. In Lindsay, E. H., Fahlbusch, V. & Mein, P. (eds.), *European Neogene Mammal Chronology*, pp. 357–74. New York: Plenum.

Rosenzweig, M. L. (1995). *Species Diversity in Space and Time*. Cambridge: Cambridge University Press.

Sepkoski, J. J., Jr. (1993). Ten years in the library: new data confirm paleontological patterns. *Paleobiology*, **19**: 43–51.

Sepkoski, J. J., Jr. (1997). Presidential address. Biodiversity: past, present, and future. *J. Paleont.* **71**: 533–9.

Steininger, F. F., Berggren, W. A., Kent, D. V., Bernor, R. L., Sen, S. & Agustí, J. (1996). Circum-Mediterranean Neogene (Miocene-Pliocene) marine-continental chronologic correlations of European mammal units. In Bernor, R. L., Fahlbusch, V. & Mittmann, H.-W. (eds.), *The Evolution of Western Eurasian Neogene Mammal Faunas*, pp. 7–46. New York: Columbia University Press.

PART II

Methods and phylogeny

3 Computer-assisted morphometry of hominoid fossils: the role of morphometric maps

Christoph P. E. Zollikofer and Marcia S. Ponce de León

Introduction

In the past decades, the rapid evolution of computer-based technologies has opened up a wide range of novel applications in the biosciences, notably in areas where large volumes of data must be acquired and analysed. Within the scope of potential computer-assisted applications, palaeoanthropology is a particularly challenging field – due to the extreme scarcity of fossil remains and the fragmentary preservation of most specimens, the data base is both limited and incomplete. It is therefore of vital interest to expand the available evidence, not only by adding recently discovered specimens to the present sample, but also by devising novel methods that permit extraction of a maximum of information from the sparse material.

Computer-based procedures have already proven especially useful in tapping new sources of information within the sample of fossil hominoids. Special-purpose computer tools for the acquisition, processing and visualisation of 3-dimensional data of fossil specimens have been integrated into a methodological framework given the label computer-assisted palaeoanthropology (CAP, Zollikofer *et al.*, 1998). CAP is based on a combination of three technologies: computer tomography, computer graphics and stereolithograpy (Figure 3.1). Computer tomography (CT) has revolutionised non-invasive data acquisition by providing X-ray based cross-sectional images of solid objects. In medical diagnostics, CT has become a standard tool and, likewise, in palaeoanthropology, CT scanning is extensively used for 'fossil diagnostics', notably to reveal internal anatomical features and regions still covered by matrix (Conroy & Vannier, 1984; Zonneveld & Wind, 1985; Conroy, 1991). Going one step beyond imaging, CT scanners also serve as powerful tools for the acquisition of 3-dimensional data – by merging a series of consecutive cross-sectional images, a graphical representation of an entire fossil specimen can be reconstructed. The subsequent application of special-purpose computer graphics procedures to these virtual objects permits non-invasive preparation, reconstruction and completion of fragmentary fossil remains on the computer screen (Zollikofer *et al.*, 1995). Furthermore, using methods from computational geometry, it is possible to quantify and visualise morphometric features that were previously accessible only to qualitative description. New morphometric evidence derived

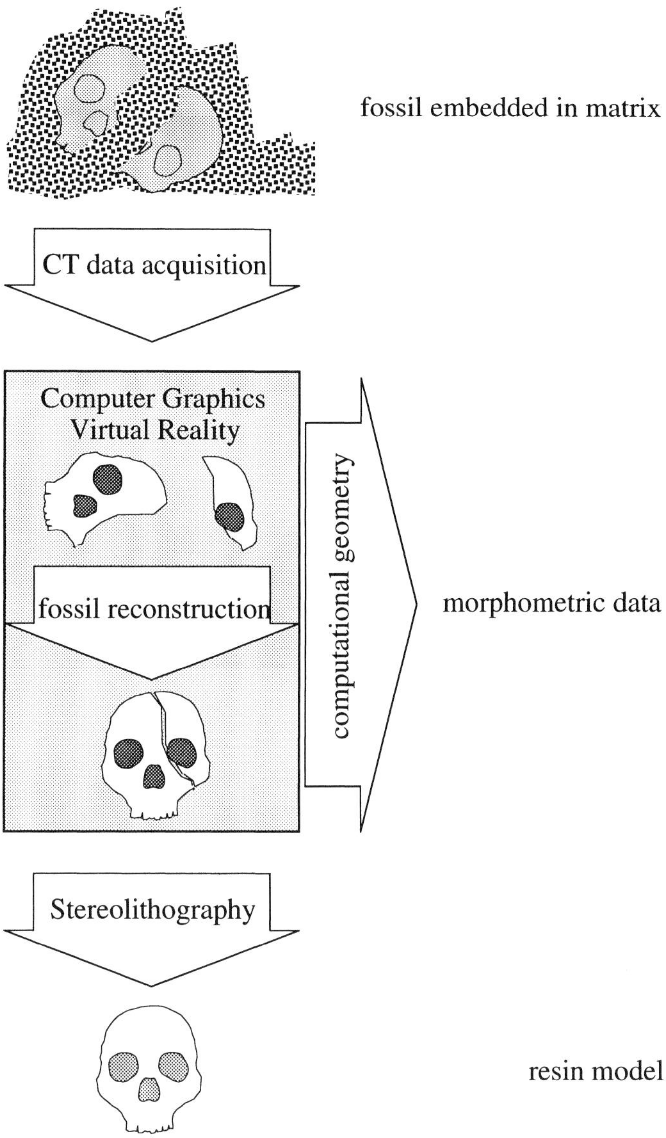

[Figure 3.1]
Computer-assisted palaeoanthropology (CAP) transforms fossil fragments into graphical object representations to permit non-invasive preparation and reconstruction. Using methods of computational geometry and computer graphics, various types of morphometric data can be derived from these virtual objects. Fossils reconstructed on the computer screen can be brought back to physical reality by means of stereolithography. CT: computer tomography.

from fossil specimens can then be compared with equivalent data obtained from extant hominoids. Following this approach, it is possible to expand considerably the amount of information that can be derived, even from fragmentary fossils.

In this study, we focus on new concepts and methods of morphometric visualisation that are designed to facilitate the exploration and interpretation of fossil morphology. Furthermore, we assess the potential of these methods for the future investigation of European hominoid fossils.

Fossil morphometry

The prevalent technique that is still used to acquire morphometric data from fossils is measurement with simple instruments such as rulers and calipers. Obviously, these tools confine sampling to a small subset of possible data types – generally distances and angles. By providing an array of computer-assisted measurement tools, it is possible to extend morphometry into the second and third dimension (i.e. area and volume data) and into the analysis of complex geometric properties. Among the wide variety of morphometric data generated by computer-assisted analyses, three principal types can be discerned.

(1) *3-dimensional object representations.* By applying image segmentation algorithms to CT data volumes, it is possible to extract object boundaries and to reconstruct 3-dimensional surfaces that provide a geometric description of the original object. These data can be rendered as graphical object representations for visualisation and manipulation on a computer screen, and for further morphometric analysis.

(2) *Landmark locations.* Landmarks represent meeting points of three or more distinct anatomical units (e.g. individual bones) or, alternatively, extreme points of morphological structures (e.g. bony spines, bulges, etc.). Landmark data are particularly relevant for comparative analyses, as they represent anatomical loci that are presumed to be homologous between the specimens under investigation. In classic morphometry, inter-landmark distances form the basis for multivariate analyses. More recently, a set of morphometric methods – known as geometric morphometrics – has been proposed based on landmark locations rather than interlandmark distances. Sampling 3-dimensional landmark coordinates on graphical object representations is straightforward, because these objects basically represent sets of interconnected 3-dimensional points.

(3) *Global and local morphometric properties.* A large amount of potentially significant morphological information remains unexplored by landmark-based methods for two main reasons. In many areas of the vertebrate skeleton, determination of homologous points is difficult or even impossible because of the lack of identifiable landmarks. Moreover, many relevant morphological traits are relatively easy to describe qualitatively but difficult to quantify. Features such as cranial capacity or cortical bone thickness cannot be identified by specifying a set of landmarks. Contrasting with the positional information provided by landmark locations, such data can be characterised as being relational. Global or local morphometric properties are expressed by geometric relations between points on the object. For example, endocranial capacity (a global property) is the volume delimited by the points defining the endocranial surface, and bone thickness (a local property) is the distance between points on opposed surfaces of the bone.

With these definitions in mind, it is possible to explore a wide variety of new morphometric parameters. We will concentrate here on issues relating to the visualisation of local morphometric properties.

The morphometric map

The procedures involved in evaluation and visualisation of local morphometric properties can be generalised by introducing the concept of a *morphometric map* (MM). An MM is defined as a 2-dimensional, planar representation of some local morphometric property of a 3-dimensional object (Figure 3.2). Constructing an MM involves two consecutive steps, each of which consists of a mapping procedure. Mapping in a mathematical sense denotes a transformation from space **A** into space **B** effected by a function F.

$$F: \mathbf{A} \rightarrow \mathbf{B}$$

In the first step, morphometric data are transformed into a visually perceivable form, for example into the values of a grey scale (or, on a computer screen, into a colour space). In the second step, the 3-dimensional geometry of the object is mapped onto a planar surface. The obvious aim of the mapping procedures is to optimise the visual representation of complex morphometric features.

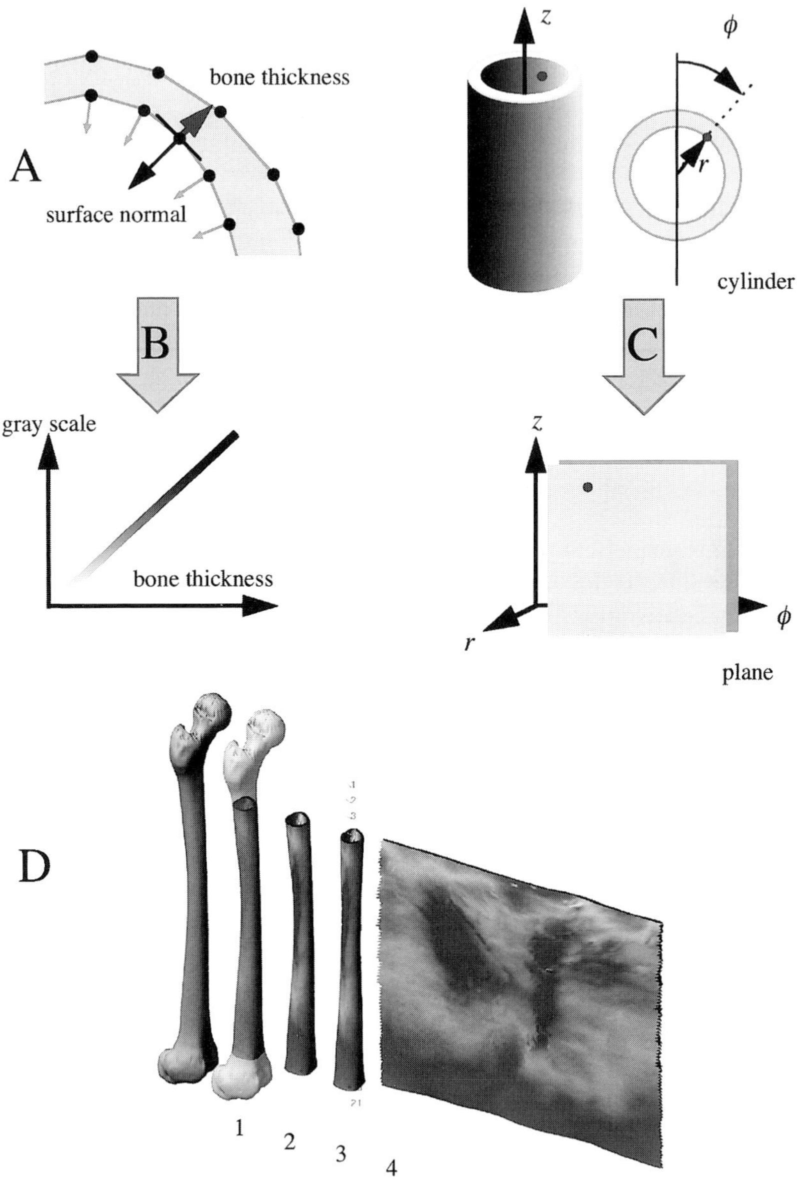

[Figure 3.2]
Construction of the morphometric map of bone thickness. Thickness is defined as the minimum normal distance from a point on the internal surface to the external surface of the bone (A). To create the morphometric map, two consecutive mapping procedures must be performed (B, C). Cortical bone thickness values are mapped onto a grey scale (B), and the diaphyseal surfaces are transformed from a cylindrical shape into planar surfaces (C). The morphometric map of a human right femur (D) is obtained by extraction of the diaphysis (1), evaluation of thickness (2), re-alignment of diaphyseal bending (3) and unrolling the shaft (4).

Bone thickness and the morphometric map

The thickness of cortical bone, notably of the cranial vault and of the diaphyses of limb bones, is the subject of studies ranging from cladistic analysis to investigation of growth processes and biomechanics (Kennedy, 1991; Runestad *et al.*, 1993; Ruff *et al.*, 1994; Lieberman, 1996; Manzi *et al.*, 1996; Trinkaus *et al.*, 1998). Using conventional measurement tools, biplanar radiography or CT imaging, quantification of thickness is restricted to cross sections through specific regions. However, thickness tends to fluctuate over the entire bone, and it is reasonable to assume that patterns of fluctuation represent data rather than noise. The acquisition and representation of comprehensive thickness data are therefore of considerable interest.

Computer tools have already proved useful in the analysis and visualisation of thickness variation. In a comparative study of cranial bone robusticity in juvenile Neanderthals and modern humans, computer-generated thickness maps were used to reveal complex patterns of thickness fluctuation and to demonstrate quantitative differences in overall parietal bone thickness between these two taxa (Zollikofer *et al.*, 1995).

Mapping femoral shaft thickness in hominoid primates

With the aim of establishing links between bone structure and locomotor habits, diaphyseal robusticity and biomechanical properties of long bones have been analysed in a wide variety of fossil and extant non-hominoid and hominoid primates (Demes & Jungers, 1993; Ruff, 1993; Trinkaus *et al.*, 1998). The primary morphometric signal – cortical bone thickness – can be directly accessed and visualised by applying the concepts of morphometric mapping. In this preliminary study, we adopt a purely descriptive approach that is devoid of functional hypotheses and aims at visualisation of a maximum of morphometric data concerning cortical diaphyseal thickness (Figures 3.2, 3.3).

We produced MMs of the femoral shafts of modern humans, a Neanderthal (Spy 2), chimpanzee and orang-utans (Figure 3.3). Volume scans of the femora were obtained on a Picker PC5000 helical CT device. Endosteal and endoperiosteal surfaces were extracted from serial cross sections reconstructed at intervals of 2 mm. The 3-dimensional object representations consisted of approximately 100,000 3-dimensional data points for each femoral shaft. Diaphyseal bone thickness was defined as the shortest distance between points on opposite surfaces of the bone and evaluated for

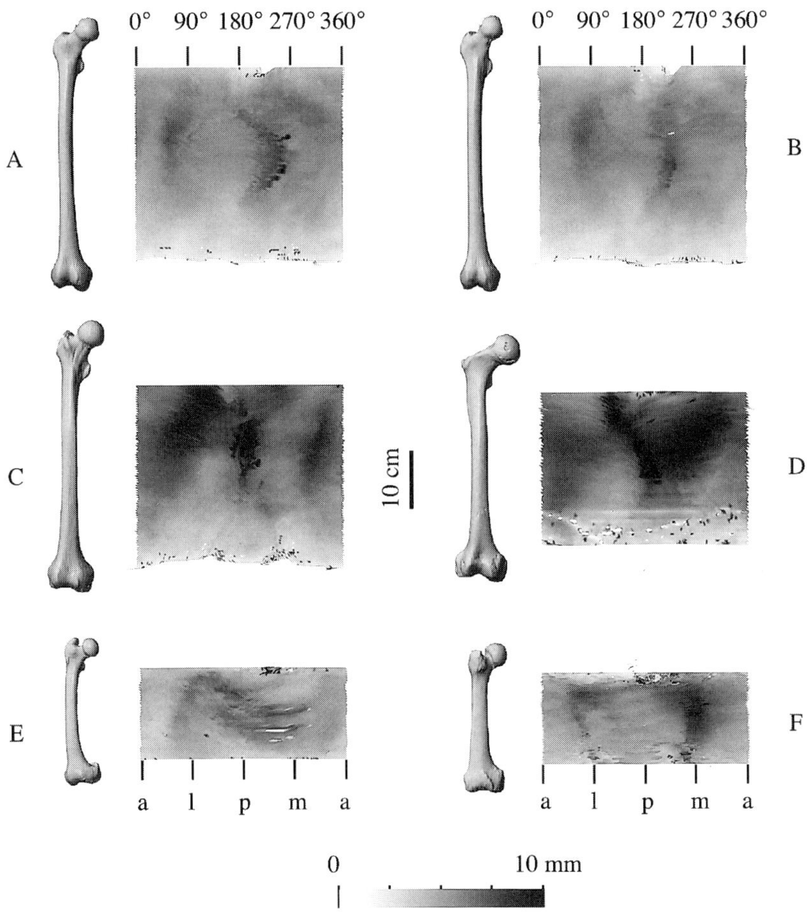

[Figure 3.3]
Cortical thickness maps of the femoral diaphysis of *Homo sapiens* (A, B: right and mirror-imaged left femora of a female; C: right femur of a male), *Homo neanderthalensis* (Spy 2, D), male *Pan troglodytes* (E) and male *Pongo pygmaeus* (F). Thickness maps represent the endosteal surface spread out from 0° to 360° (anterior → lateral → posterior → median → anterior). The grey scale represents thickness values from 0 to 10 mm, the vertical bar measures 10 cm.

each point on the endosteal surface by calculating the distance to the endoperiosteal surface along a line perpendicular to the endosteal surface (Figure 3.3A).

The first step in constructing the MM consisted of mapping the thickness data onto a grey scale (Figure 3.3B). To obtain a planar representation of the approximately cylindrical diaphysis, the basic idea was to cut open and unroll a virtual representation of the femur (Figure 3.3C). To perform unrolling in virtual reality, a cylindrical system of coordinates was superimposed onto the femoral shaft, consisting of a central axis (z) parallel to the shaft, an angle (ϕ) denoting the position around z, and an axis (r) measuring radial distance from z. Prior to spreading out strongly bent femoral shafts, the curved central axis was re-aligned using 3-dimensional thin plate spline deformation procedures (Bookstein, 1991).

Our preliminary comparative analysis of the thickness maps revealed the following characteristics.

- The maps show considerable fluctuation of femoral cortical shaft thickness, both along and around the shaft. While certain structures of the pattern can be associated with major areas of muscular insertion of the flexor and extensor complexes, various structures await further examination and interpretation.
- Comparison of the maps among human subjects reveals intra-individual as well as inter-individual variability. While the left and right femoral shafts of one and the same individual show relatively concordant patterns, inter-individual variability is conspicuous, with respect to both overall robusticity and pattern fluctuation (male versus female: Figure 3.3A, B vs. C; corresponding data for within-sex variation not shown). The causes of variability can be of an internal and/or an external nature. The areas of muscular insertion along the femoral shaft seem to be relatively variable in humans and may therefore give rise to deviating thickness patterns. Further, proximate factors such as body size and locomotor habits are known to have a major effect on the distribution of cortical bone (Trinkaus *et al.*, 1994).
- Interspecific differences are salient. With respect to their interpretation, we have to consider two points – genetics and environment. Many studies suggest that bone thickness shows a strong environmental signal, mainly because bone remodelling is a continuous process mediated by the habitual conditions of loading to which the bone is exposed (Biewener, 1990; Ruff *et al.*, 1994; Mundy, 1995). We can therefore expect that interspecific differences of the thickness patterns largely represent different loading regimes that, in turn, reflect different locomotor habits. Further analyses are necessary to specify the nature of correlation between the thickness maps and locomotor patterns, considering both the hereditary and environmental constraints of the locomotor apparatus and of locomotor behaviour.
- With respect to the analysis of fossil limb bones, thickness patterns represent a new source of morphometric data that can be used to infer information about locomotor habits. This point is of special significance, because current arguments on locomotion of fossil primates are mainly based on

the analysis of a restricted number of cross-sections and on the application of biomechanical models. Morphometric maps of diaphyseal bone thickness and of various other local morphometric parameters directly reflect local properties of the bone and can therefore be expected to yield new insights into the links between structure and function of the hominoid limb.

Acknowledgement

This study was supported by Swiss NSF grant #31-42419.94.

References

Biewener, A. A. (1990). Biomechanics of mammalian terrestrial locomotion. *Science* **250**: 1097–103.

Bookstein, F. L. (1991). *Morphometric Tools for Landmark Data.* Cambridge: Cambridge University Press.

Conroy, G. C. (1991). Enamel thickness in South African australopithecines: noninvasive evaluation by computed tomography. *Palaeontol. Afr.* **28**: 53–9.

Conroy, G. C. & Vannier, M. W. (1984). Noninvasive three dimensional computer imaging of matrix filled fossil skulls by high resolution computed tomography. *Science* **226**: 457–8.

Demes, B. & Jungers, W. L. (1993). Long bone cross-sectional dimensions, locomotor adaptations and body size in prosimian primates. *J. Hum. Evol.* **25**: 57–74.

Kennedy, G. E. (1991). On the autapomorphic traits of *Homo erectus. J. Hum. Evol.* **20**: 375–412.

Lieberman, D. E. (1996). How and why humans grow thin skulls: experimental evidence for systemic cortical robusticity. *Am. J. Phys. Anthrop.* **101**: 217–36.

Manzi, G., Vienna, A. & Hauser, G. (1996). Developmental stress and cranial hypostosis by epigenetic trait occurrence and distribution: an exploratory study on the Italian Neandertals. *J. Hum. Evol.* **30**: 511–27.

Mundy, G. R. (1995). *Bone Remodelling and Its Disorders.* London: Dunitz.

Ruff, C. B., Trinkaus, E., Walker, A. & Larsen, C. S. (1993). Postcranial robusticity in *Homo*. I. Temporal trends and mechanical interpretation. *Am. J. Phys. Anthrop.* **91**: 21–53.

Ruff, C. B., Walker, A. & Trinkaus, E. (1994). Postcranial robusticity in *Homo*. III. Ontogeny. *Am. J. Phys. Anthrop.* **93**: 35–54.

Runestad, J. A., Ruff, C. B., Nieh, J. C., Thorington, R. W. Jr. & Teaford, M. F. (1993). Radiographic estimation of long bone cross-sectional geometric properties. *Am. J. Phys. Anthrop.* **90**: 207–13.

Trinkaus, E., Churchill, S. E. & Ruff, C. B. (1994). Postcranial robusticity in Homo. II. Humeral bilateral asymmetry and bone plasticity. *Am. J. Phys. Anthrop.* **93**: 1–34.

Trinkaus, E., Ruff, C. B., Churchill, S. E. & Vandermeersch, B. (1998). Locomotion and body proportions of the Saint-Césaire 1 Chatelperronian Neandertal. *Proc. Natl. Acad. Sci.* **95**: 5836–40.

Zollikofer, C. P. E., Ponce de León, M. S., Martin, R. D. & Stucki, P. (1995). Neanderthal computer skulls. *Nature* **375**: 283–5.

Zollikofer, C. P. E., Ponce de León, M. S. & Martin, R. D. (1998). Computer-assisted paleoanthropology. *Evol. Anthrop.* **6**: 41–54.

Zonneveld, F. W. & Wind, J. (1985). High-resolution computed tomography of fossil hominid skulls: a new method and some results. In Tobias, P. V. (ed.), *Hominid Evolution: Past, Present and Future*, pp. 427–36. New York: Alan R. Liss, Inc.

4 Comparative analysis of the iliac trabecular architecture in extant and fossil primates by means of digital image processing techniques: implications for the reconstruction of fossil locomotor behaviours

Roberto Macchiarelli, Lorenzo Rook and Luca Bondioli

Introduction

Among the extant mammals, primates display a great diversity of postural and locomotor behaviours. In an evolutionary perspective, this variation relates to their adaptive ability to exploit a full range of arboreal and terrestrial substrates (Jouffroy *et al.*, 1990; Hunt *et al.*, 1996). Principal primate locomotor modes include arboreal and terrestrial quadrupedalism, vertical clinging and leaping, saltation, suspensory climbing, brachiation, knuckle and fist-walking quadrupedalism, and bipedalism (Gebo, 1996). These varied locomotor patterns result in specific types of limb use which rely on a number of anatomical and biomechanical musculoskeletal specialisations (Conroy, 1990; Fleagle, 1999).

Following the principle of uniformitarianism, the reconstruction of body postures and modes of locomotion in fossil taxa is primarily based on the comparative analysis of external bone shape, size and proportions with the respective skeletal elements of living forms. Despite its obvious – but inescapable – intrinsic limits (Preuschoft, 1975), correlations between bony morphology and positional/locomotor behaviours 'are constantly being tested and refined by experimental studies that permit a clearer understanding of the biomechanical and physiological mechanisms of primate evolution' (Fleagle, 1999: 298).

In the last few years, there has been a growing interest for potential functional analysis and interpretation of fossil trabecular bone patterns, a previously underexploited tool for various aspects of vertebrate palaeobiology. As 'living bones modify their structure at gross, tissue, and molecular levels in response to the force patterns acting on them ... the preserved structure of fossil bones may contain direct information on the forces that may have been experienced during life' (Thomason, 1995: 249). None the less, despite occasional remarks on the distribution of the natural split-lines and the radiographic appearance of fossil hominid hip bones (see Mednick, 1955; Day, 1971; Zihlman & Hunter, 1972), until now trabecular anatomy in pre-modern hominoids remains unstudied.

Since the late 1980s the research unit of the Section of Anthropology at the National Prehistoric and Ethnographic 'L. Pigorini' Museum of Rome has focused on patterning in extant and fossil trabecular bone, especially of the hip region. Amplifying some earlier work developed by, among the others, Correnti (1952a,b, 1955, 1957), Oxnard (1970, 1997) and Oxnard & Yang (1981), we have assessed the radiographic appearance of the pelvic girdle in hominoids and cercopithecoids and extant humans. In collaboration with a number of Italian and foreign scientific institutions, we have thus carried out a long-term project designed to:

- investigate the nature and extent of the relationships between trabecular architecture – especially of the ilium and proximal femur – and habitual postural/locomotor behaviours in extant human and non-human primates, according to the principle that different locomotor-related loads/stresses are responsible for different structural organisation and site-specific degrees of anisotropy of the cancellous network (see below);
- extensively document the appearance of postural/locomotion-related trabecular features/architecture in primate and non-primate mammals (Macchiarelli *et al.*, 1999a,b; for a structural description of the femoral neck in extant primates, see Rafferty, 1998; for the internal architecture of the human sacrum, see Peretz *et al.*, 1998);
- assess ontogeny and age-related intraspecific ranges of morpho-architectural variation of the cancellous network in *Homo* and non-human primate taxa (for a similar research perspective, see also Fajardo *et al.*, 1999);
- ascertain the existence of *gradiognomonic* patterns, that is of trabecular features/architecture uniquely associated with distinct locomotor modes;
- develop original investigative tools (i.e. advanced digital image processing techniques) facilitating the extraction of reliable information from fossil cancellous bone and able to enhance the strength of quantative and qualitative analyses of conventional X-ray films (Macchiarelli *et al.*, 1990, 1996a, 1998a, 1999c; Macchiarelli & Bondioli, 1994; Bondioli, 1999);
- comparatively investigate hip bone and femoral trabecular architecture in fossil primate taxa (including *Homo*) to reconstruct fossil locomotor habits (Macchiarelli *et al.*, 1995, 1996b, 1997a,b, 1998b, 1999a,c; Galichon *et al.*, 1996; Rook *et al.*, 1997, 1999);
- collect the entire body of data and electronically enhanced images of iliac trabecular patterns on a CD-ROM (Macchiarelli *et al.*, 1999b) for distribution within the monographic series *Digital Archives of Human Paleobiology* (see Geusa *et al.*, 1999; Rossi *et al.*, 1999).

Trabecular bone biomechanics and structural patterning

Cancellous bone architecture and mechanics are intimately related and the adaptation of cancellous bone to mechanical forces is well recognised (Fyhrie & Carter, 1986; Luo & An, 1998). The trabecular architecture of cancellous bone is largely determined by its mechanical environment, and the mechanical properties of cancellous bone are determined by the trabecular architecture and material properties (Odgaard *et al.*, 1997).

Meyer (1867) first noted that trajectories of bony trabeculae tend to align with the principal stress directions, and a few years later this fundamental observation was formally defined by Wolff (1870) in his 'law of bone transformation' (or 'trajectorial theory') which provides the core of much of the research on bone growth, remodelling and maintenance (Lanyon, 1974; Hayes & Snyder, 1981; Einhorn, 1996).

Bone is a 'self-optimizing' material displaying the capacity for non-destructive energy dissipation (visco-elastic behaviour) and the influence of strain rate on strength and stiffness (creep behaviour) (Linde *et al.*, 1991). Its final shape and integrity has a genetic and mechanical component but its ultimate configuration is heavily influenced by external forces and activities. Actually, bone grows in response to the daily loads applied to it and its site-specific density is largely a function of the magnitude/direction of such loads (Currey, 1984). In turn, these forces are dependent on the type of activity and the location and action of the surrounding muscular and ligamentous tissues. On the whole, more than 80% of the variance in bone biomechanical behaviour can be explained by measures of density and orientation (Goldstein *et al.*, 1993).

Along with weight transfer, peak principal stresses (mostly deriving from contraction forces exerted by muscles) determine the specific orientation of cancellous architecture (i.e. cancellous structure adapts to functional loading). Since multiaxial stresses are complicated, a complex network of cancellous bone is produced (Turner *et al.*, 1990; Turner, 1992). Anisotropy in cancellous bone is thus a consequence of the arrangement of the trabeculae themselves (so peak strains within bone are kept rather isotropic), where principal stresses determine the fabric (architecture) of cancellous bone.

The directions of the principal strains and the trabecular orientation correspond strongly, resulting in architecturally distinct patterning. Peak principal strain directions align with trabecular trajectories, and cancellous bone is more dense in regions of high shear stress. In addition, at low relative densities, the cancellous structure is composed of rod-like structures connecting to form open cells (a kind of cubic array of beam-like elements), but as the relative density increases, more material accumulates in the cell wall and the structure transforms into a more closed network of plates (a kind of closed cell cubic array of plate-like elements) (Gibson, 1985). In detail, low density, open cell, rod-like arrangements are usually found in regions where stresses are $< 0.13\,\text{g cm}^{-3}$. With stresses $> 0.20\ \text{g cm}^{-3}$, higher density, closed cell, plate-like structures occur. Finally, in sites experiencing intermediate levels of stress, trabecular structure is a combination of rod- and plate-like elements. Trabecular density is thus greatest in regions where the difference in the values of principal stresses are largest and the high density direction aligns according to the maximum strain, which is usually compressive. On the other hand, the low density direction is aligned according to the minimum stress, which could be either a low compressive or a tensile stress (Linde *et al.*, 1991).

In the vertebrate skeletal system, the hip is a key bone. In pronogrades it transmits propulsive force from the hind limbs to the trunk and part of the trunk weight to the hind limbs (*c*. 50–60% of the weight carried by obligatory bipeds). In bipeds it shifts the entire body weight from the lower lumbar vertebrae, the sacrum and the sacro-iliac joints through the ilium to the acetabulum and onto the head and neck of the femur (Reynolds, 1985). From a biomechanical perspective, the hip bone is a 'sandwich-like' construction (Currey, 1984; Dalstra & Huiskes, 1990), providing both high strength and low weight. In such kinds of 3-dimensional structures the bulk of the load is carried by a thin shell of high-modulus material (the cortical bone), while the low-weight core material (the trabecular bone) acts as a spacer in separating the outer sheets of compact bone and, more importantly, in resisting shear stresses. Lying as core material in a sandwich construction where the cortical bone transforms abruptly to cancellous bone, pelvic trabecular bone withstands predominantly shear-loading modes, against which plate-like structures form the best resistance. The iliac trabecular bone also distinguishes itself from the strictly weight-bearing trabecular bone such as that of the tibia, where prolate transverse isotropy is usually found (Dalstra *et al.*, 1993).

The trabecular pattern in bipeds directly contrasts with the quadrupedal pattern. In obligatory bipeds like humans, the major load in the pelvis is transferred through the cortical shell, where the stresses are apparently ~50 times higher than in the underlying trabecular network (Dalstra & Huiskes, 1995). Muscle forces acting on the ilium and ischium bones compensate for the forward and upward tilting action exerted by the hip joint force on the acetabulum. Because of this muscle action, stress is minimized and the pelvis is stress-relieved, thus preventing fatigue failure of the bone. During bipedal locomotion, the hip joint force is directed towards a relatively narrow strip along the anterior–superior edge of the acetabulum (Finlay *et al.*, 1986), where a high density trabecular chiasma is found. Because of load transfer, the lateral shell of the iliac cortex, just above the acetabulum and extending toward the sciatic notch, is heavily stressed (Bergmann *et al.*, 1990). Thus, the most dense trabecular bone and the thickest cortical shell are both found in these areas of high mechanical loadings (Dalstra *et al.*, 1993).

Loads across the hip bone are primarily transferred from the acetabulum through the lateral cortical shell to the sacro-iliac joint and the pubic symphysis. The stress component which actually transfers the hip joint force onto the pelvis is the radially directed component of the contact stress between acetabulum and femoral head (Dalstra & Huiskes, 1995). The highest stresses occur in the acetabular roof, where the radially directed stress component transfers the hip joint force onto the pelvis to the sacro-iliac joint and the pubic symphysis (Quesnel *et al.*, 1995).

Calcium-equivalent density estimates for the human iliac trabecular bone show that the greatest density (0.12–0.18 g cm^{-3}) occurs at the roof of the acetabulum and in the corpus of the ilium just superior to it. Conversely, the lowest densities (0.08–0.12 g cm^{-3}) are found in the blade and in the ischial bone (Oonishi *et al.*, 1983; Finlay *et al.*, 1986; Dalstra *et al.*, 1993).

The human iliac trabecular pattern

The basic comparative morphology of the human and non-human primate cancellous network of the hip bone was first detailed by Correnti (1952a, 1955; see also Latarjet & Gallois, 1910). He described a marginal (or ectochoric) and an arcuate (or endochoric) section, each consisting of a number of site-specific trabecular structures (Table 4.1).

In a two-dimensional projection, the major trabecular features distinguishing the human adult ilium (Figure 4.1d) are: a distinctive ilio-ischial (bituberal) bundle; a strong and undivided sacropubic bundle; and a diagonal (~100°) full crossing, or chiasma, over the acetabulum between the ilio-ischial and the sacropubic bundles, transversely located between the sciatic notch and the anterior–inferior iliac spine. On the whole, especially towards the body of the ilium and around the auricular region, the predominant trabecular structure is one of parallel plates, while rod-like structures mostly are found along the upper portion of the blade.

Other unique features characteristic of the so-called ‘anthropic’ pattern include: fully developed superior and anterior marginal bundles; an area of relative higher density along the superior marginal bundle, located in proximity to the anterosuperior rim of the blade; higher density supra-

Table 1. *Main trabecular structures of the human ilium (modified from Correnti, 1955; revised by Macchiarelli et al., 1999c)*

Sections	Trabecular structures	Description
Marginal (ectochoric)	superior bundle (sb)	Runs along the superior margin of the blade
	anterior bundle (ab)	Runs along the anterior margin of the blade
	posterior bundle (pb)	Runs along the posterior margin of the blade
Arcuate (endochoric)	sacropubic bundle (spb)	From the auricular surface and the posterior superior and inferior iliac spines runs diagonally downwards and forwards along the iliopectinal line as far as the pubic symphysis
	iliocotyloid bundle (icb)	Slightly differentiated from the sacropubic bundle, this thinner fan-shaped bundle runs diagonally downwards and forwards as far as the acetabulum crossing the ilioischial bundle
	ilio-ischial bundle (iib)	From the region of the tubercle of the iliac crest, posterior to the anterior spines, runs diagonally downwards and backwards with arcuate trajectory as far as the ischial tuberosity
	trabecular chiasma	High-density trabecular net transversally located at the level of the sciatic notch, defined by a distinct 'St. Andrews-like' crossing over the acetabulum between the sacropubic and the ilioischial bundles

acetabular and pericotyloid areas; and a definite distinction between the cancellous network of the supra-acetabular area and the iliac fossa.

In standard anatomical position, where the biomechanical loads are minimal, the orientation plane of the pelvis is such that the axis connecting the trabecular chiasma and the centre of the head of the femur is perpendicular to the transverse plane (Correnti, 1952b, 1955, 1957). Such an orientation plane varies continually during striding gait (Quesnel *et al.*, 1995). By means of the surrounding muscle and ligamentous tissues, the loads deriving from the sacro-iliac joint are absorbed and distributed through the chiasma by the sacropubic and the ilio-ischial bundles, and the peak strains within the cancellous bone are kept isotropic. Thus, distinct arcuate bundles – with the sacropubic more robust than the ilio-ischial one – which fully cross into a higher density trabecular area located between the superior part of the acetabulum and the corpus of the ilium are biomechanically and uniquely found in bipeds.

Our original radiographic work focused on the iliac cancellous network of infant and juvenile humans ($N = 200$). This work shows that, concurrent with the adoption of bipedal gait as an obligatory locomotion mode, the typical bipedal pattern is established early in childhood through progressive

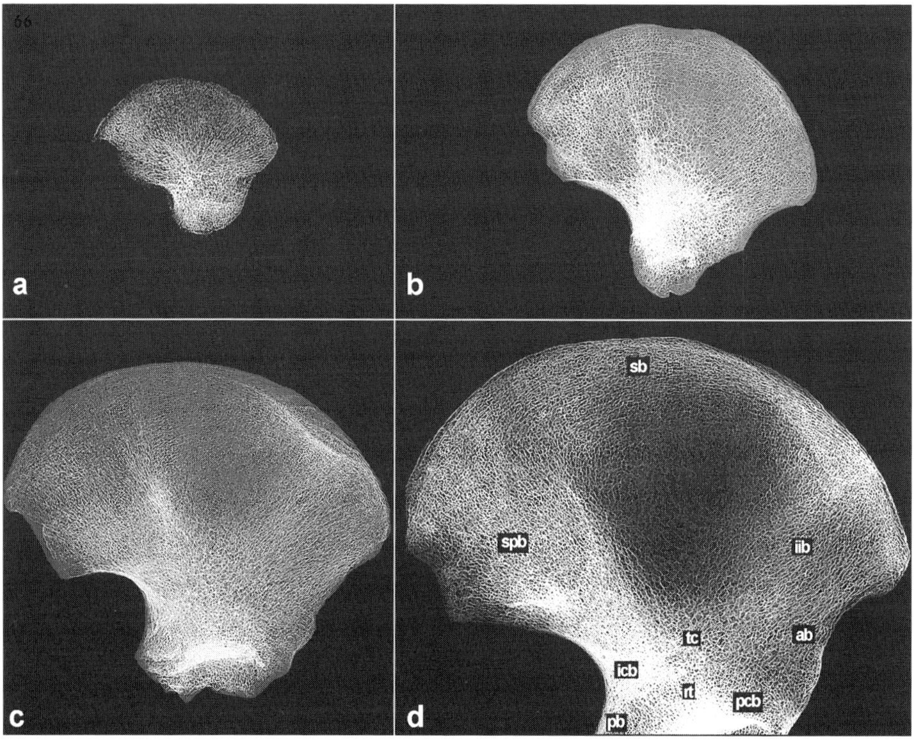

[Figure 4.1]
Age-related structural patterning of the iliac cancellous network in *Homo sapiens*. (a) newborn (SCR 696); (b) ~ 5 years old (SCR 741); (c) ~ 11 years old (SCR 290); (d) ~ 40 years old (SCR 252). Digitally processed images; dimensions not to scale.

Trabecular features labeled in the adult specimen (d) (in alphabetical order): ab: anterior bundle; icb: iliocotyloid bundle; iib: ilioischial bundle; pcb: pericotyloid bundle; pb: posterior bundle; rt: radial trabeculae; sb: superior bundle; spb: sacropubic bundle; tc: trabecular chiasma (see Table 4.1 for description).

strengthening of the arcuate section from a previously poorly differentiated ('isotropic') cancellous network (Figure 4.1). As biomechanical strains increase in frequency and magnitude, trabeculae undergo site-specific thickening (especially along the sacropubic bundle running from the auricular surface and the posterior–superior and inferior iliac spines towards the pubic symphysis) and the degree of anisotropy of the network as a whole progressively increases (the so-called 'Poisson's ratio' of the human adult cancellous network is ~0.2; Dalstra *et al.*, 1993). However, this pattern is only maintained if an individual walks. For example, in phocomelics who never walked, a typical human-like trabecular architecture is never established. Rather, the trabecular pattern is weak and unorganized.

Furthermore, in bedridden patients with greatly reduced mobility and weight bearing, the iliac network consistently shows substantial asymmetry and morpho-functional deviations from the 'normal' pattern (Geraets *et al.*, 1998).

On the whole, these observations point to the importance of environment (as opposed to genetics) in maintaining hip bone trabecular patterning. However, while the relevance of locomotion-related strains in structuring, differentiating and maintaining cancellous networks is readily apparent, we also note that 'the load-transfer mechanism and the stress patterns of the pelvic bone under normal physiological conditions are still not well understood' (Dalstra & Huiskes, 1995: 715).

The iliac trabecular pattern in extant non-human primates

As previously noted, primates probably exhibit greater locomotion-related diversity in external hip morphology and function than any other mammalian order (MacLatchy & Bossert, 1996). Our extensive research shows that, to some extent, similar activity patterns yield similar variation in the structural organisation and degree of anisotropy of the iliac cancellous network in non-human primates.

The primates currently comprising our radiographic sample include a hundred specimens representing prosimians (*Propithecus, Perodicticus*), New World monkeys (*Lagothrix, Alouatta*), Old World monkeys (*Colobus, Presbytis, Theropithecus, Papio, Macaca, Cercopithecus*), lesser (*Hylobates*) and great (*Pongo, Gorilla, Pan*) apes. Besides these primate taxa, for comparative purposes we have also investigated the iliac cancellous patterning in a number of non-primate mammals, including marsupials (*Macropus*), sloths (*Bradypus, Choloepus*), rodents (*Lepus*), carnivores (*Canis, Ursus, Martes, Meles, Lutra, Felis, Panthera*) and artiodactyls (*Dama, Capreolus, Ovis, Capra, Rupicapra*). The digital elaboration and careful comparative analysis of this growing set of radiographic images offers a unique opportunity to assess the nature of the relationships between postural/locomotor behaviours and hip bone trabecular architecture across a wide variety of small, medium and large body-sized primates and non-primate mammals.

Four examples of iliac trabecular patterns in prosimians (*Propithecus*) and Old World monkeys (*Colobus, Macaca, Papio*) are shown in Figure 4.2. These are for the most part single specimens whose pattern we assume represents the specific taxon's typical trabecular pattern.

In *Propithecus verreauxi* (Figure 4.2a), a vertical clinger and leaper (85% of leaping; in Gebo, 1996) member of the Indriidae (sifaka) weighing ~3 kg, the

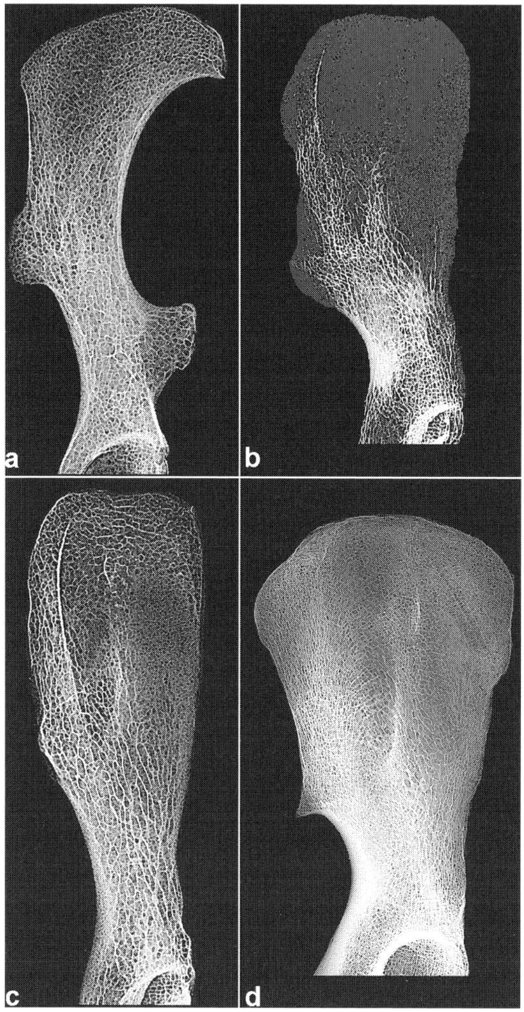

[Figure 4.2]
Architectural variation of the iliac cancellous network in adult extant prosimians and Old World monkeys. (a) *Propithecus verreauxi* (AIZIU 7255); (b) *Colobus badius* (TRV 33105); (c) *Macaca fascicularis* (WFU 364); (d) *Papio sphinx* (AIZIU PAL 109). Digitally processed images; contours approximate; dimensions not to scale.

cancellous network of the corpus of the ilium is slightly more structured than the blade. None the less, a relatively developed ilio-ischial-like bundle – which is absent in cercopithecoids – runs in parallel to the anterior iliac margin towards the acetabulum. Apparently, the least stressed portion of the ilium corresponds to the posterosuperior aspect of the blade, where thinner and fewer trabeculae are found. It is perhaps interesting to note that, when on the ground, *P. verreauxi* progresses using bipedal hops (Fleagle, 1999).

Colobus badius (Figure 4.2b) and *Macaca fascicularis* (Figure 4.2c) represent arboreal quadrupedalists displaying different amounts of overall climbing. In *C. badius* (~8 kg; 35–42% of the time is quadrupedal and 37–15% of the time is climbing; in Gebo 1996) both the marginal and arcuate sections of the cancellous network are poorly developed. However, the contrast between loaded and poorly loaded (or even unloaded) iliac sites is quite evident. In the red colobus most of the blade does not show structured trabeculae. Trabecular thickening and patterning is mostly found along the sacropubic axis towards the upper margin of the acetabulum, as well as around the sciatic notch, where a relatively developed fan-shaped iliocotyloid bundle intersecting the sacropubic bundles is detectable. Conversely, both the anterior and the ilio-ischial bundles are virtually absent. In *M. fascicularis*, the crab-eating macaque (3–5 kg; 60–65% of time is quadrupedal and 27–23% of time is climbing; in Gebo 1996), the iliac trabecular structure consists of a rather homogeneously scattered network characterized by extremely loose meshes. In this case, a modestly developed superior and anterior bundle can be observed, while the most dense site showing relatively thicker trabeculae again is found in the supra-acetabular region and along the sciatic notch margin.

Papio sphinx, the mandrill (Figure 4.2d), is a 13–30 kg almost exclusively ground quadrupedalist (99%; in Gebo, 1996). Similar to *Colobus* and *Macaca*, its 'gait-related' trabecular system (Correnti, 1955) is represented only by the marginal section, whereas a true arcuate section is only poorly developed. The posterior and the (very poorly structured) anterior bundles run nearly parallel along the free margins of the hip bone without crossing or fully converging over the acetabulum. The cancellous network as a whole appears poorly anisotropic, but some site-specific trabecular thickening and patterning is detectable around the auricular surface area, along the sciatic notch margin, and towards the upper acetabular rim.

Examples of variation of the iliac trabecular pattern in extant hominoids (*Hylobates, Pongo, Pan*) are shown in Figure 4.3.

By moving almost exclusively in a two-armed brachiation and slower quadrumanous climbing, gibbons (Figure 4.3a and 4.3b) are the most suspensory among all primates (Fleagle, 1980; Gittins, 1983; Srikosamatara, 1984; Cannon & Leighton, 1994). The iliac trabecular patterns shown by *Hylobates syndactylus* and *H. lar* substantially overlap, but, particularly along the anterior free margin of the blade, the larger siamang (10–11 kg) displays a more structured frame (Figure 4.3a) compared to the 5–6 kg white-handed gibbon (Figure 4.3b). While both use a variety of seated and

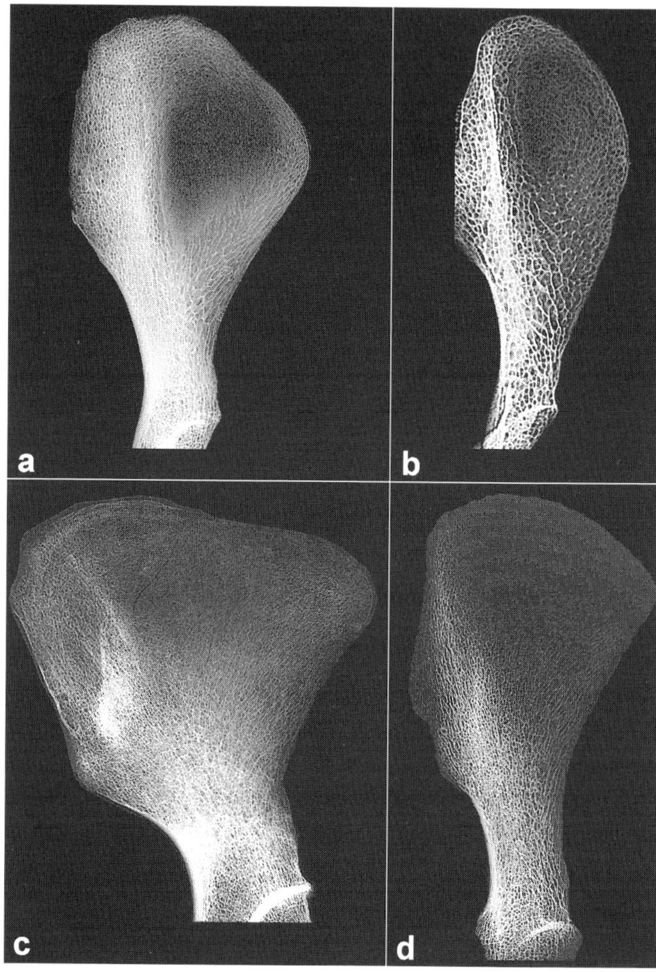

[Figure 4.3]
Architectural variation of the iliac cancellous network in adult extant lesser and great apes. (a) *Hylobates syndactylus* (AIZIU 1726); (b) *Hylobates lar* (AIZIU 10786); (c) *Pongo pygmaeus* (AIZIU 8609); (d) *Pan paniscus* (TRV 29058). Digitally processed images; contours approximate; dimensions not to scale.

suspensory feeding postures, siamangs mainly move by slow, pendulum-like arm-over-arm brachiation, whereas the smaller *H. lar* uses more rapid ricocheting brachiation (Fleagle, 1999). In some ways, their iliac trabecular architecture is qualitatively intermediate between cercopithecoids and great apes. The cancellous network is relatively structured and reflects biomechanical loads rather homogeneously distributed across both the blade and the corpus of the ilium. However, compared to the great ape condition, its meshes appear more open (even in sites of higher density) and the general appearance of the cancellous bone as a whole is substantially less anisotropic. This is likely related to the gibbons lighter weight and greater arboreality compared to the other apes.

A more structured trabecular pattern where the distinction between features belonging respectively to the marginal and to the arcuate sections appears rather well defined is shown by *Pongo pygmaeus* (Figure 4.3c). In the trees, orang-utans move almost exclusively by slow quadrumanous climbing, while on the ground they are quadrupedal. Orang-utans use both seated and suspensory postures when feeding (Sugardjito & van Hooff, 1986; Cant, 1987). In this large Asian ape the sacropubic and the ilio-ischial-like bundles grow away from the trabecular frame and flow downwards and inwards as far as the body of the ilium, where higher density is found in the area where these bundles merge (located far above the acetabular upper rim). Like gorillas, but different from chimpanzees, this area shows a number of thicker trabeculae tranversely crossing the lower blade with an arcuate trajectory. Increased functional density is also found in the iliocotyloid bundle, while most of the iliac fossa is poorly loaded. Compared to the human condition, no substantial distinction in relative density can be traced between the sacropubic and the ilio-ischial-like bundle, and the predominant trabecular structure is one of rather homogeneous rod-like structures. Occasionally, a small area of increased density is found along the superior marginal bundle, in proximity to the posterosuperior rim of the blade.

Pan paniscus, the bonobo, moves mainly on the ground by knuckle-walking, yet it is the most bipedal of all the extant apes. When feeding in trees, the pygmy chimpanzee performs a variety of quadrupedal, bipedal and, mostly, suspensory locomotor and postural activities (Susman, 1984; Hunt, 1991a,b; Doran, 1993a,b, 1997). Among the great apes its iliac trabecular pattern (Figure 4.3d) is rather characteristic, with a quite marked distinction in relative density between the sacropubic and the ilio-ischial-like bundle. However, even within the sacropubic bundle, the trabeculae are thinner and much less structured than in humans. The poorly developed anterior and posterior marginal bundles partially fade into the arcuate section bundles. Over the acetabulum, the sacropubic bundle divides into a ventral branch running towards the pubis, and a dorsal branch bending backwards and joining the posterior marginal bundle. A true human-like ilio-ischial bundle is absent, whereas a weak spinocotyloid bundle (comparable to an extension of the human anterior marginal one) runs in an oblique direction approximately from the anterior–superior iliac spine towards the body of the ilium, where it joins the peripheral trabeculae of the sacropubic bundle. As observed in orang-utans, the spinocotyloid and the sacropubic bundles do not fully cross into a true chiasma (i.e. they do not form a 'St. Andrews-like' cross as seen in humans), but only partially flow into a slightly higher density confluence of trabeculae located high above the acetabular upper rim.

In extant primates the functionally-related structural arrangement of the pelvic cancellous network thus shows substantial differences which are correlated to a variety of habitual postural and locomotor repertoires. A human, an 'ape' and a 'monkey' pattern are unmistakably distinguishable. However, within the investigated taxa, architectural variation also exists, so, an invariable one-to-one relationship between a specific positional/locomotor mode and a site-specific trabecular arrangement is difficult to show. This does not mean there are no fundamental functionally-related differences among these taxa, only that there are obvious biomechanical limits to the expression of the trabecular patterning according to specific locomotor/positional behaviours. Furthermore, hip bone developmental patterns and age-, sex- (perhaps), size- and shape-related architectural variation of the primate cancellous network are still poorly known (even in *Homo*). Yet in humans at least, sex is not a significant variable (human males and females have indistinguishable patterns, but this may not be true, for instance, in orang-utangs), and size does not appear to be a major factor, since in humans the typical trabecular pattern begins early, when body size is small. Also, original evidence from our radiographic record shows that: (1) basically identical iliac trabecular architectures are found in various species which exhibit substantial variation in body size, but share similar locomotor behaviours; and (2) for extant hominoids, the fundamental architectural differences between *Homo* and all the other great apes (*Pongo, Gorilla, Pan*) are positional/locomotor-related, not body-size dependent.

Analysis of the fossil cancellous network

Many sophisticated analytical approaches have been used for 3-dimensional imaging studies of the cancellous bone architecture. Serial reconstruction, micro-computerised tomography (μ-CT), quantitative computerised tomography (QCT), X-ray tomographic microscopy (XTM), nuclear magnetic resonance (NMR) are, among others, the most frequently used methods (for a review, see Odgaard, 1997). Unfortunately, all of these techniques are problematic (or unsuitable) for large-scale projects of data-gathering from fossil specimens, which are difficult to transport, may have sizes exceeding those allowed for micro-analytical studies, and often are kept in countries where access to high technology equipment is problematic or impossible.

In spite of its lower analytical power, conventional radiography remains the primary tool used so far for imaging trabecular architectures, and it is expected that it will remain useful for studying bone in the next decade (Geraets, 1998). Although conventional plane-film radiography offers only two-dimensional projections of the bone network, this technique has many advantages. Clinical radiography is available almost everywhere, the analysis itself is non-invasive and non-destructive, spatial resolution is high and results are permanently recorded on easy-to-transport and durable media.

For the purposes of our investigations, the external (gluteal) face of the ilium of each hip bone is placed in contact with the radiographic plate. The radiological parameters adopted (distance from the source of radiation, kV, mA, time exposure) obviously vary. They are systematically calibrated in order to obtain the optimum exposure relative to each bone's density. This variable is a function of the degrees of integrity and fossilisation, size and age at death of the specimens. Also, since bone thickness varies topographically across the ilium, a minimum of three different individual exposures are commonly made. The first exposure is calibrated to the lower density trabecular net usually surrounding the iliac fossa and the peripheral portion of the marginal bundles. The second focuses on the higher density inner structures (arcuate bundles) and the third concerns the highest density trabecular area found in the thicker supra-acetabular region.

However, despite these precautions, the extraction of reliable information from fossil cancellous bone is a rather complex task. A remarkable degree of variation in the general preservation conditions and radiographic appearance of the cancellous network (unrelated to the external bone morphology) is usually found in fossil specimens, sometimes also between different regions within the same specimen. Consequently, because of site-specific taphonomic dynamics and diagenetic microchanges, radiopacity does not represent a reliable indicator of relative trabecular densities.

In order to minimise the effects on cancellous bone of post-mortem disturbance factors and to allow the extraction from calibrated X-ray films of fossil specimens of reliable structural information, we have experienced a number of advanced digital image processing (DIP) techniques (Figure 4.4). According to our results derived from a variety of applications in palaeobiological research (e.g. Macchiarelli *et al.*, 1990, 1996a, 1998a; Macchiarelli & Bondioli, 1994; Geusa *et al.*, 1999; Rossi *et al.*, 1999; Bondioli, 1999), DIP represents an investigative tool capable of accurately enhancing the strength of qualitative and quantitative analyses. Our work convinces us of the reliability and replicability of the estimates.

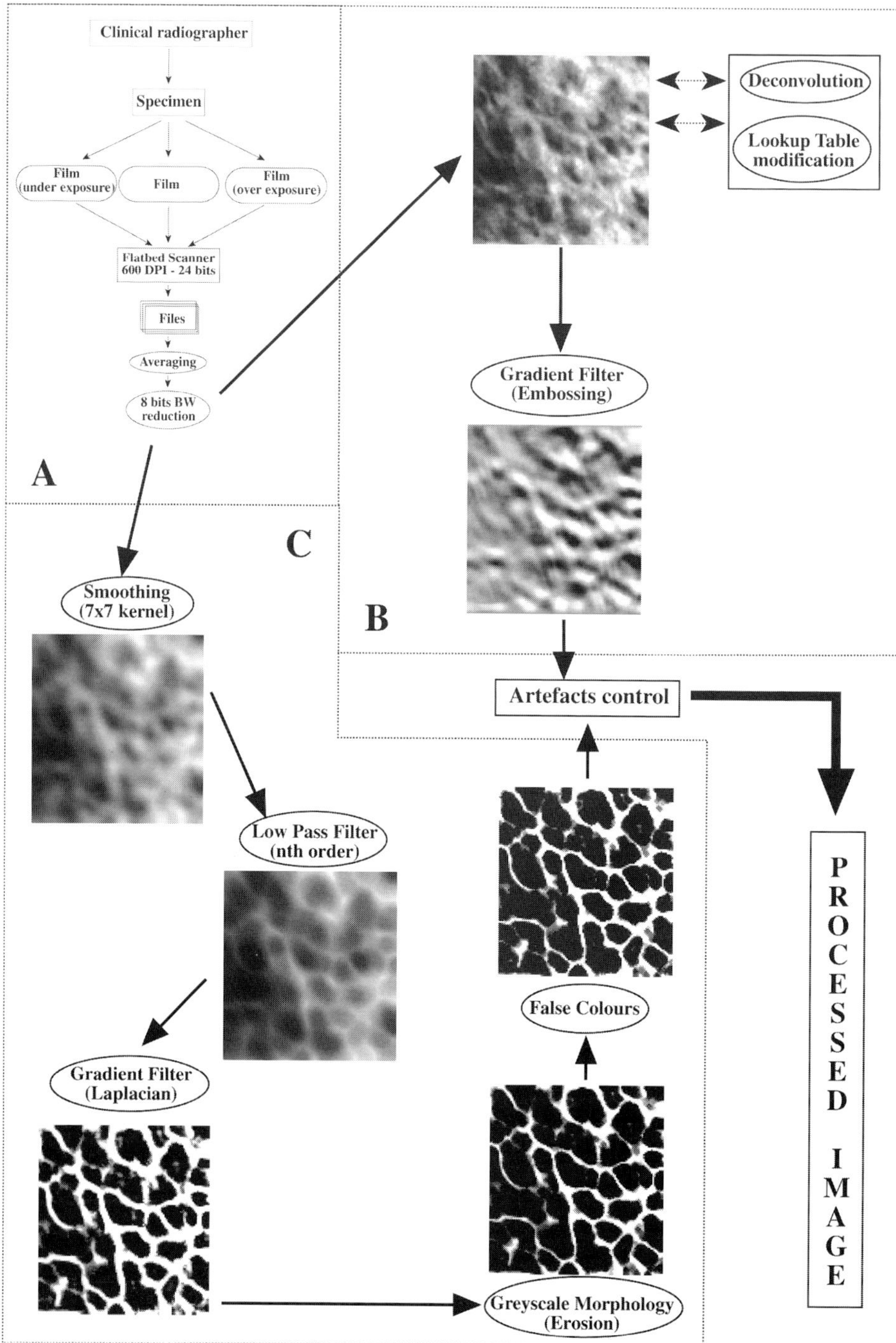

[Figure 4.4]
Flow chart of the main phases of digital image processing of fossil bone X-ray films. (A): acquisition; (B): pre-processing; (C): segmentation. See the text for explanations.

Digital image acquisition (Figure 4.4A)

Calibrated original films are transferred into a numerical format with an AGFA ARCUS 2 transparency scanner in a resolution of 600 DPI in true colours (24 bit plane). This resolution has been selected as a compromise between the sometimes conflicting needs for fine detail retrieval and global analysis of the structures. In fact, the 5×5 and 7×7 pixel-wide kernels used for image filtering encompass a distance ranging between 0.21 and 0.29 mm that approximates the mean thickness of the trabeculae (at 600 DPI resolution each pixel represents 0.042 mm of the original film size).

The scanner is linked with an Apple Power Macintosh 8100 PC running an Adobe Photoshop 5.0 plug-in as scanning software. In order to standardise the images and to keep the instrumental parameters constant, the program is set without tone and brightness corrections (tone variation being dependent only upon the film optical density). Images are recorded in TIFF format without compression. To allow easier processing, the 24 bit original images ($>$ 16 million colours) are converted into a greyscale format (256 grey levels). The original greyscale dominion of the film is shifted so that its mid-scale fits that of the digital image.

Image pre-processing (Figure 4.4B)

The software used for DIP is a combination of three different packages: National Institutes of Health NIH Image v. 1.61b; Graftek Optilab Pro 2.5; and the deconvolution routines elaborated by Watkins *et al.* (1993).

Changes in the Look-Up Table function (LUT) are performed to increase site-specific contrasts of intensity profiles. In particular, the use of the 'Inverse Power' and the 'Square Root' LUTs improves image brightness and increases contrast in dark areas, resulting in better image definition. For light areas in the radiographs, the use of their inverse functions reduces brightness and increases contrast.

To limit the effects of fossilisation, which causes the fading of the cancellous gradients and a general hazy appearance of the trabeculae, a specific deconvolution technique widely used in CCD image astronomy is adopted (Watkins *et al.*, 1993). This technique compensates for signal deterioration and restores the image as close as possible to its original appearance. This iterative algorithm uses the Gaussian function as a point spread function responsible for signal deterioration. Preliminary elaborations aimed at highlighting specific directional alignments of the bundles are performed by the application of directional embossing filters and by pseudocolor transformations.

Segmentation (Figure 4.4C)

In order to enhance the bone-marrow threshold, a non-binary segmentation of the radiographic images is obtained by using three analytical tools run sequentially. The first consists of a 7×7 smoothing convolution filter (low pass), with the central coefficient equal to zero and the remaining values set to 1. By fading local anisotropy but still maintaining the macroscopic features unaltered, this filter has the effect of smoothing the overall shape of objects and reducing noise produced by the film grain.

The order-N filter used as a second analytical tool belongs to the class of so-called non-linear filters. It replaces each pixel value with a non-linear function of its surrounding neighbours. To each pixel is assigned the Nth value of the $F \times F-1$ neighbours sorted in ascending order (N and F being specified by the user; in our analysis, the best results were obtained by using $F=5$ and $N=3$). If $N < (F \times F-1)/2$, the order-N filter has a tendency to erode bright regions and, consequently, to dilate the dark ones. On the whole, this tool reduces the noise generated by the fact that a radiographic image results from the two-dimensional projection of a three-dimensional object, thus enhancing the spatial concordances (the brighter features) and reducing the discordances (darker regions) among the different trabecular layers. However, it has to be stressed that, in the case of low-contrast uniform areas, this filter may produce local artefacts, where random arrangements of clear and dark spots are tendentially transformed into a 'lattice of vacuoles'.

The high-pass omnidirectional Laplacian filter used as a third processing tool highlights the variation of the light intensity surrounding a pixel. It mostly contributes to the extraction of the contours and outlines details. A 7×7 kernel with the central value set to 49 is usually applied in order to add the contours extracted by the Laplacian to the source image. As a final result of the non-binary segmentation, the elaborated images show distinct enhancement of the cancellous network, even in highly fossilised specimens, with unambiguous distinction between bone and no-bone.

Further image processing foresees the use of greyscale erosion morphological operators in order to compensate for the size expansion of the trabeculae (an effect resulting from the use of the Laplacian kernel), and the application of threshold operators capable of transforming the trabecular network into a binary image which can be measured for orientation (Geraets *et al.*, 1997; Geraets, 1998) and quantified by means of particle analysis tools.

Concordance between the trabecular morphology extracted by DIP and the network shown in the original films is tested by the use of the so-called 'skeleton' morphology analysis function (for technical details, see the Graftek Optilab Pro 2.5 user manual, 1994: 128).

The fossil evidence

To predict the postural/locomotor modes in fossil primates (including hominids), we comparatively investigated the appearance of the hip (and femoral) cancellous architecture of Plio-Pleistocene Old World monkeys (*Paracolobus chemeroni, Theropithecus oswaldi, Macaca majori, M. sylvana*), Miocene hominoids (*Proconsul heseloni, P. nyanzae, Limnopithecus legetet, Pliopithecus vindobonensis, Oreopithecus bambolii*), and Plio-Pleistocene hominids (*Australopithecus africanus, A. robustus, Homo* spp.). In detail, our currently available human fossil record includes: KNM-ER 1808, 3228 (East Turkana, Kenya); KNM-WT 15000 (West Turkana, Kenya); OH 28 (Olduvai Gorge, Tanzania); Krapina 207, 209, 211 (Croatia); Kebara 2 (Israel); Mladeč 21, 22 (Czech Republic). Beside this growing primate sample, we have also considered a number of non-primate fossil specimens, representing Plio-Pleistocene carnivores (*Canis etruscus, Homotherium crenatidens*) and Mio-Pliocene perissodactyls (*Chalicotherium grande, Tapirus arvernensis*).

The entire body of original and processed images of fossil and extant primate and non-primate specimens is available at the Section of Anthropology of the National Prehistoric Ethnographic 'L. Pigorini' Museum (Rome). In order to assure full interactive access to this original scientific documentation and to stimulate further independent observations, analyses, interpretations, all the images will be made available for free distribution in CD-ROM format.

On the whole, our preliminary radiographic survey of the fossil remains gave satisfactory results, proving trabecular structure preserved enough in most (despite not in total) of the specimens examined to allow site-specific textural analysis and architectural reconstruction. However, although conventional radiographic investigations have been successfully carried out on extinct reptile and mammal specimens (e.g. Thomason, 1985a,b; Alexander, 1989), it is certainly true that a variety of taphonomic disturbance factors – including crushing, distortion, mineralisation and percolation of mineral matrix through the trabeculae – may affect, or even inhibit, the structural analysis of fossil cancellous bone. In our investigative routine we thus try to minimise (or, whenever possible, even to avoid) the effects of these disturbance factors through the systematic use of digital technology (see above).

Following a summary of the evidence previously derived from the image processing of calibrated hip bone X-ray films (and CT-scans) for *Oreopithecus bambolii* and South African australopithecines, we briefly document some cases selected from our digital record – i.e. the Middle Miocene *Pliopithecus vindobonensis*, two Pleistocene macaques, and the

OH 28 *Homo erectus* – showing different preservation conditions and degrees of integrity, fossilisation and general appearance of the cancellous network.

Oreopithecus bambolii (IGF 11778)

Oreopithecus bambolii is a Late Miocene (9 to 7 Ma) endemic primate named by Gervais (1872a,b) from finds in the lignite mines at Monte Bamboli, in the province of Grosseto, Italy (Azzaroli *et al.*, 1986). Since the early discoveries, its affinities were the subject of lively discussion (in Delson, 1986). Gervais himself thought *Oreopithecus* mostly resembled gorilla, and thus placed the new genus among the great apes (Hominoidea), though he also pointed to a certain convergence with the Cercopithecidae on the strength of some features of its molars. This opinion was accepted by some students, but it was not long before other opinions appeared. Rütimeyer (1876) found a relationship with the gibbons, while Forsyth Major (1880) even saw some similarities with *Homo*. Schlosser's classification of *Oreopithecus* as cercopithecoid (Schlosser, 1887) was soon criticised by Ristori (1890), whose examination of the material from Monte Bamboli and from two other Grosseto mines (Casteani and Monte Massi) led him to reinstate this primate among the apes.

Interest in this purely Italian fossil then waned for over 40 years. Its revival was sparked by the work of Hürzeler and his personal finds in the Baccinello mine, once again in the province of Grosseto. His examination of this material and the other finds in Italian museums (Hürzeler, 1949, 1954a, 1958, 1960, 1968; Hürzeler & Engesser, 1976) led him to the conclusion that *Oreopithecus* should be placed on a collateral phyletic branch close to the Hominidae *s.l.*. A lively debate thus ensued. In the 1970s, the tendency was to reallocate *Oreopithecus* among the Cercopithecoidea (Szalay & Berzi, 1973; Szalay, 1975; Delson, 1979, 1986; Szalay & Delson, 1979). Today, however, while it cannot be said that a total consensus has been reached, current evaluations connect *Oreopithecus* to certain primitive African hominoids that subsequently broke off from the lines leading to the extant apes (Harrison 1986a,b, 1991; Sarmiento, 1987, 1998; Benefit & McCrossin, 1997). Recent studies (Rook, 1993; Harrison & Rook, 1994, 1997; Rook *et al.*, 1996; Moyà-Solà & Köhler, 1997) suggest a phylogenetic relation between *Oreopithecus* and the European Late Miocene *Dryopithecus* on the strength of a number of craniodental and postcranial characters. Accordingly, *Oreopithecus* should be seen as an endemic, highly specialised member of the Dryopithecinae.

This controversial question of taxonomy has arisen not only from the fact that *Oreopithecus* displays specific adaptations formed under very marked conditions of endemism, but also and above all, because of the previous shortage of fossil evidence for the Miocene hominoids in Europe and Asia. The latest discoveries, particularly the very recent recovery of a postcranial skeleton of *Dryopithecus* at Can Llobateres, Spain, have provided a more distinct picture of the wide morphological variety displayed by the Miocene hominoids (Moyà-Solà & Köhler, 1993, 1995, 1996).

Besides its taxonomic placement, the locomotor mode of *Oreopithecus* has been long debated. The primary issues concern some morphological features which were interpreted as possibly related to bipedality. Today it is generally assumed that *Oreopithecus* has a postcranial skeleton functionally related to climbing and suspension (Harrison, 1991), but this argument has been recently reconsidered by Köhler & Moyà-Solà (1997) and Moyà-Solà *et al.* (1999). According to them, *Oreopithecus* shows a number of ape-like features (including wide thorax, short trunk, high intermembral index and an abducting hallux) combined with characteristics such as lumbar lordosis, a hominid-like very large ischial spine, an extremely short pubic symphysis, an epicondylar angle of the femur, a peculiar foot morphology and hominid-like precision grip capability in the hand. On the whole, this complex of morphological features suggests a significant bipedal component in the *Oreopithecus* habitual locomotor behaviour (but see Wunderlich *et al.*, 1999).

The claim that the postural and locomotor behaviour of this Late Miocene ape included bipedality has been corroborated by the structural analysis of its hip bone cancellous network (Rook *et al.*, 1999).

Available *Oreopithecus* hip bones consist of two specimens, BAC 76 and IGF 11778, recovered in late 1950s at Baccinello. BAC 76 represents a fragmentary blade and corpus of a right ilium, while IGF 11778 – which belongs to a young adult articulated skeleton – includes the incomplete right and left ilia, the left ischium, pubis and the sacrum. A series of CT-scans and calibrated radiographs revealed cancellous bone preserved enough for digital image enhancement in both the right and left IGF 11778 fragments, but not in BAC 76.

In order to reconstruct a virtually complete *Oreopithecus* ilium suitable for structural analysis and functional interpretation, enhanced X-ray films of the right IGF 11778 incomplete blade have been electronically mirrored and superimposed on the left specimen. Following a number of preliminary attempts performed on both high-quality screens and printed images (Macchiarelli *et al.*, 1997a,b; Rook *et al.*, 1997), a satisfactory result (in Rook *et al.*,

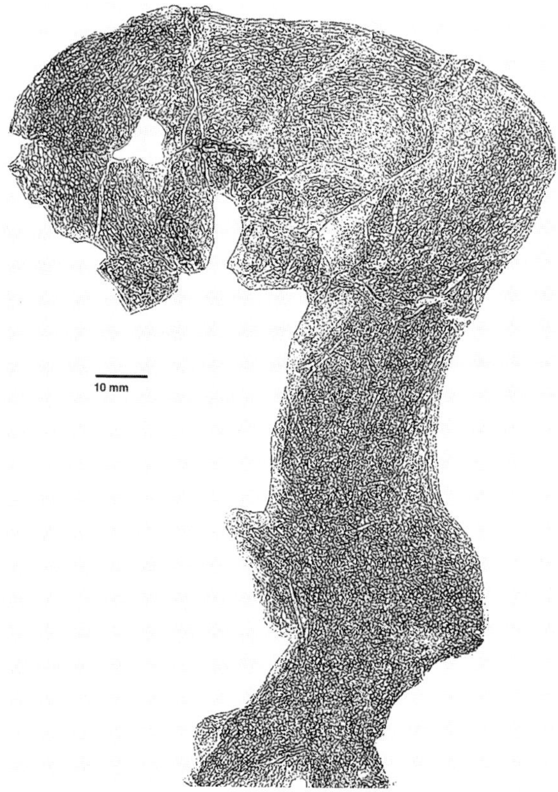

[Figure 4.5]
Cancellous network of the IGF 11778 *Oreopithecus bambolii* young adult ilium (Baccinello, Italy, Late Miocene). The specimen has been electronically reconstructed by superimposition of the mirrored right blade to the incomplete left hip bone. Digitally processed image; contour approximate.

1999) has been obtained by means of the non-binary segmentation procedure previously described. This procedure was run in order to enhance the directionality pattern of the pixels and to guarantee the careful alignment of the trabecular structures.

The original image in Figure 4.5 shows a further elaboration of the *Oreopithecus* iliac cancellous pattern which results from binary thresholding of the segmented image. In such a case, the network is depicted in its directional and textural components.

Comparative site-specific structural analysis of the *Oreopithecus* posterosuperior, anterosuperior, and antero-inferior margins of the iliac blade (where the cancellous network of a marginal trabecular bundle is well defined) and of the supra-acetabular region (where substantial portions of the sacropubic and the ilioischial bundles distinctly cross) with those of extant cercopithecoids, apes and humans, demonstrates that *Oreopithecus* habitually supported and transmitted a considerable amount of vertical weight to the lower limbs (Rook *et al.*, 1999). These data do not prove that *Oreopithecus* was bipedal, but its overall similarity to bipedal hominids makes full scale bipedalism more likely than chimp-like partial uprightness.

The South African australopithecines

Since late 1994, in collaboration with V. Galichon (Collège de France, Paris) and P.V. Tobias (University of the Witwatersrand, Johannesburg), two of us (L.B. and R.M.) have investigated the iliac trabecular structure preserved in eight South African australopithecines belonging to the taxa *A. africanus* and *A.* (*P.*) *robustus* (Macchiarelli *et al.*, 1995, 1996b, 1998b, 1999c; Galichon, 1997). The sample represents two juvenile (MLD 7, MLD 25), two adolescent (SK 3155, TM 1605) and four adult individuals (Sts 14, Stw 65, Stw 431, SK 50). A detailed critical description of the collected evidence is reported in Macchiarelli *et al.* (1999c).

Given previous work on australopithecine locomotion (e.g. Lovejoy, 1988; Ohman *et al.*, 1997; Latimer & Ward, 1998; but see also Spoor *et al.*, 1994; Clarke & Tobias, 1995; Berger & Tobias; 1996; Spoor & Zonneveld, 1997), it is not surprising that DIP of the adolescent and adult ilia revealed a number of human-like trabecular features. These features include: a relatively well developed ilio-ischial bundle, a distinct sacropubic bundle; a supra-acetabular crossing, an area of relative higher density along the superior marginal bundle located in proximity to the anterosuperior rim of the blade; and a thin fan-shaped iliocotyloid bundle arising from the greater sciatic notch margin. Further, as detected by DIP and confirmed by CT-scans (Galichon, 1997; Galichon & Thackeray, 1997; Coppens *et al.*, 1998), the trabecular structures of the inner (arcuate) section were moderately differentiated from the peripheral (marginal) ones.

None the less, as reconstructed in the form of a mosaic of electronically elaborated radiographic images, the australopithecine trabecular pattern also showed a unique set of morphological features. With special regard to the modern adult human condition, these include less developed superior and anterior marginal bundles, a less distinct ilio-ischial bundle, lower density supra-acetabular and pericotyloid areas, a fan-shaped rather than diagonal full crossing between the ilioischial and the sacropubic bundles, minor distinctions between the cancellous network of the supra-acetabular area and the iliac fossa, a higher proportion of low density, open-cell, rod-like structures and a less anisotropic appearance of the entire cancellous network.

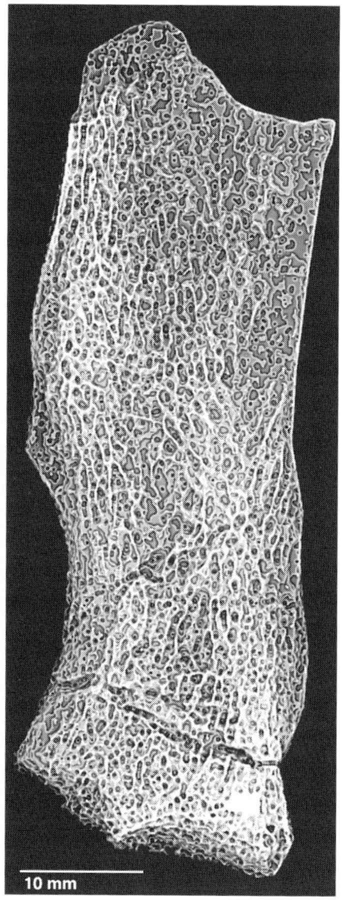

[Figure 4.6]
Cancellous network of the NHMW 1970/1397/53 *Pliopithecus vindobonensis* adult right ilium (Neudorf an der March, Czech Republic, middle Miocene). Digitally processed image; contour approximate. Original X-ray film courtesy of G. Höck and M. Teschler-Nicola (Naturhist. Mus., Wien, Austria).

On the whole, these features point to qualitatively and quantitatively different peak principal strains imposed upon the australopithecine hip cancellous bone during growth, possibly related to disequilibrium in the use of the hip joint muscles (Galichon & Thackeray, 1997), and/or to different mechanisms for channeling vertical forces through the vertebral column than humans (Sanders, 1998: 249). Whatever biomechanical factors might be responsible, available evidence shows that in cancellous bone the strength and direction of the postural/locomotion-related peak strains imposed upon the australopithecine pelvis differed from those acting daily on extant humans and apes. This suggests a unique suite of habitual behaviours characterising these early hominids.

Support for this conclusion was provided by the investigation of the extraordinarily well-preserved trabecular structure of the human-like shape MLD 7 specimen from Makapansgat (3.2–3.0 Ma). The iliac trabecular patterns of this juvenile *A. africanus* and that of an extant human counterpart of similar age (~6.6–11.3 years old; Conroy & Vannier, 1991a,b) share a number of important features. These include thinning of the superior bundle in postero-anterior direction, increased density of the most anterior portion of the iliac crest and of the greater sciatic notch margin, sub-perpendicular alignment of the upper trabeculae of the sacropubic bundle with respect to the posterior portion of the superior bundle, relative higher density of the sacropubic bundle with respect to the ilio-ischial one, general morphology of the posterior marginal bundle extension and fan-shaped morphology of the iliocotyloid arcuate bundle and relatively distinct radial trabeculae. In addition, MLD 7 uniquely shows less distinct superior and anterior marginal bundles, a weakly developed ilio-ischial bundle with minor textural distinction from the superior and anterior marginal bundles, a less well-developed arcuate sacropubic bundle progressively weakening towards the acetabulum, partial rather than full crossing between the ilio-ischial and the sacropubic bundles, lower trabecular density of the body of the ilium and of the supra-acetabular and pericotyloid areas, a lower proportion of higher density, closed cell, plate-like structures, a honeycomb appearance of the cancellous network as a whole, and a substantially lower degree of textural anisotropy (for description and images, see Macchiarelli *et al.*, 1998b, 1999c).

Since this australopithecine child surely had already adopted the characteristic suite of postural/locomotion behaviours typical of its taxon, it seems reasonable to postulate that its most frequent locomotion was bipedalism, but with a gait somewhat different from a modern human child.

Pliopithecus vindobonensis (NHMW 1970/1397/53)

Pliopithecids are a conservative group of small- to medium-sized catarrhines widely distributed in Eurasia since the end of early Miocene to the late Miocene. Although no clear direct ancestor has been identified outside Eurasia in the Paleogene or early Miocene fossil record, it is currently assumed that pliopithecids originated in Africa during the Oligocene and dispersed into Eurasia at the time of the collision between the Afro-Arabian and the Eurasian plates (Harrison, 1987; Bernor *et al.*, 1988; Harrison *et al.*, 1991; Andrews *et al.*, 1996).

Pliopithecids are a quite diverse family, including two subfamilies and a dozen of different species. The inferred differences in estimated body size range, the reconstructed ecological associations and the differences in dietary behaviours all together testify a wide range of adaptive diversity (Andrews *et al.*, 1996).

In the literature there has been a broad consensus on a close relationship between pliopithecids and extant gibbons (Gervais, 1849; Hürzeler, 1954b; Simons, 1972; Simons & Fleagle, 1973), although it has been also stressed that a number of pliopithecid resemblances to hylobatids are due to similarity in overall size and retention of plesiomorphic features (Groves, 1972; Delson & Andrews, 1975; Szalay & Delson, 1979; Harrison, 1987).

The postcranial anatomy of most pliopithecids is relatively poorly known, since most of the species are known from jaw fragments or isolated teeth. A remarkable exception is represented by the Middle Miocene *Pliopithecus vindobonensis* from Neudorf an der March, Czech Republic (Zapfe & Hürzeler, 1957; Zapfe, 1958, 1960). In this type locality, the species is represented by three partial skeletons which preserve morphological details of unique relevance for taxonomic and phylogenetic reconstruction, as well as for positional/locomotion reconstruction.

Morphofunctional analysis of the *P. vindobonensis* postcranial skeleton suggests that this gibbon-size fossil primate (~7 kg) was an arboreal quadruped with suspensory abilities like those of the larger platyrrhines (Fleagle, 1986, 1999; Bacon, 1994). Unfortunately, its pelvis is too poorly represented for a detailed comparative analysis. However, DIP of the X-ray film of the NHMW 1970/1397/53 specimen (labelled as C 39 in Zapfe, 1960) has revealed interesting structural details (Figure 4.6).

The specimen consists of an incomplete adult right ilium including the upper portion of the acetabulum. Its maximum height is 69.0 mm; the minimum breadth of the corpus is 18.5 mm. Despite the high degree of fossilisation and two major cracks running obliquely across the iliac body, its cancellous network is still preserved in the blade, the entire corpus and around the preserved acetabular margin. As effect of the application of the *N*th order filter to the digital image (see under 'Analysis of the fossil cancellous network'), some structural artefacts in form of 'lattice of vacuoles' appear in the superior portion of the blade.

The cancellous pattern as a whole is formed by low to middle density rod-like structures. A proportionally higher density sacropubic-like bundle runs from the anterior margin of the auricular surface downwards and forwards as far as the anterosuperior rim of the acetabulum. This bundle-like trabecular structure is oblique in its first portion, then curves downwards and becomes sub-vertical across the corpus of the ilium. While a pericotyloid-like bundle may be present, no evidence of radial trabeculae exists around the acetabulum.

Interestingly, just above the major crack, the apparent cross between the higher density sacropubic-like bundle and some ilioischial-like thicker trabeculae originating from the antero-inferior margin of the blade (reaching the posterior margin of the corpus with an oblique course) delimits a lower density pointed arch-shaped supra-acetabular area, which cancellous weft is closer to that of the upper blade.

When compared to the generalized 'monkey' pattern (see under 'The iliac trabecular pattern in extant non-human primates'), the *P. vindobonensis* network appears more structured. Despite the limits implicit in structural evaluations based on the digital processing of incomplete and highly fossilised specimens and the necessity of a certain degree of unavoidable extrapolation required to electronically fill in network discontinuities, its trabecular pattern is apparently closer to *Hylobates syndactylus* (Figure 4.3A). Also, although the skeleton of *Pliopithecus* is morphologically much like that of a large living platyrrhine such as *Lagothrix* (Fleagle, 1999), according to our record their iliac trabecular architecture differs considerably.

Macaca majori (Ty 12548) and *M. sylvana* (PFT 1653 [6])

Fossil remains from Plio-Pleistocene Italian deposits attributable to the genus *Macaca* have long been known (Cocchi, 1872; Forsyth Major 1872; Ristori, 1890; Portis 1917; Rook, 1997). However, with the exception of some specimens from Capo Figari, Sardinia (Azzaroli, 1946), until recent times most of the recovered fossils were highly fragmentary and, like other Western European sites (Ardito & Mottura, 1987), primarily comprised of mandibles and teeth. This scarcity of specimens and their extremely conservative morphological features makes it difficult to assess the diversity of taxa proposed for the European fossil samples and to distinguish them from the extant *M. sylvana* Linnaeus, 1758. Accordingly, some authors (Szalay & Delson, 1979; Delson, 1980a,b) regard the European fossil record as simply representing temporal-geographical subspecies of *M. sylvana*.

Most of the Italian finds are from the early and middle Pleistocene and two forms of macaques have been so far recognised: *M. majori*, a species endemic to Sardinia, and *M. sylvana (florentina)*, from northern and central Italy.

The first indication of a primate occurring in the fossiliferous breccias at

Capo Figari, northeastern Sardinia, was given by Dehaut in 1911. Dehaut (1911) described a new primate species, *Ophtalmomegas lamarmorae*, based on cranial fragments that subsequently were identified as having belonged to a large bird (see Vigne, 1989: p. 20). Later on, he attributed to the same species an M_3 of macaque origin (Dehaut, 1914). Since 1910, systematic excavations at the same site were succesfully carried out by Forsyth Major (1913).

The first detailed study of the *Macaca* fossil sample from Capo Figari stored at the University of Florence was by Azzaroli (1946), who erected the name of *M. majori* and described it as an endemic, dwarf species. However, there is still no consensus about either the species designation for these fossil macaques, nor for the supposed insular dwarfism condition (for discussion, see Szalay & Delson, 1979; Delson, 1980a,b; Ardito & Mottura, 1987).

Overall, the latest early to early middle Pleistocene *M. majori* from Capo Figari is represented by an extraordinarily large sample including more than 250 specimens stored at the University of Florence, the Natural History Museum (London) and at the Naturhistorisches Museum (Basel).

A total of 78 specimens from eight individuals of *M. sylvana (florentina)* currently stored at the University of Perugia come from lignite excavations begun in 1986 at the Pietrafitta mine in Umbria (Ambrosetti *et al.*, 1987, 1992a; Gentili *et al.*, 1998). Located just south of the Lake Trasimene basin, Pietrafitta is an intramontane, tectonic, sedimentary basin indirectly linked to the opening of the Tyrrhenian sea. It is filled with approximately 100 m of pelite, the upper portion of which consists of clayey deposits about 8 m thick, with lignite interbeddings, and topped by weakly stratified, lacustrine, silty clays (Ambrosetti *et al.*, 1989, 1992b). The late early Pleistocene lignites have yielded abundant vertebrates, molluscs, insects, macroflora and pollens. It was formed in a fen or marsh with abundant organic matter, as well as sporadic, thin clay intercalations with intraformational clasts and molluscs.

Figure 4.7 shows the digitally processed X-ray films of two fossil macacques from Capo Figari and Pietrafitta, respectively. The Ty 12548 specimen (Figure 4.7a) is still an unpublished and incomplete adult right hip bone of *M. majori* from the Swiss collection, extracted in recent years from a bone-breccia block originally collected in 1913–14 by Forsyth Major. The distance between the upper margin of the acetabulum and the most distant point of the blade is 61.5 mm; the minimum breadth of the corpus is 17.5 mm. The PFT 1653 (6) *M. sylvana* specimen (Figure 4.7b) is an unpublished incomplete adult left hip bone. The distance between the upper margin of the acetabulum and the most distant point of the blade is 82.5 mm; the minimum breadth of the corpus is 24.4 mm.

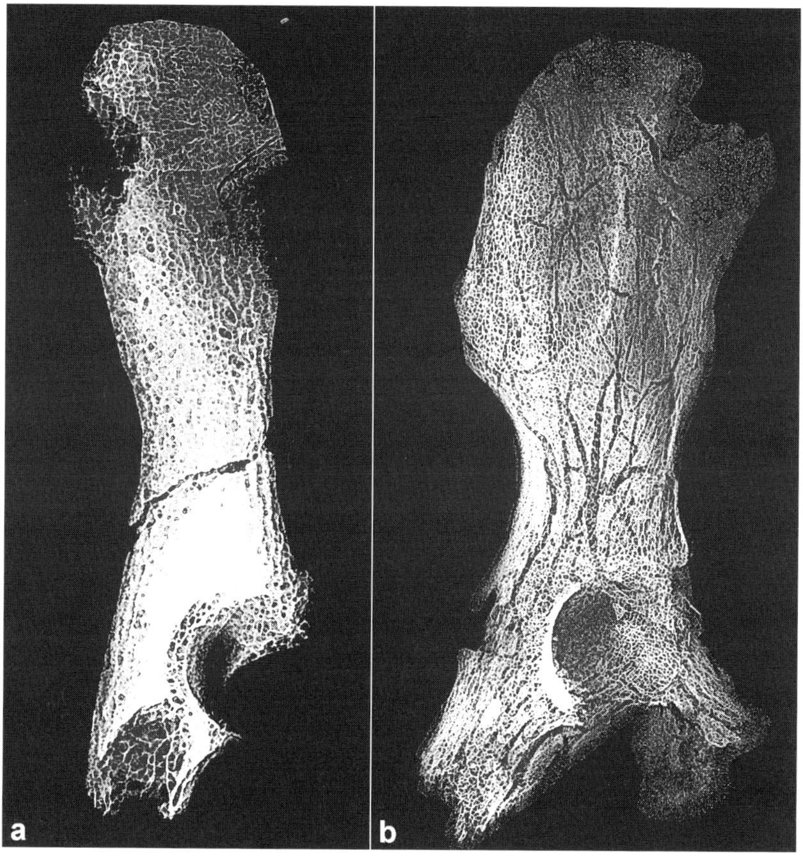

[Figure 4.7]
Iliac cancellous network in two fossil macaque adult specimens: (a) Ty 12548 *M. majori* (Capo Figari, Italy, latest Early to early Middle Pleistocene); (b) PFT 1653 (6) *M. sylvana* (*florentina*) (Pietrafitta, Italy, late Early Pleistocene). False colours digitally processed images; contours approximate; dimensions not to scale. Original X-ray films courtesy of B. Engesser (Naturhist. Mus., Basel, Switzerland) and S. Gentili (Univ. of Perugia, Italy).

Compared to PFT 1653 (6), Ty 12548 (Figure 4.7a) is less complete and shows a higher degree of fossilisation. Because of the specific type of elaboration, as in the case of *Pliopithecus vindobonensis* (Figure 4.6), its enhanced image evidences a 'lattice of vacuoles', mostly concentrated in the corpus, around and over the acetabular rim. However, despite some obvious site-specific discontinuities in the network (in part related to the transverse fracture of the corpus), the general cancellous pattern of the blade is rather distinct.

Compared to the *M. sylvana* specimen (Figure 4.7b), the *M. majori* trabecular network shows much wider cells. Tentatively, this may reflect different ages at death, but a size-related, at least partial effect, cannot be excluded. Conversely, no obvious structural difference is discernable in their overall monkey-like patterns. Both cancellous networks show sub-vertical trabecular thickening and patterning along the sacropubic axis and in correspondence of the posterior marginal frame, while the antero-superior portion of the blades shows extremely loose meshes reflecting substantially lower loads. Like extant adult monkeys (see Figure 4.2b,c,d), *M. sylvana* has the most dense iliac bone at the sciatic notch.

Homo erectus (OH 28)

OH 28 is a ~0.7 Ma associated part of a left hip bone and a femur shaft discovered in 1970 at the surface in the upper part of Bed IV (locality WK), at Olduvai Gorge, Tanzania (Leakey, 1971: 234). Although the hip bone had been chewed (perhaps by a crocodile), it preserves the lower part of the ilium and some of the ischium. All of the iliac crest is missing, and cracking has caused some slight lateral displacement of the anterior part of the blade, with an irregular depression of the surface (Day, 1971).

The ilium is very large. It shows marked flare and a thickened vertical pillar extending from the acetabulum to the top of the bone; correspondingly, a very stout horizontal bar runs posteriorly from the acetabulum above the greater sciatic notch to the posterior iliac spine, along the lower margin of the iliac fossa. A wide sciatic notch suggests that this is from a female. The obliquely placed sacral articulation is relatively small, while the acetabulum is very large in vertical diameter and deep, with a thickened rim (for general description and comparisons, see Day, 1971, 1973, 1978, 1982, 1986; Rightmire, 1993: 85). Recently estimated body mass for this *Homo erectus* is 66.6 kg (Ruff *et al.*, 1997).

The original radiographic analysis of the pelvic fragment (Day, 1971) disclosed a number of interesting additional morphological details. Current electronic enhancement of a new set of calibrated X-ray films allows a rather detailed investigation of the highly structured OH 28 iliac cancellous network (Figure 4.8).

Despite the incompleteness and high degree of fossilisation of the specimen (which also shows a number of deep cracks), its architectural pattern substantially overlaps the modern human adult (see Figure 4.1d). With special reference to the so-called arcuate section (see Table 4.1), a sacropubic bundle, a ilioischial bundle, and their supra-acetabular diagonal crossing, transversally located at the sciatic notch level, are quite distinct

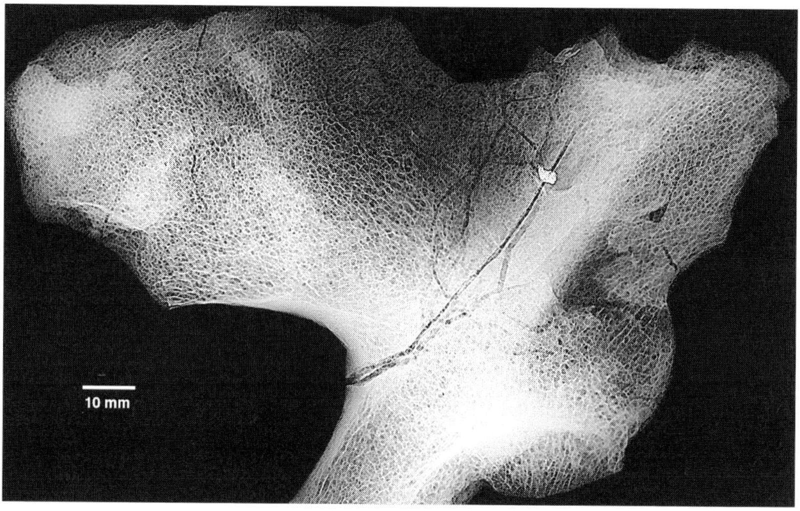

[Figure 4.8]
Cancellous network of the OH 28 *Homo erectus* adult left ilium (Olduvai Gorge, Tanzania, ~ 0.7 Ma). Digitally processed image; contour approximate. Original X-ray films courtesy of A. Walker (Pennsylvania State Univ., U.S.A.) and C.V. Ward (Univ. of Missouri, U.S.A).

trabecular features. In the blade, the textural distinction between the arcuate trabecular section and the iliac fossa is also definite. At the preserved upper part of the auricular surface and at the sciatic notch area, respectively, the root of a rather developed superior bundle (marginal section) and a fan-shaped iliocotyloid bundle are traceable. As expected because of the significant amount of radially directed stresses occurring at the acetabular roof during bipedal locomotion, the supra-acetabular and pericotyloid areas show very high relative density, with distinct radial trabeculae arranged around the thick acetabular rim.

Interestingly, the orientation of the acetabulocristal buttress, which in OH 28 is more prominent than on any modern human ilium (Day, 1971, 1986), and that of the (incomplete) ilioischial bundle apparently do not fully coincide – the angle between the sacropubic bundle and the iliac pillar is ~85°, while that one between the sacropubic and the ilioschial bundle, i.e. the trabecular chiasma, is ~96°. Also, despite the fact that in DIP analyses of fossil specimens radiopacity does not represent a reliable indicator of relative trabecular densities, it is important to notice the exceptionally high trabecular density characterising the OH 28 ilioischial bundle, which, along the pillar, even exceedes the strong sacropubic bundle. Compared to the extant human condition, both bundles display a higher proportion of high density, closed cell, plate-like structures.

 According to our currently available digital record, among the fossil hominids, similarly developed trabecular features can be observed in the ~1.95 Ma KNM-ER 3228 adult right hip bone from Koobi Fora, Kenya (Rose, 1984), but not, for example, in the ~1.53 Ma KNM-WT 15000 juvenile ilia from Nariokotome, Kenya (Walker & Ruff, 1993).

Acknowledgements

This study is part of an ongoing research project supported by the Italian National Research Council to the Research Unit *Methodologies and techniques for the analysis of the human paleobiological heritage: procedures and digital archives*, within the 'Cultural Heritage' Project (www.culturalheritage.cnr.it).

For having granted access to osteological (fossil and recent specimens) and/or original radiographic material in their care, we thank the following scientific Institutions: Anthrop. Inst., Zurich-Irchel Univ. (Switzerland); Cleveland Mus. Nat. Hist. (USA); Croatian Nat. Hist. Mus. (Croatia); Dept. Anat. Sci., Univ. of the Witwatersrand (South Africa); Dept. Anat. & Anthrop., Sackler Faculty of Med., Tel-Aviv Univ. (Israel); Dept. Anthrop., Pennsylvania State Univ. (USA); Dept. Anthrop., Univ. of Michigan (USA); Dept. Anthrop., Univ. Toronto (Canada); Dept. Anthrop., Wake Forest Univ. (USA); Dip. Anat., Farmacol. e Med. Leg., Univ. Torino (Italy); Dip. Biol. Anim., Univ. Pavia (Italy); Dip. Sci. Terra, Univ. Perugia (Italy); Dip. Sci. Terra & Museo Storia Nat., Univ. Firenze (Italy); Lab. Phys. Anthrop., Kyoto Univ. (Japan); Mus. Royal de l'Afrique Centr. (Belgium); Mus. Civ. Zool. Roma (Italy); Mus. St. Nat., Univ. Pavia (Italy); Mus. Naz. Preist. Etnogr. 'L. Pigorini' (Italy); Nat. Museums of Kenya (Kenya); Naturhist. Mus. (Switzerland); Naturhist. Mus. (Austria); Primate Res. Inst., Kyoto Univ. (Japan); Transv. Mus. (South Africa).

Since the very beginning of this project, a number of colleagues provided us substantial help, technical support, scientific collaboration. For their generous contribution to the development of the original radiographic record, special thanks are due to: G. Abungu (Kenya), D. R. Begun (Canada), L. R. Berger (South Africa), R. Carlini (Italy), M. Cereza (Italy), E. Cioppi (Italy), R. J. Clarke (South Africa), Y. Coppens (France), B. Engesser (Switzerland), G. Ficcarelli (Italy), D. Formenti (Italy), V. Galichon (France), S. Gentili (Italy), G. Giacobini (Italy), G. Höck (Austria), F. M. Kirera (Kenya), M. Köhler (Spain), H. Kritscher (Austria), B. Latimer (USA), M. G. Leakey (Kenya), R. D. Martin (Switzerland), S. Moyà-Solà (Spain), M. Nakatsukasa (Japan), R. Orban-Segebarth (Belgium), J. Radovčič (Croatia), Y. Rak (Israel), C. Rovati (Italy), C. Savorè (Italy), H. Seidler (Austria), A. Tagliacozzo (Italy),

M. Teschler-Nicola (Austria), J. F. Thackeray (South Africa), P. V. Tobias (South Africa), D. Torre (Italy), V. Vomero (Italy), A. Walker (USA), C. V. Ward (USA), D. S. Weaver (USA), M. H. Wolpoff (USA). For their valuable scientific contribution in terms of discussion, criticism, suggestions, we are also particularly indebted to: L. Aiello (UK), G. C. Conroy (USA), R. S. Corruccini (USA), D. W. Frayer (USA), R. L. Lee (UK), O. C. Lovejoy (USA), P. O'Higgins (UK), D. S. Weaver (USA), G. W. Weber (Austria), T. D. White (USA).

Essential technical assistance during various phases of the work (graphic, photographic, radiographic record of the specimens; transformation of the original X-ray films into a numerical format; digital image processing) was kindly provided by: G. Calandra (Italy), D. Capannolo (Italy), M. Dazzi (Italy), V. Galichon (France), G. Geusa (Italy), K. Isler (Switzerland), F. M. Kirera (Kenya), N. Malit (Kenya), G. Pedicelli (Italy), W. Recheis (Austria), F. Salomone (Italy), C. Savorè (Italy), A. Sperduti (Italy), R. van De Ritt (South Africa).

Radiographic investigation and scientific interpretation of the South African australopithecine specimens were developed between 1994 and 1996 jointly with V. Galichon (supported by the Collège de France) and P. V. Tobias (South Africa), to which L.B. and R.M. wish to express the deepest gratitude for their collaborative spirit and substantial contribution to this pilot study.

Finally, we wish to recognize the support provided by the European Science Foundation to the Network on Hominoid Evolution & Environmental Change in the Neogene of Europe, and particularly to thank L. de Bonis and G. Koufos for their kind invitation to contribute the 3rd Workshop on Phylogeny of Eurasian Neogene Hominoid Primates (Nikiti, Greece, 21-25/10/1998).

Without the original stimulus provided by V. Correnti (1909–91) to L.B. and R.M., this project would not have been initiated.

References

Alexander, R. (1989). *Dynamics of Dinosaurs & Other Extinct Giants*. New York: Columbia University Press.

Ambrosetti, P., Argenti, A., Basilici, G., Gentili, S. & Ikome, F. E. (1992b). The fossil vertebrates of the Pietrafitta basin (Perugia, Umbria-Italy): preliminary analysis of the processes responsable for 'paleontological production'. In (Gall, J. C. & Grauvogel-Stamm, L. (eds.), *Taphonomy: Processes and Products*, pp. 3–4 (abstract). Strasbourg: Europ. Palaeont. Ass. Workshop.

Ambrosetti, P., Capurso, R., Covino, R., Gallo, G., Monicchia, R. & Salvatici, B. (1992a). *Storia, Energia, Ambiente: Un Modello di Centrale a Pietrafitta*. Perugia: Protagon.

Ambrosetti, P., Cattuto, C. & Gregori, L. (1989). Geomorfologia e neotettonica nel bacino di Tavernelle/Pietrafitta (Umbria). *Quaternario* **2**: 57–64.

Ambrosetti, P., Faraone, A. & Gregori, L. (1987). Pietrafitta: un museo di paleontologia in Umbria. *Museol. Sci.* **4**: 99–118.

Andrews, P., Harrison, T., Delson, E., Bernor, R. & Martin, L. (1996). Distribution and biochronology of European and Southwest Asian Miocene Catarrhines. In R. L. Bernor, & V. Fahlbusch (eds.), *Evolution of Neogene Continental Biotypes in Central Europe and the Eastern Mediterranean*, pp. 167–207. New York: Columbia University Press.

Ardito, G. & Mottura, A. (1987). An overview of the geographic and chronologic distribution of West European cercopithecoids. *Hum. Evol.* **1**: 29–45.

Azzaroli, A. (1946). La scimmia fossile della Sardegna. *Riv. Sc. Preist.* **1**: 68–76.

Azzaroli, A., Boccaletti, M., Delson, E., Moratti, G. & Torre, D. (1986). Chronological and paleogeographical background to the study of *Oreopithecus bambolii. J. Hum. Evol.* **15**: 533–40.

Bacon, A. M. (1994). Nouvelles perspectives sur la locomotion de Pliopithecus vindobonensis (Zapfe et Hürzeler, 1957). *C.R. Acad. Sci. Paris* **318**: 259–66.

Benefit, B.R. & McCrossin, M.L. (1997). New fossil evidence bearing on the relationships of *Nyanzapithecus* and *Oreopithecus*. *Am. J. Phys. Anthrop.* (Suppl.) **24**: 74 (abstract).

Berger, L. R. & Tobias, P. V. (1996). A chimpanzee-like tibia from Sterkfontein, South Africa and its implications for the interpretation of bipedalism in *Australopithecus africanus*. *J. Hum. Evol.* **30**: 343–8.

Bergmann, G., Graichen, F. & Rohlmann, A. (1990). Instrumentation of a hip joint prosthesis. In G. Bergmann, F. Graichen & A. Rohlmann (eds.), *Implantable Telemetry in Orthopaedics*, pp. 35–63. Berlin: Freie Universität.

Bernor, R. L., Flynn, L. J., Harrison, T., Hussain, S. T. & Kelley, J. (1988). *Dionysopithecus* from southern Pakistan and the biochronology and biogeography of early Eurasian catarrhines. *J. Hum. Evol.* **17**: 339–58.

Bondioli, L. (1999). Investigating trabecular architectures by means of advanced digital image processing techniques. *IVth Int. Anthrop. Congress of Aleš Hrdlička*, Workshop on *Computed-Assisted Palaeoanthropology & Palaeoprimatology*, Prague. (Abstract.)

Cannon, C. H. & Leighton, M. (1994). Comparative locomotory ecology of gibbons and macaques: selection of canopy elements for crossing gaps. *Am. J. Phys. Anthrop.* **93**: 505–24.

Cant, J. G. H. (1987). Positional behavior of female Bornean orangutans (*Pongo pygmaeus*). *Am. J. Primatol.* **12**: 71–90.

Clarke, R. J. & Tobias, P. V. (1995). Sterkfontein Member 2 foot bones of the oldest South African hominid. *Science* **269**: 521–4.

Cocchi, I. (1872). Su di due scimmie fossili italiane. *Boll. Comit. Geol. Ital.* **3**: 59–71.

Conroy, G. C. (1990). *Primate Evolution*. New York: W. W. Norton & Co.

Conroy, G. C. & Vannier, M. W. (1991a). Dental development in South African australopithecines. Part I: Problems of pattern and chronology. *Am. J. Phys. Anthrop.* **86**: 121–36.

Conroy, G. C. & Vannier, M. W. (1991b). Dental development in South African australopithecines. Part II: Dental stage assessment. *Am. J. Phys. Anthrop.* **86**: 137–56.

Coppens, Y., Thackeray, J. F. & Galichon, V. (1998). Ilio-ischiatic and sacro-pubic structures recognized from CT-scans of South African hominid fossils. *Dual Congress 1998*, Sun City. (Abstract.)

Correnti, V. (1952a). Sulla disposizione delle trabecole dell'osso dell'anca in relazione alla loro morfodinamica (studio radiologico). *Atti. Acc. Naz. Lincei*, **12**: 518–23.

Correnti, V. (1952b). Osservazioni morfofisiologiche sull'osso dell'anca e sul bacino secondo un nuovo piano di orientamento. *Atti. Acc. Naz. Lincei*, **12**: 729–37.

Correnti, V. (1955). Le basi morfomeccaniche della struttura dell'osso iliaco. *Riv. Antrop.* **42**: 289–336.

Correnti, V. (1957). L'architettura del bacino umano ed il suo piano di orientamento fisiologico. *Riv. Antrop.* **44**: 3–68.

Currey, J. (1984). *The Mechanical Adaptations of Bones*. Princeton: Princeton University Press.

Currey, J. (1986). Power law models for the mechanical properties of cancellous bone. *Eng. Med.* **15**: 53–154.

Dalstra, M. & Huiskes, R. (1990). The pelvic bone as a sandwich construction: a three dimensional finite element study. *Proc. ESB* **7**: B32.

Dalstra, M. & Huiskes, R. (1995). Load transfer across the pelvic bone. *J. Biomech.* **28**: 715–24.

Dalstra, M., Huiskes, R., Odgaard, A. & Van Erning, L. (1993). Mechanical and textural properties of pelvic trabecular bone. *J. Biomech.* **26**: 523–35.

Day, M. H. (1971). Postcranial remains of *Homo erectus* from Bed IV, Olduvai Gorge, Tanzania. *Nature* **232**: 383–7.

Day, M. H. (1973). Locomotor features of the lower limb in hominids. *Symp. Zool. Soc. London* **33**: 29–51.

Day, M. H. (1978). Functional interpretation of the morphology of postcranial remains of early African hominids. In C. Jolly (ed.), *Early Hominids of Africa*, pp. 311–45. London: Duckworth.

Day, M. H. (1982). The *Homo erectus* pelvis: Punctuation or gradualism? *Congr. Int. Pal. Hum.*, pp. 411–421. Nice.

Day, M. H. (1986). Bipedalism: pressures, origins and modes. In B. Wood, L. Martin & P. Andrews (eds.), *Major Topics in Primate and Human Evolution*, pp. 188–202. Cambridge: Cambridge University Press.

Dehaut, E. G. (1911). Animaux fossiles du Cap Figari. *Mat. servir Hist. Zool. Paléont. Iles Corse Sardaigne* (Paris) **3**: 53–9.

Dehaut, E. G. (1914). Animaux fossiles du Cap Figari. *Mat. servir Hist. Zool. Paléont. Iles Corse Sardaigne* (Paris) **5**: 77.

Delson, E. (1979). *Oreopithecus* is a cercopithecoid after all. *Am. J. Phys. Anthrop.* **50**: 431–2.

Delson, E. (1980a). Fossil macaques, phyletic relationships and a scenario of deployment. In D. E. Lindburg (ed.), *The Macaques, Studies in Ecology, Behaviour and Evolution*, pp. 10–30. New York: Van Nostrand.

Delson, E. (1980b). A study of fossil Old World monkeys of the circum-Mediterranean region. *Nat. Geogr. Soc.* **944**: 125–8.

Delson, E. (1986). An anthropoid enigma: Historical introduction to the study of *Oreopithecus bambolii*. *J. Hum. Evol.* **15**: 523–31.

Delson, E. & Andrews, P. (1975). Palaeoecology and zoogeography of the Old World monkeys. In R. Tuttle (ed.), *Primate Functional Morphology and Evolution*, pp. 37–64. The Hague: Mouton.

Doran, D. M. (1993a). Comparative locomotor behavior of chimpanzees and bonobos: the influence of morphology on locomotion. *Am. J. Phys. Anthrop.* **91**: 83–98.

Doran, D. M. (1993b). Sex differences in adult chimpanzee positional behavior: The influence of body size on locomotion and posture. *Am. J. Phys. Anthrop.* **91**: 99–115.

Doran, D. M. (1997). Ontogeny of locomotion in mountain gorillas and chimpanzees. *J. Hum. Evol.* **32**: 323–44.

Einhorn, T. A. (1996). Biomechanics of bone. In J. P. Bilezikian, L. G. Raisz & G. A. Rodan (eds.), *Principles of Bone Biology*, pp. 25–35. San Diego: Academic Press.

Fajardo, R. J., Ryan, T. M. & Kappelman, J. (1999). Quantitative analysis of primate trabecular bone architecture: comparison of High Resolution X-ray Computed Tomography and histologic sections. *Am. J. Phys. Anthrop.* **28** (Suppl.): 125 (abstract).

Finlay, J. B., Bourne, R. B., Landsberg, P. D. & Andreae, P. (1986). Pelvic stresses in vitro-I. Malsizing of endoprostheses. *J. Biomech.* **19**: 703–14.

Fleagle, J. G. (1980). Locomotion and posture. In D. J. Chivers (ed.), *Malayan Forest Primates*, pp. 191–207. New York: Plenum Press.

Fleagle, J. G. (1986). The fossil record of early catarrhine evolution. In B. Wood, L. Martin & P. Andrews (eds.), *Major Topics in Primate and Human Evolution*, pp. 130–49. Cambridge: Cambridge University Press.

Fleagle, J. G. (1999). *Primate Adaptation and Evolution*. San Diego: Academic Press.

Forsyth Major, C. J. (1872). Notes sur des singes fossiles trouvés en Italie, précédées d'un apercu sur les quadrumanes fossiles en général. *Atti. Soc. It. Sci. Nat.* **15**: 79–95.

Forsyth Major, C. J. (1880). Beiträge zur Geschichte der fossilen Pferde inbesondere Italiens. *Schweiz. Paläont. Abhandl.* **7**: 1–154.

Forsyth Major, C. J. (1913). Observations sur la faune des Mammiferes quaternaires de la Corse et de la Sardaigne. *Natura* **4**: 594.

Fyhrie, D.P. & Carter, D.R. (1986). A unifying principle relating stress to trabecular bone morphology. *J. Orthop. Res.* **4**: 304–17.

Galichon, V. (1997). *Etude de la Structure Trabeculaire Interne de l'Ilium des Premiers Hominides d'Afrique du Sud (Analyse d'Image Digitales)*. Ph.D. Thesis, Museum d'Histoire Naturelle, Paris.

Galichon, V. & Thackeray, J.F. (1997). CT scans of trabecular bone structures in the ilia of Sts 14 (*Australopithecus africanus*), *Homo sapiens* and *Pan paniscus*. *South Afr. J. Sci.* **93**: 179–80.

Galichon, V., Bondioli, L. & Macchiarelli, R. (1996). Trabecular architecture of the hominid pelvis. *Am. J. Phys. Anthrop.* **22** (Suppl.): 108–9 (abstract).

Gebo, D. L. (1996). Climbing, brachiation, and terrestrial quadrupedalism: Historical precursors of hominid bipedalism. *Am. J. Phys. Anthrop.* **101**: 55–92.

Gentili, S., Mottura, A. & Rook, L. (1998). The Italian fossil primate record: Recent finds and their geological context. *Geobios* **31**: 675–86.

Geraets, W. G. M. (1998). Comparison of two methods for measuring orientation. *Bone* **23**: 383–8.

Geraets, W. G. M., Van der Stelt, P. F., Lips, P., Elders, P. J. M., Van Ginkel, F. C. &

Burger, E. H. (1997). Orientation of the trabecular pattern of the distal radius around the menopause. *J. Biomech.* **30**: 363–70.

Geraets, W. G. M., Van der Stelt, P. F., Lips, P. & Van Ginkel, F. C. (1998). The radiographic trabecular pattern of hips in patients with hip fractures and in elderly control subjects. *Bone* **22**: 165–73.

Gervais, P. (1849). *Zoologie et Paleontologie Française.* Paris: Bertrand.

Gervais, P. (1872a). Coup d'oeil sur les mammiféres d'Italie, suivie de la description d'une espèce nouvelle de singe provenant des lignites du Monte Bamboli. *J. Zool.* **1**: 219–35.

Gervais, P. (1872b). Sur un singe fossile, d'espèce non encore décrite, qui a été découverte au Monte Bamboli. *C. R. Hebdom. Sean. Acad. Sci. Paris* **74**: 1217–23.

Geusa, G., Bondioli, L., Capucci, E., Cipriano, A., Grupe, G., Savorè, C. & Macchiarelli, R. (1999). Osteodental biology of the people of *Portus Romae* (Necropolis of Isola Sacra, 2nd-3rd Cent. AD). II. Dental cementum annulations and age at death estimates. *Digital Archives of Human Paleobiology* **2** (L. Bondioli & R. Macchiarelli, Eds.). Roma: Museo Nazionale 'L. Pigorini' (CD-ROM; ISBN 88-87563-01-2).

Gibson, L. J. (1985). The mechanical behaviour of cancellous bone. *J. Biomech.* **18**: 317–28.

Gittins, S. P. (1983). Use of the forest canopy by the agile gibbon. *Folia Primat.* **40**: 134–44.

Goldstein, S. A., Goulet, R. & McCubbrey, D. (1993). Measurement and significance of three-dimensional architecture to the mechanical integrity of trabecular bone. *Calcif. Tissue Int.* **53** (Suppl. 1): 127–33.

Groves, C. (1972). Systematics and phylogeny of gibbons. *Gibbon and Siamang* **1**: 1–89.

Harrison, T. (1986a). A reassessment of the phylogenetic relationships of *Oreopithecus bambolii* Gervais. *J. Hum. Evol.* **15**: 541–83.

Harrison, T. (1986b). New fossil anthropoids from the middle Miocene of East Africa and their bearing on the origin of the Oreopithecidae. *Am. J. Phys. Anthrop.* **71**: 265–84.

Harrison, T. (1987). The phylogenetic relationships of the early catarrhine primates: a review of the current evidence. *J. Hum. Evol.* **16**: 41–80.

Harrison, T. (1991). The implication of *Oreopithecus bambolii* for the origin of bipedalism. In Y. Coppens & B. Senut (eds.), *Origine(s) de la Bipédie chez les Hominidés,* pp. 235–44. Paris: CNRS.

Harrison, T. & Rook, L. (1994). The taxonomic, phylogenetic and biochronological implications of new *Oreopithecus* specimens from Baccinello V2. *Am. J. Phys. Anthrop.* **18** (Suppl.): 102 (abstract).

Harrison, T. & Rook, L. (1997). Enigmatic anthropoid or misunderstood ape? The phylogenetic status of *Oreopithecus bambolii* reconsidered. In D. R. Begun, C. V. Ward & M. D. Rose (eds.), *Miocene Hominoid Origins and Adaptations,* pp. 327–62. New York: Plenum Press.

Harrison, T., Delson, E. & Jian, G. (1991). A new species of *Pliopithecus* from the middle Miocene of China and its implications for early catarrhine zoogeography. *J. Hum. Evol.* **21**: 329–61.

Hayes, W. C. & Snyder, B. (1981). Toward a quantitative formulation of Wolff's Law in trabecular bone. In S. C. Cowin (ed.), *Mechanical Properties of Bone*, pp. 43–68. New York: ASME.

Hunt, K. D. (1991a). Positional behavior in the Hominoidea. *Int. J. Primatol.* **12**: 95–118.

Hunt, K. D. (1991b). Mechanical implications of chimpanzee positional behavior. *Am. J. Phys. Anthrop.* **86**: 521–36.

Hunt, K. D., Cant, J. G. H., Gebo, D. L., Rose, M. D., Walker, S. E. & Youlatos, D. (1996). Standardized descriptions of primate locomotor and postural modes. *Primates* **37**: 363–87.

Hürzeler, J. (1949). Neubeschreibung von *Oreopithecus bambolii* Gervais. *Schweiz. Paläont. Abhandl.* **66**: 1–20.

Hürzeler, J. (1954a). Zur systematischen Steleung von *Oreopithecus*. *Verh. Naturf. Ges. Basel*, **65**: 88–95.

Hürzeler, J. (1954b). Contribution à l'odontologie et à la phylogénèse du gendre *Pliopithecus* Gervais. *Ann. Paléont.* **40**: 1–63.

Hürzeler, J. (1958). *Oreopithecus bambolii* Gervais: A preliminary report. *Verh. Naturf. Ges. Basel*, **69**: 1–47.

Hürzeler, J. (1960). The significance of *Oreopithecus* in the genealogy of man. *Triangle* **4**: 164–74.

Hürzeler, J. (1968). Questions et reflexions sur l'histoire des Anthropomorphes. *Ann. Paléont.* **54**: 195–233.

Hürzeler, J. & Engesser, B. (1976). Les faunes de mammifères néogènes du Bassin de Baccinello (Grosseto, Italie). *C.R. Acad. Sci. Paris* **296**: 497–503.

Jouffroy, F. K., Stack, M. H. & Niemitz, C., Eds. (1990). *Gravity, Posture and Locomotion in Primates*. Firenze: Il Sedicesimo.

Köhler, M. & Moyà-Solà, S. (1997). Ape-like or hominid-like? The positional behavior of *Oreopithecus bambolii* reconsidered. *Proc. Nat. Acad. Sci. USA* **94**: 11747–50.

Lanyon, L. E. (1974). Experimental support for the trajectorial theory of bone structure. *J. Bone Joint Surg.* **56 B**: 160–6.

Latarjet, A. & Gallois, A. (1910). L'architecture intérieure de l'os iliaque et de la ceinture pelvienne. *Bibliograph. Anat.* **20**: 55–69.

Latimer, B. M. & Ward, C. V. (1998). Locomotion in *Australopithecus*: Evidence from the leg and foot. *Dual Congress 1998*, Sun City. (abstract.)

Leakey, M. D. (1971). *Olduvai Gorge. Volume 3. Excavations in Beds I and II, 1960–1963*. Cambridge: Cambridge University Press.

Linde, F., Nørgaard, P., Hvid, I., Odgaard, A. & Søballe, K. (1991). Mechanical properties of trabecular bone. Dependency on strain rate. *J. Biomech.* **24**: 803–9.

Lovejoy, C. O. (1988). Evolution of human walking. *Sc. Am.* **259**: 118–25.

Luo, Z. P. & An, K. N. (1998). A theoretical model to predict distribution of the fabric tensor and apparent density in cancellous bone. *J. Math. Biol.* **36**: 557–68.

Macchiarelli, R. & Bondioli, L. (1994). Linear densitometry and digital image processing of proximal femur radiographs: implications for archaeological and forensic anthropology. *Am. J. Phys. Anthrop.* **93**: 109–22.

Macchiarelli, R., Bondioli, L., Coppens, Y. & Galichon, V. (1995). L'architettura trabecolare dell'osso dell'anca negli ominidi. In C. Peretto & S. Milliken (eds.), *11th Congr. Antrop. Ital. L'Adattamento Umano all'Ambiente. Passato e Presente*, pp. 45–6. Isernia: C. Iannone Ed. (Abstract.)

Macchiarelli, R., Bondioli, L., Galichon, V. & Tobias, P. V. (1999c). Hip bone trabecular architecture shows uniquely distinctive locomotor behaviour in South African australopithecines. *J. Hum. Evol.* **36**: 211–32.

Macchiarelli, R., Bondioli, L. & Rook, L. (1999a). Locomotion-related iliac trabecular patterns in fossil and extant primates. *Am. J. Phys. Anthrop.* **28** (Suppl.): 187–8 (abstract.)

Macchiarelli, R., Galichon, V., Bondioli, L. & Tobias, P. V. (1996b). Hip bone trabecuar architecture and locomotor behaviour in South African australopithecines. In C. Peretto (ed.) *XIII Int. Congr. Prehist. Protohist. Sc. The Sections, Abstracts*, vol. 1, pp. 109–10. Forlì: ABACO. (Abstract.)

Macchiarelli, R., Galichon, V., Bondioli, L. & Tobias, P.V. (1998b). Hip bone trabecular architecture and locomotor behaviour in South African australopithecines. In F. Facchini, A. Palma di Cesnola, M. Piperno & C. Peretto (eds.), *Proceedings of the XIII International Congress of Prehistoric and Protohistoric Sciences*, vol. 2, pp. 35–41. Forlì: ABACO.

Macchiarelli R., Galichon V., Geusa G., Rossi P.F., Sperduti A. & Bondioli L. (1996a). Digital image analysis in human paleobiology and paleoanthropology. *10th Congress of the European Anthropological Association*, Brussels. (Abstract.)

Macchiarelli, R., Geusa, G., Rossi, P.F., Salomone, F., Sperduti, A & Bondioli, L. (1998a). Tecnologie avanzate in paleobiologia umana: verso un accesso pubblico alle collezioni odontoscheletriche. In Consiglio Nazionale delle Recherche (ed.) *Science and Technology for the Safeguard of Cultural Heritage in the Mediterranean Basin*, vol 2, pp. 1411–23. Palermo: Tipolitografia Luxograph.

Macchiarelli, R., Köhler, M., Moyà-Solà, S., Rook, L. & Bondioli, L. (1997a). The *Oreopithecus* hip bone trabecular pattern reconstructed and investigated by means of digital image processing techniques. Comparative analysis with extant hominoids. *Int. Symp. Humans from the Past: Advancement in Research and Technology*, Rome. (Poster.)

Macchiarelli, R., Rook, L. & Bondioli, L. (1997b). Reconstruction of the locomotor behaviour in fossil primates – with special reference to *Oreopithecus* – by means of digital image processing of the hip bone trabecular architecture. *12th Congr. Ass. Antrop. It.*, Palermo, *Antrop. Cont.* (Abstract.)

Macchiarelli, R., Rook, L. & Bondioli, L. (1999b). Architectural Variation of the Iliac Trabecular Structure in Extant and Fossil Primates, Including *Homo*. Atlas of Electronically Processed Radiographic Images. *Digital Archives of Human Paleobiology* **3** (L. Bondioli & R. Macchiarelli, Eds.). Roma: Museo Nazionale 'L. Pigorini' (CD-ROM, in preparation).

Macchiarelli, R., Sperduti, A. & Bondioli, L. (1990). L'indagine radiografica dello scheletro nella attribuzione dell'età alla morte. II. Analisi sperimentale dei corpi vertebrali. *Riv. Antrop.* **68**: 103–27.

MacLatchy, L. M. & Bossert, W. H. (1996). An analysis of the articular surface distribution of the femoral head and acetabulum in anthropoids, with implications for hip function in Miocene hominoids. *J. Hum. Evol.* **31**: 425–53.

Mednick, L. W. (1955). The evolution of the human ilium. *Am. J. Phys. Anthrop.* **13**: 203–16.

Meyer, G. H. (1867). Die architektur der spongiosa. *Arch. Anat. Physiol. wiss. Med.* **34**: 615–28.

Moyà-Solà, S. & Köhler, M. (1993). Recent discoveries of *Dryopithecus* shed new light on evolution of great apes. *Nature* **365**: 543–5.

Moyà-Solà, S. & Köhler, M. (1995). New partial cranium of *Dryopithecus* Lartet, 1863 (Hominoidea, Primates) from the Upper Miocene of Can Llobaretes, Barcelona, Spain. *J. Hum. Evol.* **29**: 101–39.

Moyà-Solà, S. & Köhler, M. (1996). A *Dryopithecus* skeleton and the origins of great-ape locomotion. *Nature* **379**: 156–9.

Moyà-Solà, S. & Köhler, M. (1997). The phylogenetic relationship of *Oreopithecus bambolii* Gervais 1872. *C.R. Acad. Sci. Paris* **324**: 141–8.

Moyà-Solà, S., Köhler, M. & Rook, L. (1999). Evidence of hominid-like precision grip capability in the hand of the Miocene ape *Oreopithecus. Pro. Nat. Acad. Sci. USA* **96**: 313–17.

Odgaard, A., (1997). Three-dimensional methods for quantification of cancellous bone architecture. *Bone* **20**: 315–28.

Odgaard, A., Kabel, J., van Rietbergen, B., Dalstra, M. & Huiskes, R. (1997). Fabric and elastic principal directions of cancellous bone are closely related. *J. Biomech.* **30**: 487–95.

Ohman, J. C., Krochta, T. J., Lovejoy, C. O., Mensforth, R. P. & Latimer, B. (1997). Cortical bone distribution in the femoral neck of hominoids: Implications for the locomotion in *Australopithecus afarensis. Am. J. Phys. Anthrop.* **104**: 117–31.

Oonishi, H., Isha, H. & Hasegawa, T. (1983). Mechanical analysis of the human pelvis and its application to the artificial hip joint-by means of the three dimensional finite element method. *J. Biomech.* **16**: 427–44.

Oxnard, C. E. (1970). The use of optical data analysis in functional morphology: Investigation of vertebral trabecular patterns. *Burg Wartenst. Symp.* **48**: 1–30.

Oxnard, C. E. (1997). From optical to computational Fourier transforms: the natural history of an investigation of the cancellous structure of bone. In P. E. Lestrel (ed.), *Fourier Descriptors and their Applications in Biology*, pp. 379–408. Cambridge: Cambridge University Press.

Oxnard, C. E. & Yang, H. C. L. (1981). Beyond biometrics: Studies of complex biological patterns. *Symp. Zool. Soc. London* **46**: 127–67.

Peretz, A. M., Hipp, J. A. & Heggeness, M. H. (1998). The internal bony architecture of the sacrum. *Spine* **23**: 971–4.

Portis, A. (1917). I primi avanzi di quadrumani del suolo di Roma. *Boll. Soc. Geol. It.*, **35**: 239–78.

Preuschoft, H. (1975). Body posture and mode of locomotion in fossil primates. Method and example: *Aegyptopithecus zeuxis*. In S. Kondo, M. Kawai, A. Ehara & S. Kawamura (eds.), *Proceedings from the Symposia of the Fifth Congress of the International Primatological Society*, pp. 345–59. Tokyo: Japan Science Press.

Quesnel, T., Gueritey, P.M. & Gonon, G.P. (1995). Biomechanics of the hip: forces exerted during walking. *Surg. Radiol. Anat.* **17**: 249–53.

Rafferty, K. L. (1998) Structural design of the femoral neck in primates. *J. Hum. Evol.* **34**: 361–83.

Reynolds, T. R. (1985). Mechanics of increased support of weight by the hindlimbs. *Am. J. Phys. Anthrop.* **67**: 335–49.

Rightmire, G. P. (1993). *The Evolution of* Homo erectus. *Comparative Anatomical Studies of an Extinct Human Species.* Cambridge: Cambridge University Press.

Ristori, G. (1890). Le scimmie fossili italiane. *Boll. Regio Com. Geol. Italia.* **1**: 178–96.

Rook, L. (1993). A new find of *Oreopithecus* (Mammalia, Primates) in the Baccinello Basin (Grosseto, southern Tuscany). *Riv. It. Paleont. Stratigr.* **99**: 255–62.

Rook, L. (1997). *Macaca sylvanus florentina* (Cocchi, 1872) from the Pleistocene Mugello basin (Northern Apennines, Italy). *Boll. Soc. Paleont. It.* **35**: 357–9.

Rook, L., Bondioli, L., Köhler, M., Moyà-Solà, S. & Macchiarelli, R. (1999). *Oreopithecus* was a bipedal ape after all: Evidence from the iliav cancellous architecture. *Proc. Natl. Acad. Sci. USA* **96**: 8795–9.

Rook, L., Harrison, T. & Engesser, B. (1996). The taxonomic status and biochronological implications of new finds of *Oreopithecus* from Baccinello (Tuscany, Italy). *J. Hum. Evol.* **30**: 3–27.

Rook, L., Macchiarelli, R., & Bondioli, L. (1997). Digital image processing of the hip bone trabecular architecture for locomotor behaviour reconstruction in fossil primates. *Geoitalia 1997.* (Abstract.)

Rose, M. D. (1984). A hominine hip bone, KNM-ER 3228, from East Lake Turkana, Kenya. *Am. J. Phys. Anthrop.* **63**: 371–8.

Rossi, P. F., Bondioli, L., Geusa, G. & Macchiarelli, R. (1999). Osteodental Biology of the People of *Portus Romae* (Necropolis of Isola Sacra, 2nd–3rd Cent. AD). I. Enamel Microstructure and Developmental Defects of the Primary Dentition. In (Bondioli, L. & Macchiarelli, R. (eds.), *Digital Archives of Human Paleobiology*, vol. 1. Roma: Museo Nazionale 'L. Pigorini' (CD-ROM; ISBN 88-87563-00-4).

Ruff, C. B., Trinkaus, E. & Holliday, T. W. (1997). Body mass and encephalization in Pleistocene *Homo. Nature* **387**: 173–6 (and relevant supplementary information on www.nature.com).

Rütimeyer, L. (1876). *Über Pliocen und eisperiode auf beiden Seiten der Alpen.* Basel: Georg Publishing House.

Sanders, W. J. (1998). Comparative morphometric study of the australopithecine vertebral series Stw-H8/841. *J. Hum. Evol.* **34**: 249–302.

Sarmiento, E. E. (1987). The phylogenetic position of *Oreopithecus* and its significance in the origin of the Hominoidea. *Am. Mus. Novit.* No. 2881, pp. 1–44.

Sarmiento, E. E. (1998). Generalized quadrupeds, committed bipeds, and the shift to open habitats: An evolutionary model of hominid divergence. *Am. Mus. Novit.* No. 3250, pp. 1–78.

Schlosser, M. (1887). Die Affen, Lemuren, Chiropteren usw. aus dem europaischen Teriär. *Beitr. Paläont. Geol. Inst. Hung.* **6**: 1–162.

Simons, E. G. (1972). *Primate Evolution: An Introduction to Man's Place in Nature.* New York: MacMillian.

Simons, E. G. & Fleagle, J. G. (1973). The history of extinct gibbon-like primates. *Gibbon and Siamang* **2**: 121–48.

Spoor, F. & Zonneveld, F. (1997). The bony labyrinth of Sts 5 ('Mrs Ples'). *South Afr. J. Sci.* **93**: 180.

Spoor, F., Wood, B. A. & Zonneveld, F. (1994). Implications of early hominid labyrinthine morphology for evolution of human bipedal locomotion. *Nature* **369**: 645–8.

Srikosamatara, S. (1984). Ecology of pileated gibbons in South-East Thailand. In H. Preuschoft, D. J. Chivers, W. Y. Brockelman & N. Creel (eds.), *The Lesser Apes*, pp. 242–57. Edinburgh: Edinburgh University Press.

Sugardjito, J. & van Hoof, J. (1986). Age-sex class differences in the positional behaviour of the Sumatran Orang-utan (*Pongo pygmaeus abelii*) in the Gunung Leuser National Park, Indonesia. *Folia Primat.* **47**: 14–25.

Susman, R. L. (1984). The locomotor behavior of *Pan paniscus* in the Lomako Forest. In R. L. Susman (ed.), *The Pygmy Chimpanzee*, pp. 369–93. New York: Plenum Press.

Szalay, F. S. (1975). Haplorhine relationships and the status of the Anthropoidea. In R. Tuttle (ed.), *Primate Functional Morphology and Evolution*, pp. 3–22. The Hague: Mouton.

Szalay, F. S. & Berzi, A. (1973). Cranial anatomy of *Oreopithecus*. *Science* **180**: 183–6.

Szalay, F. S. & Delson, E. (1979). *Evolutionary History of the Primates*. New York: Academic Press.

Thomason, J. J. (1985a). Estimation of the locomotory forces and stresses acting on the limb bones of recent and extinct equids. *Paleobiol.* **11**: 287–98.

Thomason, J. J. (1985b). The relationship of trabecular architecture to inferred loading patterns in the third metacarpals of the extinct equids *Merychippus* and *Mesohippus*. *Paleobiol.* **11**: 323–35.

Thomason, J. J. (1995). To what extent can the mechanical environment of a bone be inferred from its internal architecture? In J. J. Thomason (ed.), *Functional Morphology in Vertebrate Paleontology*, pp. 249–63. Cambridge: Cambridge University Press.

Turner, C. H. (1992). On Wolff's law of trabecular architecture. *J. Biomech.* **25**: 1–9.

Turner, C. H., Cowin, S. C., Rho, J. Y., Ashman, R. B. & Rice, J. C. (1990). The fabric dependence of the orthotropic elastic constants of cancellous bone. *J. Biomech.* **23**: 549–61.

Vigne, J. D. (1989). Les Mammiferes Post-Glaciaires de Corse. Etude Archeozoologique. *Gallia Préhist* (Suppl.), **26**: 1–377.

Walker, A. & Ruff, C. B. (1993). The reconstruction of the pelvis. In A. Walker & R. E. F. Leakey (eds.), *The Nariokotome* Homo erectus *Skeleton*, pp. 221–33. Cambridge: Harvard University Press.

Watkins, C., Sadun, A. & Marenka, S. (1993). *Modern Image Processing*. Boston: Academic Press.

Wolff, J. (1870). Uber die innere architektur der knochen unk ihre bedeutung fur die frage vom knochenwachstum. *Arch. Pathol. Anat. Physiol. Klin. Med. Virchows Arch.* **50**: 389–453.

Wunderlich, R. E., Walker, A. & Jungers, W. L. (1999). Rethinking the positional repertoire of *Oreopithecus*. *Am. J. Phys. Anthrop.* (Suppl.), **28**: 282 (abstract.)

Zapfe, H. (1958). The skeleton of *Pliopithecus* (*Epipliopithecus*) *vindobonensis* Zapfe and Hürzeler. *Am. J. Phys. Anthrop.* **16**: 441–58.

Zapfe, H. (1960). *Die Primatenfunde aus der miozänen Spaltenfüvon Neudorf an der March (Děvínská Nová Ves), Tschechoslowakei. Schweiz. Paläont. Abhandl.* **78**. Basel: Birkhäuser Verlag.

Zapfe, H. & Hürzeler, J. (1957). Die fauna der Miozänen Spaltenfüllung von Neudorf an der March (CSR). Primates. *Sitz. Osterr. Akad. Wiessen.* **166**: 113–23.
Zihlman, A. L. & Hunter, W. S. (1972). A biomechanical interpretation of the pelvis of *Australopithecus*. *Folia Primatologica*, **18**: 1–19.

5 Dental microwear and diet in Eurasian Miocene catarrhines

Tania King

Introduction

Diet is a key factor in primate evolution. It affects adaptation, behaviour and morphology. An understanding of dietary preferences allows inferences to be made about how fossil species interacted with their environment. While the study of dental morphology and gross wear can provide much information about diet, the examination of microscopic wear on the surfaces of teeth can provide more subtle information relating to dietary regime and feeding behaviour in fossil primates. The aim of this chapter is to bring together, from the literature, what is known about the diets of Eurasian catarrhines from the dental microwear evidence. Research to date has focused on the diets and feeding adaptations of Middle and Late Miocene catarrhines from sites in Western and Central Europe and Southern Asia.

Evidence for the diets of catarrhines belonging to the subfamilies Pliopithecinae and Crouzeliinae has been investigated. Pliopithecine species examined include *Pliopithecus vindobonensis* from Neudorf in Slovakia (MN 6), *Pliopithecus platyodon* from the Austrian site of Göriach (MN 6), as well as pliopithecid specimens from Castell de Barbera, in the Vallès de Penedès in northeastern Spain (MN 8), whose taxonomic assignment is less certain. One crouzeliine species, *Anapithecus hernyaki,* from the site of Rudabánya, Hungary (late MN 9), has been analysed to date. The hominoid taxa examined represent four subfamilies of the Hominidae – Dryopithecinae (comprising both the kenyapithecini and the dryopithecini, as well as *Oreopithecus*), Ponginae and the Homininae. The kenyapithecini are represented by *Griphopithecus alpani* from Paşalar, Turkey (MN6). Dryopithecine taxa include *Dryopithecus brancoi* from the Hungarian site of Rudabánya (MN 9), *Dryopithecus crusafonti* from Can Ponsic in northeastern Spain (MN 9) and *Dryopithecus laietanus* from Can Llobateres (MN 9), also located in northeastern Spain. Ponginae specimens examined comprise one species – *Sivapithecus indicus* from the Siwaliks, Pakistan (the equivalent of MN 11 – MN 7/8). *Oreopithecus bambolii* specimens from the sites of Baccinello, Monte Bamboli and Ribolla (MN12–13) have been the subject of dental microwear analysis, and represent the Oreopithecinae. Dietary reconstruction using dental microwear analysis has also been carried out for the hominine *Ouranopithecus macedoniensis* from the localities of Ravin de la Pluie, Xirochori and Nikiti in Macedonia, Greece (MN 10).

Dental microwear

Dental microwear analysis has been routinely used in the investigation of diet in extinct species for more than 20 years. Early research showed that dietary type and seasonal variation in diet could be detected in sympatric hyraxes (Walker *et al.*, 1978). Molar position, facet type, age and sex were found to influence microwear patterns in *Pan troglodytes*, and these variations were related to the interaction of two mechanical factors (shear and compression) during chewing (Gordon, 1980, 1982). Examination of the microwear patterns of a number of extant primate species demonstrated that folivores could be distinguished from frugivores, and that soft fruit eaters could be differentiated from hard-object feeders on the basis of differences in dental microwear patterns (Teaford & Walker, 1984). Seasonal and ecological differences in diet were also found to affect dental microwear patterns, while other research showed that subtle variations in diet may not be detectable in museum specimens (Teaford, 1985; Teaford & Robinson, 1989; Teaford & Glander, 1991; Teaford & Runestad, 1992). Much research has been conducted to examine the causes of microwear formation. Experiments have shown that some food types result in the formation of similar microwear features and that the relative hardness of food items and abrasives produce different effects on enamel (Covert & Kay, 1981; Peters, 1982; Teaford & Oyen, 1989a,b; Lucas, 1994). Attention has focused on the role of grit and plant phytoliths in the formation of dental microwear (Covert & Kay, 1981; Peters, 1982; Ungar *et al.*, 1995). Research has also been carried out to examine the effect of enamel structure and the role different masticatory forces on microwear feature formation (Maas, 1991, 1994).

Analysis of microwear on the anterior dentition also provides information about diet and feeding behaviour. Research on both primates and human populations has shown that it is possible to distinguish between the husking of fruit with an abrasive skin, and leaf and pith stripping in chimpanzees, the ingestion of grit-covered seeds, rhizomes and roots in baboons, and leaf and pith stripping behaviours in gorillas (Ryan, 1981). Microwear present on the anterior dentitions of Inuits has been related to the ingestion of gritty meat and fish, as well as non-dietary activities (Ryan & Johanson, 1989). Kelley (1990) found variations in incisor microwear between species with the same broad dietary preferences and suggested that these differences relate to variations in feeding behaviour and/or the physical properties of certain food items. Other research has shown that the microwear patterns of folivore–frugivores are distinct from those of frugivore–insectivores (Ungar, 1990a,b). Examination of the incisor microwear of wild-shot museum primate specimens has been compared with the feeding behaviour observed in the same species in the wild (Ungar, 1992, 1994). Results indicate that scratch density is linked to the amount of anterior tooth use when abrasive foods are consumed. Striation width was related to the proportion of phytoliths and grit ingested, and the orientation of striations was associated with the direction in which food items were pulled across the teeth.

Dental microwear studies of extant and fossil primate dentitions have generally centered on the examination of wear patterns on incisor and molar teeth. In the case of molar teeth, research has focused on occlusal wear facets. The chewing cycle, in simple terms, consists of puncture crushing – where food is broken down so that molar surfaces can approach each other closely enough for chewing to take place – and two chewing phases (Kay & Hiiemae, 1974; Kay, 1977). Phase I occurs when the mandible moves upwards towards the maxilla and into central occlusion where cusps interlock. During this phase of the chewing cycle food is sheared and forces are exerted parallel to the occlusal surfaces. During Phase II of the chewing cycle the mandible moves out of central occlusion. Forces applied include elements of both shearing and crushing – both parallel and perpendicular to the direction of jaw movement. Ten molar wear facets have been described – eight Phase I facets and two Phase II facets (Butler, 1952; Kay & Hiiemae, 1974). An additional Phase II facet was subsequently identified by Kay (1977).

Microwear features are usually classified as pits and scratches on the basis of a length to width ratio of 4 : 1 (e.g. Solounias *et al.*, 1988; Teaford, 1988; Teaford & Runestad, 1992; Ungar, 1996; King *et al.*, 1999). That is, all features with a length to width ratio below 4 : 1 are classified as pits, and all those with ratios equal to or above 4 : 1 are defined as scratches. Research has shown that for molar teeth pit percentages (the proportion of pits relative to total microwear features) is a good indicator of dietary behaviour. This microwear variable distinguishes primate folivores and frugivores on the one hand, and primate soft fruit eaters and hard object feeders on the other (Teaford & Walker, 1984; Teaford, 1988). Folivores have lower percentages of pits than frugivores, and primate hard object feeders have higher percentages of pits than soft fruit eaters do.

Results

The dental microwear of a number of Eurasian catarrhines has been examined by several researchers. Ungar (1996) has investigated the incisor and molar microwear of *Pliopithecus platyodon* and *P. vindobonensis*, the pliopithecid specimens from Castell de Barbera, Spain, and *Anapithecus*

Table 5.1. *Miocene Eurasian catarrhine taxa which have been the subject of dental microwear analysis*

Teaford & Walker (1984); Teaford (1988)	Ungar (1996)	King *et al.* (1999)
Sivapithecus indicus	*Anapithecus hernyaki*	*Griphopithecus alpani*
	Dryopithecus brancoi	
	D. crusafonti	
	D. laietanus	
	Oreopithecus bambolii	
	Ouranopithecus macedoniensis	
	Pliopithecus platyodon	
	P. vindobonensis	
	pliopithecid specimens from Castell de Barbera	

hernyaki (Table 5.1). The microwear patterns of *Dryopithecus brancoi, D. crusafonti, D. laietanus, Oreopithecus bambolii* and *Ouranopithecus macedoniensis* were also examined in this study. For the study of molar microwear, facet 9 (Kay & Hiiemae, 1974; Kay, 1977), a Phase II facet, was examined. Results from the molar microwear analysis indicate that percentages of pits range from 17% in *Oreopithecus bambolii* to 58% in *Ouranopithecus macedoniensis* (Figure 5.1). *Dryopithecus* spp. displayed 38% pits, *Pliopithecus* spp. possessed 37% pits and *A. hernyaki* displayed 45% pits, and these taxa are intermediate to *O. macedoniensis* and *Oreopithecus bambolii* in this microwear variable. The pliopithecid specimens from Castell de Barbera had 17% pits and are similar to *O. bambolii* in this respect. Significant differences were found between *O. bambolii* and *Ouranopithecus macedoniensis* in percentage pitting, while no significant differences were indicated for the remaining taxa.

These results suggest that *Oreopithecus bambolii* and the pliopithecids from Castell de Barbera are more similar to extant primate folivores in the percentages of pits they display. *Dryopithecus* spp., *Pliopithecus* spp., and *A. hernyaki* and are similar to primate soft fruit eaters in percentage pitting. *Ouranopithecus macedoniensis* displays high percentages of pits and this indicates that it may have been ingesting hard objects habitually.

Microwear on the labial surfaces of upper central incisors was also examined in the study by Ungar (1996). Due to a lack of suitable specimens *Oreopithecus bambolii* and the Castell de Barbera specimens could not be included in the analysis. Results indicated that these Miocene catarrhines varied significantly according to the density of striations they displayed. *Ouranopithecus macedoniensis* and *P. platyodon* had significantly more

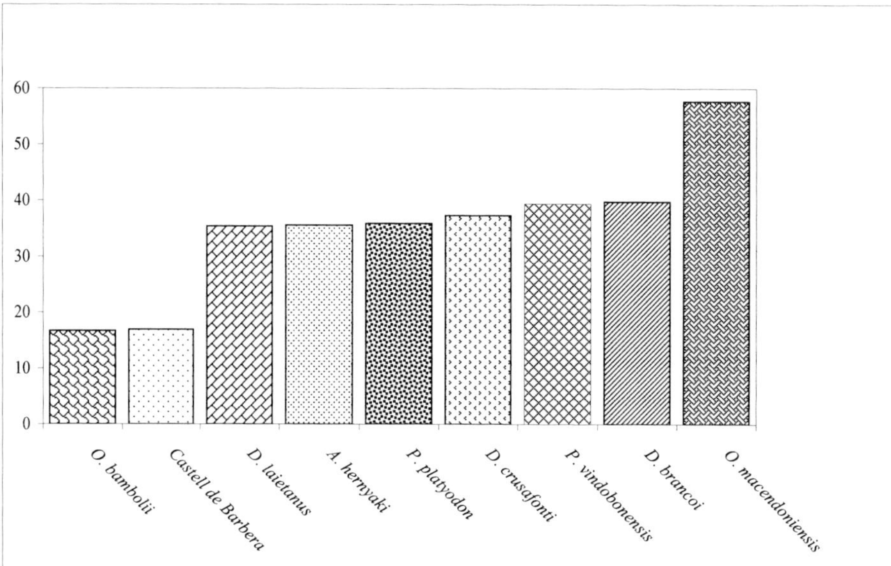

[Figure 5.1]
Percentage pitting on molar Phase II facets in European Miocene catarrhine taxa. Data from Ungar (1996). (*Oreopithecus bambolii; Dryopithecus laietanus; Anapithecus hernyaki; Pliothecus platyodon; Dryopithecus crusafonti; Pliothecus vindobonensis; Dryopithecus brancoi;* Ouranopithecus macedoniensis.)

striations than *Dryopithecus* spp. *Dryopithecus* spp. and *A. hernyaki* displayed similar numbers of scratches.

Species also varied according to their striation dimensions. *O. macedoniensis* displayed narrower striations than all other taxa (Figure 5.2). Statistical tests revealed that *O. macedoniensis* had narrower scratches than *Dryopithecus* spp. *P. platyodon* also had narrower striations than all other taxa apart from *O. macedoniensis*, although this was not a significant difference. The results suggest that these taxa were using their anterior dentitions to ingest small or angular objects more regularly than the other species. Feature orientation also varied among these catarrhine taxa. *O. macedoniensis* had significantly more striations that were mesiodistally oriented than the other species examined. This may indicate that *O. macedoniensis* used its anterior teeth more often for the stripping of foods laterally than the other taxa included in the study.

The microwear patterns of *Sivapithecus indicus* has been examined by Teaford & Walker (1984). This data was later re-analysed using a scratch: pit ratio of 4: 1 (Teaford, 1988). The microwear patterns of a number of extant primates which included both hard and soft object feeders were also investigated in this study. *S. indicus* displayed similar percentages of pits (35%) to *Pan troglodytes* and was intermediate to primate leaf-eaters and hard object

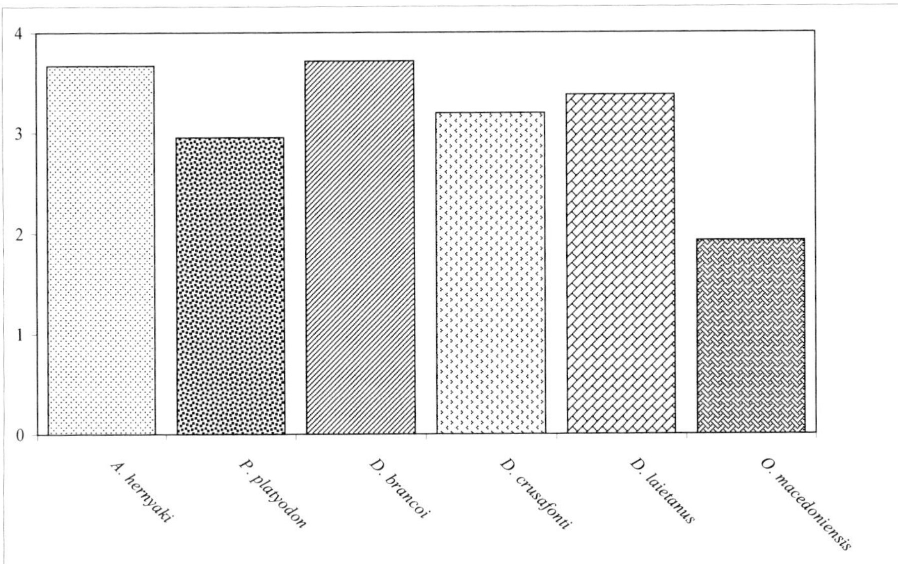

[Figure 5.2]
Striatum widths (μm) on incisor surfaces of European Miocene catarrhine taxa. Data from Ungar (1996). (Species list as for Figure 5.1.)

feeders for this microwear variable. The authors concluded that *S. indicus* had a diet that was similar to that of the chimpanzee.

Research has also focused on the microwear patterns of *Griphopithecus alpani* from the site of Paşalar in northwestern Turkey (King *et al.*, 1999). The molar microwear of *G. alpani* was compared with that of three extant hominoid taxa – *Gorilla gorilla gorilla, Pan troglodytes verus* and *Pongo pygmaeus pygmaeus*. All three extant taxa are predominantly frugivorous. *G. g. gorilla* – the western lowland gorilla – has a diet which, in contrast to that of the mountain gorilla, *Gorilla gorilla berengei*, is comprised largely of fruit. Research at Lopé and Belinga in Gabon has shown that fruit forms between 45% and 79% of the diet of this sub-species (Tutin & Fernandez, 1985, 1987; Tutin *et al.*, 1991). The remainder of the diet is comprised of young and mature leaves, seeds, pith, bark, insects, and roots, gall and fungi. Research at Gombe, Tanzania, and also in Gabon has shown that fruit comprises between 48% and 68% of all dietary items ingested by *Pan troglodytes verus* (Hladik, 1977; Goodall, 1986). The rest of the diet consists of leaves and leaf buds, seeds, blossoms, stems, pith, bark and resin. The chimpanzees at Gombe also supplement their diet with birds, and small and medium-sized mammals, as well as insects (Goodall, 1986). Fruit forms between 54% and 85% of the diet of *Pongo p. pygmaeus* (see references in Rodman, 1988). In contrast to *Gorilla gorilla gorilla* and *Pan troglodytes verus, Pongo pygmaeus pygmaeus* ingests hard and very large fruits, and

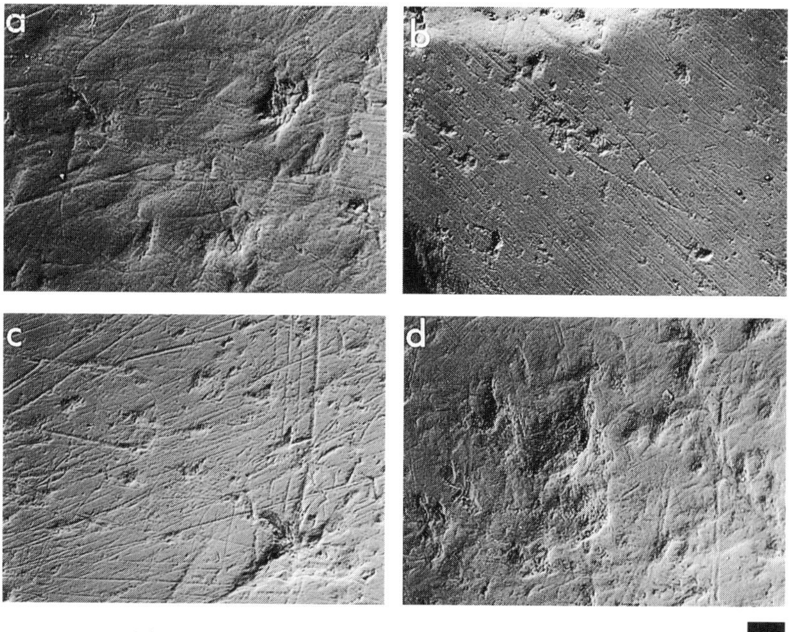

[Figure 5.3]
Interspecific variation in the Phase II (facet 9) dental microwear patterns of extant hominoids and *Griphopithecus alpani*. (a) *Gorilla gorilla gorilla*, (b) *Pan troglodytes verus*, (c) *Pongo pygmaeus pygmaeus* and (d) *G. alpani*. Scale bar represents 50 μm. Published in King *et al.* (1999).

these items form about 35% of all fruits consumed (MacKinnon, 1977). It has also been found that *P. p. pygmaeus* ingests fruits with harder husks and more fleshy pericarps, and also consumes unripe fruit more frequently than other sympatric primates (Ungar, 1992, 1995). *P. p. pygmaeus* supplements its diet with leaves, shoots, bark, wood, insects, eggs, fungi, honey, soil, lianas and epiphytes (MacKinnon, 1977).

Microwear features on three molar occlusal facets were investigated by King *et al.* (1999). One Phase I facet – facet 1 – and two Phase II facets – facets x and 9 – were examined. Analysis of variance revealed significant differences for several microwear variables among the extant apes and *Griphopithecus alpani* (Figure 5.3). Both *P. p. pygmaeus* and *G. alpani* displayed greater frequencies of microwear features than *Pan troglodytes verus* and *Gorilla gorilla gorilla*, and the difference between *Griphopithecus alpani* and *Gorilla gorilla gorilla* was significant (Figure 5.4). Pit density varied significantly according to taxa with *Griphopithecus alpani* and *Pongo pygmaeus pygmaeus* displaying greater numbers of pits than *Gorilla gorilla gorilla*. In addition, *Griphopithecus alpani* displayed significantly larger frequencies of pits than *Pan troglodytes verus* (Figure 5.5). Percentage pitting (the proportion of pits relative to total frequencies of features) was higher in *G. alpani* (55%) and *Pongo pygmaeus pygmaeus* (43%) than in *Gorilla gorilla gorilla* (33%) and *Pan troglodytes verus* (40%), with *Griphopithecus alpani* having significantly greater percentages of pits than *Gorilla gorilla gorilla* (Figure 5.6). Lastly, striation width varied according to taxa. The analysis revealed that *Gorilla gorilla gorilla* and *Pan troglodytes verus* displayed significantly narrower scratches than do *Pongo pygmaeus pygmaeus* and *Griphopithecus alpani* (Figure 5.7).

Research by Teaford (1988) has shown that it is possible to relate dental microwear in living primates to broad dietary categories. The diet of three groups of extant hominoid examined by King *et al.* (1999) is mainly comprised of fruit and the percentages of pits displayed by these taxa is within the range described by Teaford (1988) for living primates which are frugivorous. Of the living great apes only *P. pygmaeus pygmaeus* is known to ingest hard objects on occasion. This difference in the diets of the extant hominoids is reflected in their microwear patterns, with *P. p. pygmaeus* displaying significantly higher densities of features – i.e. pits – and higher percentages of pits (although this was not a significant result) than *Gorilla gorilla gorilla* and *Pan troglodytes verus*.

The results of the study by King *et al.* (1999) indicate that *Griphopithecus alpani* was predominantly frugivorous. Of the extant great apes, *G. alpani* was more similar in its microwear to *Pongo pygmaeus pygmaeus*. This suggests that *G. alpani* had a diet which was more similar to that of *P. p. pygmaeus* than to that of *Gorilla gorilla gorilla* or *Pan troglodytes verus*. The high percentages of pits displayed by *Griphopithecus alpani* compared to *Pongo pygmaeus pygmaeus* may indicate that it ingested harder fruits and/or objects than does the orang-utan.

Clear differences between the patterns of microwear on Phase I and Phase II facets, irrespective of species, also emerged from the study by King *et al.* (1999). The Phase II facets examined were very similar in their microwear patterns and only one significant difference was indicated between facets 9 and x – facet x displayed more striations than facet 9. Both Phase II facets differ from the Phase I wear surface examined in having more microwear features, greater frequencies of pits and striations and higher percentages of pits than facet 1. The lack of difference between the two Phase II facets is important in terms of sampling in microwear studies. Dietary reconstructions based on molar occlusal microwear are generally made using Phase II wear facets. Where one Phase II facet may be damaged by postmortem processes, sample sizes may be increased by substituting it with another Phase II facet.

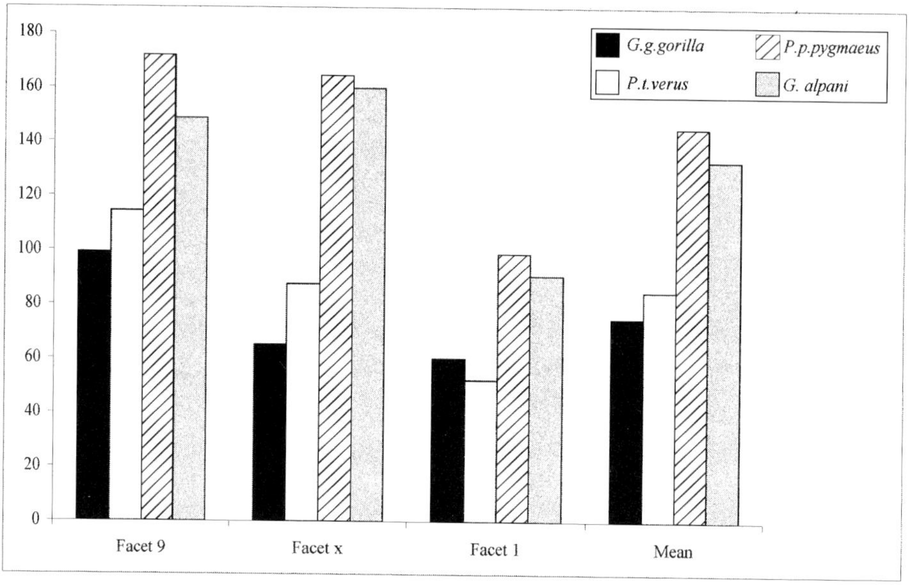

[Figure 5.4]
Total microwear frequency comparisons for *Gorilla gorilla gorilla*, *Pan troglodytes verus*, *Pongo pygmaeus pygmaeus*, and *Griphopithecus alpani*. Data from King *et al*. (1999).

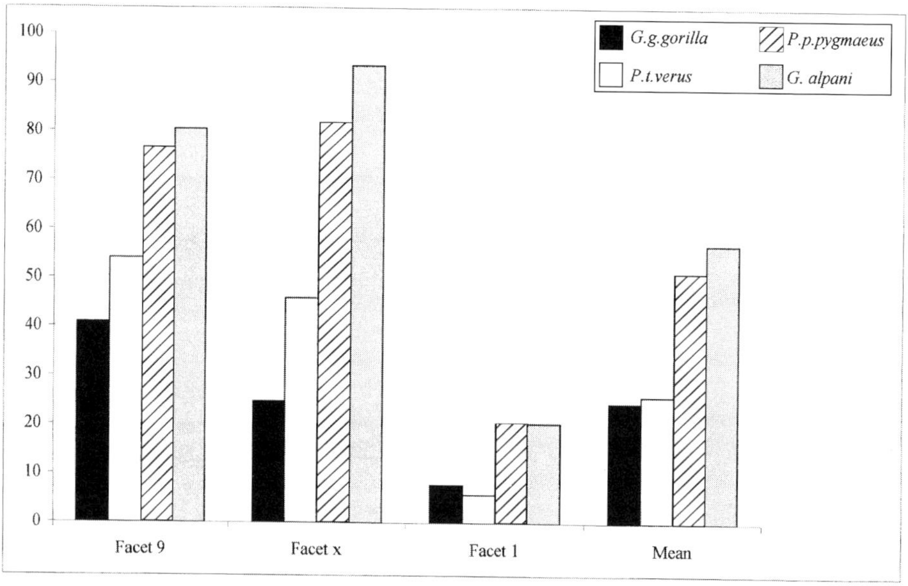

[Figure 5.5]
Pit frequency comparisons on molar facets for *Gorilla gorilla gorilla*, *Pan troglodytes verus*, *Pongo pygmaeus pygmaeus*, and *Griphopithecus alpani*. Data from King *et al*. (1999).

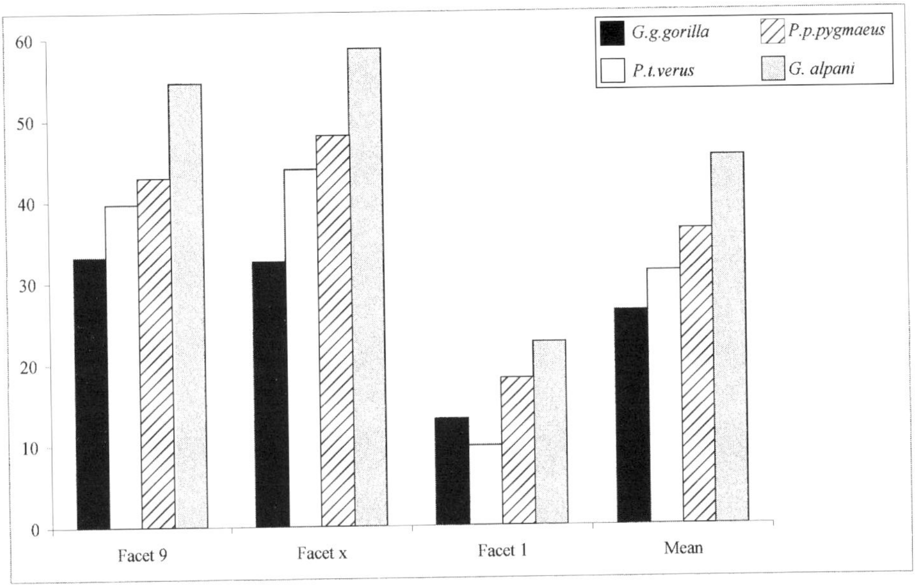

[Figure 5.6]
Pit percentage comparisons on molar facets for *Gorilla gorilla gorilla*, *Pan troglodytes verus*, *Pongo pygmaeus pygmaeus*, and *Griphopithecus alpani*. Data from King *et al*. (1999).

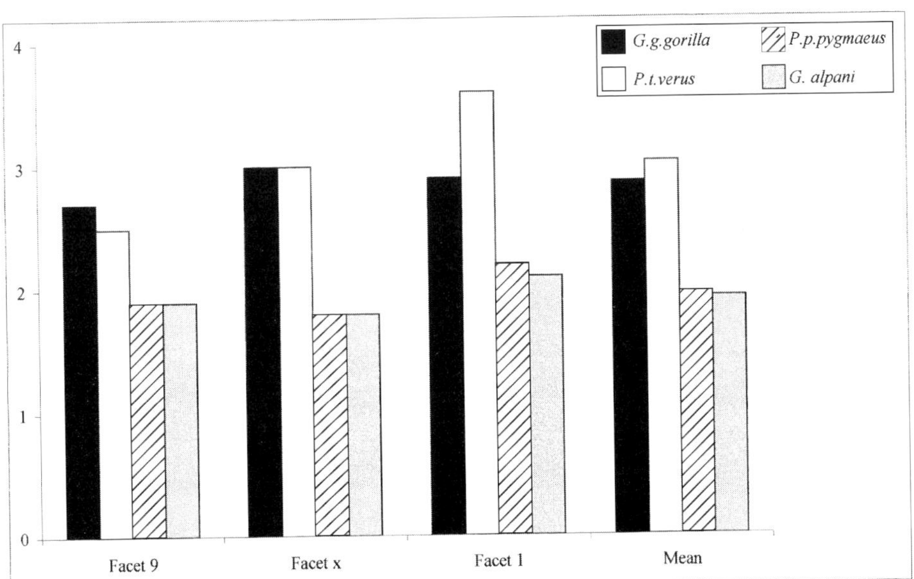

[Figure 5.7]
Striation width comparisons on molar facets for *Gorilla gorilla gorilla*, *Pan troglodytes verus*, *Pongo pygmaeus pygmaeus*, and *Griphopithecus alpani*. Striation widths are in μm. Data from King *et al*. (1999).

Discussion and conclusions

The following conclusions can be made about the dietary regimes of a number of Miocene Eurasian catarrhines.

Ouranopithecus macedoniensis

Large percentages of pits have been found on the molar teeth of *O. macedoniensis* and it has been suggested that this hominoid ingested a significant amount of hard objects (Figure 5.8) (Ungar, 1996). It has been postulated that the palaeoenvironment of *O. macedoniensis* was essentially open and had two distinct seasons (Bonis & Koufos, 1994). These authors also proposed that this hominid exploited items that would be available during the dry season, such as nuts, seeds, roots and tubers. The extremely thick molar enamel displayed by *O. macedoniensis*, as well as the microwear patterns present on its molar teeth are compatible with this suggestion. In addition, the narrow striations on the incisor teeth of *O. macedoniensis* may indicate that it was feeding on or near to the ground (Ungar, 1996). Recent research on Sumatran anthropoid incisor microwear has demonstrated that those primates which feed on or near to the ground have narrower striations than those which feed higher up in the canopy. This has been linked to size differences between soil particles and plant phytoliths (Ungar *et al.*, 1995).

Griphopithecus alpani

G. alpani displays high percentages of pits indicating that it was a frugivore and it may also have consumed a certain amount of hard dietary items (Figure 5.8) (King *et al.*, 1999). Evidence from palaeosols (Bestland, 1990), stable isotope analysis of enamel and soils (Quade *et. al.*, 1995) and mammalian community structure (Andrews, 1990) suggest that the palaeoenvironment at Paşalar was one of strongly seasonal sub-tropical deciduous closed forest/woodland. *G. alpani*, like *Ouranopithecus macedoniensis*, possessed thick molar enamel and this may have enabled it to exploit hard objects, such as hard fruits and nuts, during the dry season.

Oreopithecus bambolii and pliopithecid specimens from Castell de Barbera

Folivory is suggested by the low percentages of pits found on the molars of *Oreopithecus bambolii* and Pliopithecid specimens from Castell de Barbera (Figure 5.8) (Ungar, 1996). Evidence from Baccinello suggests that the

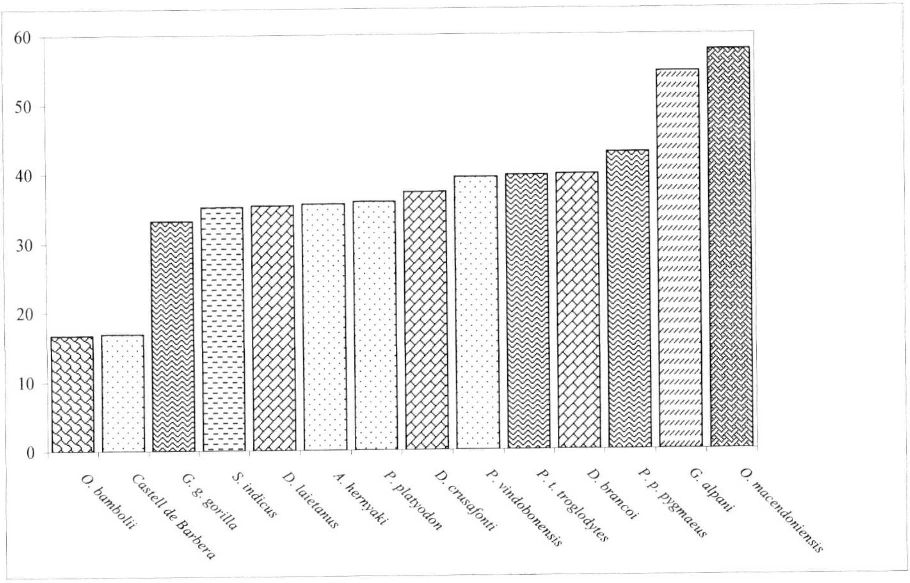

[Figure 5.8]
Percentage pitting comparisons for molar Phase II surface in a range of Eurasian Miocene hominoid taxa and three extant ape taxa. Data from King *et al.* (1999); Teaford (1988) and Ungar (1996). (Species list as for Figure 5.1 plus *Gorilla gorilla gorilla*; *Sivapithecus indiscus*, *Pan troglodytes troglodytes*, *Pongo pygmaeus pygmaeus*; *Griphopithecus alpani*.)

habitat of *O. bambolii* was wet mixed mesophytic forest (Harrison & Harrison, 1989). It has been suggested that the environment of the Castell de Barbera specimens was wet closed canopy forest (Köhler, 1993).

Dryopithecus spp

Intermediate percentages of pits on the dentitions of *Dryopithecus brancoi*, *D. crusafonti*, *D. laietanus* suggest that these thin-enamelled hominoids were soft fruit eaters (Ungar, 1996). Palaeoenvironmental evidence has shown that these hominoids inhabited equable sub-tropical seasonal forests (Andrews, 1996).

Anapithecus hernyaki and *Pliopithecus platyodon* and *Pliopithecus vindobonensis*

These taxa all display similar percentages of pits to those of extant primate soft frugivores (Ungar, 1996). These catarrhines inhabited sub-tropical woodland forests with a type of swamp vegetation indicated for Rudabánya (Kretzoi *et al.*, 1974; Kordos, 1982; Nagatoshi, 1986; Andrews, 1996).

Sivapithecus indicus

Soft fruit eating is also suggested for *S. indicus* which displays intermediate percentages of pits to primate folivores and hard objects specialists (Teaford, 1988). Palaeoenvironmental evidence indicates that this hominoid lived in sub-tropical forests which although seasonal, were wetter than those inhabited by *Griphopithecus alpani* (Andrews, 1996). Scott *et al.* (1999) suggest that the habitat of *S. parvarda,* also from the Siwaliks, Pakistan, was predominantly forested with some less densely covered areas also present.

The diversity in Eurasian catarrhine diets is largely due to the fact that *Griphopithecus alpani* and *Ouranopithecus macedoniensis* may have ingested large amounts of hard objects. None of the living great apes are hard object specialists despite the fact that *Pongo pygmaeus pygmaeus* does ingest these items occasionally (Ungar, 1995). Among modern anthropoids *Lophocebus albigena,* which has 55% pits (Teaford, 1988) on its molar teeth and which ingests hard objects habitually, may be a better dietary analogue for *Griphopithecus alpani* and *O. macedoniensis* (Figure 5.8).

Acknowledgements

I thank the organizers for inviting me to participate in the workshop and contribute to this volume. I am grateful to Peter Andrews and an anonymous reviewer for helpful comments on an earlier version of the chapter.

References

Andrews, P. (1990). Palaeoecology of the Miocene fauna from Paşalar, Turkey. *J. Hum. Evol.* **19**: 569–82.

Andrews, P. (1996). Palaeoecology and hominoid palaeoenvironments. *Biol. Rev.* **71**: 257–300.

Bestland, E. A. (1990). Sedimentology and paleopedology of Miocene alluvial deposits at the Paşalar Hominoid site, Western Turkey. *J. Hum. Evol.* **19**: 363–77.

Bonis, L. de & Koufos, G. D. (1994). Our ancestors' ancestor: *Ouranopithecus* is a Greek link in human ancestry. *Evol. Anthrop.* **3**: 75–84.

Butler, P. M. (1952). The milk-molars of Perissodactyla, with remarks on molar occlusion. *Proc. Zool. Soc. (London)* **121**: 777–817.

Covert, H. H. & Kay, R. F. (1981). Dental microwear and diet: implications fordetermining the feeding behaviour of extinct primates, with a comment on the dietary pattern of *Sivapithecus. Am. J. Phys. Anthrop.* **55**: 331–6.

Goodall, J. (1986). *The Chimpanzees of Gombe: Patterns of behaviour.* Cambridge, MA: Harvard University Press.

Gordon, K. D. (1980). *Dental Attrition in the Chimpanzee*. Ph.D. Dissertation, Yale University.

Gordon, K. D. (1982). A study of microwear on chimpanzee molars: implications for dental microwear analysis. *Am. J. Phys. Anthrop.* **59**: 195–215.

Harrison, T. S. & Harrison, T. (1989). Palynology of the late Miocene Oreopithecus-bearing lignite from Baccinello, Italy. *Palaeogeogr. Palaeoclimatol. Palaeoecol.* **76**: 45–65.

Hladik, C. M. (1977). Chimpanzees of Gombe and the chimpanzees of Gabon: some comparative data on the diet. In Clutton-Brock, T. H. (ed.), *Primate Ecology: Studies of Feeding and Ranging Behaviour in Lemurs, Monkeys and Apes*. London: Academic Press.

Kay, R. F. (1977). The evolution of molar occlusion in *Cercopithecoidea* and early catarrhines. *Am. J. Phys. Anthrop.* **46**: 327–52.

Kay, R. F. & Hiiemae, K. M. (1974). Jaw movement and tooth use in recent and fossil primates. *Am. J. Phys. Anthrop.* **40**: 227–56.

Kelley, J. (1990). Incisor microwear and diet in three species of *Colobus*. *Folia Primatol.* **55**: 73–84.

King, T., Aiello, L. C. & Andrews, P. (1999). Dental microwear of Griphopithecus alpani. *J. Hum. Evol.* **36**: 3–31.

Köhler, M. (1993). Skeleton and habitat of recent and fossil ruminants. *Munch. Geowissen. Abhandl.* **25**: 1–88.

Kordos, L. (1982). The prehominid locality of Rudabánya (NE Hungary) and its neighbourhood: A palaeogeographic reconstruction. *Mag. All. Foldt. Intez. Evi. Jel.* **1980**: 395–406.

Kretzoi, M., Krolopp, E., Lörincz, H. & Pálfalvy, I. (1974). A Rudabanyai Alsopannoniai Prehominidas Lelohely Floraja Faunaja es Retegtani Helyzete. *M. All. Foldtani Intezet Evi Jelentense* pp. 365–93.

Lucas, P. (1994). Fundamental physical properties of fruits and seeds in primate diets. In (Ehara, A., Kimura, T., Takenaka,O. & Iwamoto, M. (eds.), *Primatology Today: proceedings of the XIIIth Congress of the International Primatological Society, Nagoya and Kyoto, 18th–24th July, 1990*, pp. 125–8. Amsterdam: Elsevier Science.

Maas, M. C. (1991). Enamel structure and microwear: an experimental study of the response of enamel to shearing forces. *Am. J. Phys. Anthrop.* **85**: 31–50.

Maas, M. C. (1994). A scanning electron-microscopic study of *in vitro* abrasion of mammalian tooth enamel under compressive loads. *Archs. Oral Biol.* **39**: 1–11.

MacKinnon, J. (1977). A comparative ecology of Asian apes. *Primates* **18**: 747–72.

Nagatoshi, K. (1986). Miocene hominoid environments of Europe and Turkey. *Palaeogeogr. Palaeoclimatol. Palaeoecol.* **61**: 145–54.

Peters, C. R. (1982). Electron-optical microscopic study of incipient dental microdamage from experimental seed and bone crushing. *Am. J. Phys. Anthrop.* **57**: 283–301.

Quade, J., Cerling, T. E., Andrews, P. & Alpagut, B. (1995). Paleodietary reconstruction of Miocene faunas from Paşalar, Turkey using stable carbon and oxygen isotopes of fossil tooth enamel. *J. Hum. Evol.* **28**: 373–84.

Rodman, P. S. (1988). Diversity and consistency on ecology and behaviour. In (Schwartz, J. H. (ed.), *Orang-utan Biology*, pp. 31–51. New York: Oxford University Press.

Ryan, A. S. (1981). Anterior dental microwear and its relationship to diet and feeding behavior in three African primates (*Pan troglodytes, Gorilla gorilla*, and *Papio hamydryas*). *Primates* **22**: 533–50.

Ryan, A. S. & Johanson, D. C. (1989). Anterior dental microwear in *Australopithecus afarensis*: comparisons with human and non-human primates. *J. Hum. Evol.* **18**: 235–68.

Scott, R. S., Kappelman, J. & Kelley, J. (1999). The paleoenvironment of *Sivapithecus parvarda. J. Hum. Evol.* **36**: 245–74.

Solounias, N., Teaford, M. F. & Walker, A. (1988). Interpreting the diet of extinct ruminants: the case of a non-browsing giraffid. *Palaeobiol.* **14**: 287–300.

Teaford, M. F. (1985). Molar microwear and diet in the genus Cebus. *Am. J. Phys. Anthrop.* **66**: 363–70.

Teaford, M. F. (1988). A review of dental microwear and diet in modern mammals. *Scanning Microsc.* **2**: 1149–66.

Teaford, M. F. & Glander, K. E. (1991). Dental microwear in live, wild-trapped *Alouatta palliata* from Costa Rica. *Am. J. Phys. Anthrop.* **85**: 313–19.

Teaford, M. F. & Oyen, O. J. (1989a). Differences in the rate of molar wear between monkeys raised on different diets. *J. Dent. Res.* **68**: 1513–18.

Teaford, M. F. & Oyen, O. J. (1989b). *In vivo* and *in vitro* turnover in dental microwear. *Am. J. Phys. Anthrop.* **80**: 447–60.

Teaford, M. F. & Robinson, J. G. (1989). Seasonal or ecological differences in diet and molar microwear in *Cebus nigrivittatus. Am. J. Phys. Anthrop.* **80**: 391–401.

Teaford, M. F. & Runestad, J. A. (1992). Dental microwear and diet in Venezuelan primates. *Am. J. Phys. Anthrop.* **88**: 347–64.

Teaford, M. F. & Walker, A. (1984). Quantitative differences in dental microwear between primate species with different diets and a comment on the presumed diet of *Sivapithecus indicus. Am. J. Phys. Anthrop.* **64**: 191–200.

Tutin, C. & Fernandez, M. (1985). Foods consumed by sympatric populations of *Gorilla gorilla gorilla* and *Pan troglodytes troglodytes* in Gabon: some preliminary data. *Int. J. Primatol.* **6**: 27– 43.

Tutin, C. & Fernandez, M. (1987). Sympatric gorillas and chimpanzees in Gabon. *Anthroquest* No. 37: 3–6.

Tutin, C. E. G., Fernandez, M., Rogers, M. E., Williamson, E. A. & McGrew, W. C. (1991). Foraging profiles of sympatric lowland gorillas and chimpanzees in the LopeReserve, Gabon. *Phil. Trans. R. Soc. Lond. B* **334**: 179–86.

Ungar, P. S. (1990a). Incisor microwear and feeding behavior in *Alouatta seniculus* and *Cebus olivaceus. Am. J. Phys. Anthrop.* **20**: 43–50.

Ungar, P. S. (1990b). A preliminary analysis of incisor microwear and feeding behavior in two platyrrhine species. *Am. J. Phys. Anthrop.* **81**: 310.

Ungar, P. S. (1992). *Incisor Microwear and Feeding Behavior of Four Sumatran Anthropoids.* Ph.D. Dissertation, State University of New York at Stony Brook.

Ungar, P. S. (1994). Incisor microwear of Sumatran anthropoid primates. *Am. J. Phys. Anthrop.* **94**: 339–63.

Ungar, P. S. (1995). Fruit preferences of four sympatric primate species at Ketambe, Northern Sumatra, Indonesia. *Int. J. Primatol.* **16**: 221–45.

Ungar, P. S. (1996). Dental microwear of European Miocene catarrhines: evidence for diets and tooth use. *J. Hum. Evol.* **30**: 335–66.

Ungar, P. S, Teaford, M. F., Glander, K. E. & Pastor, R. F. (1995). Dust accumulation in the canopy: a potential cause of dental microwear in primates. *Am. J. Phys. Anthrop.* **97**: 93–9.
Walker, A., Hoeck, H. N. & Perez, L. (1978). Microwear of mammalian teeth as an indicator of diet. *Science* **201**: 908–10.

6 How reliable are current estimates of fossil catarrhine phylogeny? An assessment using extant great apes and Old World monkeys

Mark Collard and Bernard Wood

Introduction

Cladistic analysis has been used for more than 20 years to reconstruct the phylogenetic relationships of fossil catarrhine species and genera (e.g. Delson & Andrews, 1975; Eldredge & Tattersall, 1975; Delson, 1977; Delson *et al.*, 1977; Tattersall & Eldredge, 1977; Andrews, 1978, 1992; Corruccini & McHenry, 1980; Harrison, 1982; Skelton & McHenry, 1986; Wood & Chamberlain, 1986, 1987; Andrews & Martin, 1987; Chamberlain & Wood, 1987; Strasser & Delson, 1987; Stringer, 1987; Wood, 1988, 1991, 1992; Skelton & McHenry, 1992; Lieberman *et al.*, 1996; Begun *et al.*, 1997; Cameron, 1997; Rae, 1997; Strait *et al.*, 1997). However, it is now apparent that, in contrast to the situation with higher-level primate taxa (Harrison, 1993), few of the relationships supported by these analyses can be considered to be reliable. This is demonstrated by the small increases in length required to alter the topologies of the most parsimonious cladograms. For example, the addition of only one step converts the *Homo* monophyly seen in Wood's (1991) most parsimonious cladogram into *Homo* paraphyly, as well as altering the relationships of *A. africanus* (Wood, 1992). Likewise, the addition of two steps to the cladogram preferred by Strait *et al.* (1997) results in *Homo* paraphyly (Wood & Collard, 1999). These examples are taken from the hominin palaeontological literature, but they could easily have been taken from studies of Miocene hominoids, Eurasian pliopithecids, or fossil Old World monkeys (e.g. Harrison, 1993; Rae, 1997). The unreliability of the most parsimonious cladograms is also illustrated by the results of Corruccini's (1994) bootstrap re-analysis of hominin data from Wood & Chamberlain (1986), Skelton *et al.* (1986), Chamberlain & Wood (1987) and Skelton & McHenry (1992). He found the relationships of most of the species and genera to be ambiguous. The only statistically significant result he obtained was that *Paranthropus robustus* and *P. boisei* are more closely related to each other than they are to any other species.

Our inability to reliably reconstruct the phylogenetic relationships of fossil catarrhine species and genera has frequently been attributed to faulty alpha taxonomy, the choice of characters examined or to the way in which the cladistic methodology has been implemented (Chamberlain & Wood, 1987; Skelton & McHenry, 1992; Strait *et al.*, 1997; Skelton & McHenry, 1998;

Strait & Grine, 1998). Recently, however, it has been suggested that the problem may lie with the data on which we normally rely (Hartman, 1988; Lieberman, 1995, 1997, 1999; Lieberman *et al.*, 1996). Unlike the investigation of the relationships between living taxa, in which any available evidence, be it anatomical, biochemical, genetic or behavioural, can be used to establish relationships, studies involving fossil taxa are limited to those parts of the phenotype that are commonly preserved in the fossil record. As far as the fossil catarrhines are concerned, this means that cladistic studies are mostly based on evidence that can be gleaned from the various hard tissues that make up the bones and teeth. Thus, most studies have been based upon dental, cranial, mandibular and, to a lesser extent, postcranial characters. This is certainly so for the fossil hominins (e.g. Eldredge & Tattersall, 1975; Tattersall & Eldredge, 1977; Delson *et al.*, 1977; Corruccini & McHenry, 1980; Skelton *et al.*, 1986; Wood & Chamberlain, 1986, 1987; Chamberlain & Wood, 1987; Arsuaga *et al.*, 1991; Wood, 1991, 1992; Skelton & McHenry, 1992; Lieberman *et al.*, 1996; Strait *et al.*, 1997), and perusal of published cladograms suggest that this is also the case for investigations of the evolutionary relationships of other fossil catarrhines (e.g. Harrison, 1982, 1989; Andrews & Martin, 1987; Strasser & Delson, 1987; Andrews, 1992; Rose *et al.*, 1992; Benefit, 1993; Moyà-Solà & Köhler, 1993, 1995; Kelley *et al.*, 1995; Begun *et al.*, 1997; Cameron, 1997; McCrossin & Benefit, 1997; Rae, 1997).

How can we assess the reliability of catarrhine craniodental evidence for reconstructing the phylogenetic relationships of species and genera? One approach is to analyse comparable evidence from closely-related extant taxa whose relationships have been established using molecular techniques and judge the resulting morphology-based hypotheses against the molecular phylogeny (Hartman, 1988). Congruence between the morphological and molecular phylogenies for the extant taxa indicates that the fossil evidence can be reasonably assumed to be reliable for phylogenetic reconstruction, whereas incongruence suggests the converse.

This approach, which assumes that molecular data are superior to morphological data for phylogenetic reconstruction, is rejected by some cladists, who deny that some classes of data are more reliable than others for the purposes of phylogenetic reconstruction, and argue that cladistic analyses should be based on all the available evidence (e.g. Smith, 1994; Kluge, 1998). We understand why these workers take this view, but believe they are mistaken. There are several reasons why, when a conflict occurs between molecular and hard tissue-based phylogenies, the former should be favoured, at least at the low taxonomic levels being considered here. First, phylogenetic relationships are genetic relationships. It is genes that are

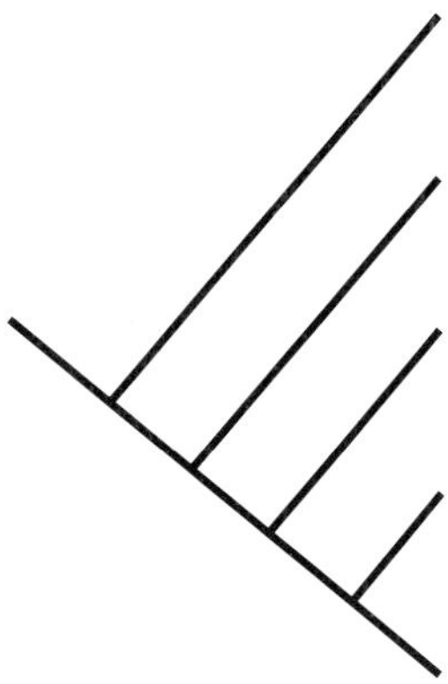

[Figure 6.1]
Hominoid molecular relationships.

passed between generations, not morphological characters. Thus, in phylogenetics, morphology can never be more than a proxy for molecular data. Secondly, it is well documented that many reproductively-defined species are genetically distinct, but dentally and osteologically indistinguishable. Since speciation events create phylogenetic relationships, there is thus an a priori expectation that characters of the teeth and skeleton will be less useful for phylogeny estimation than genetical characters. Thirdly, because many osseous and other morphological characters are clearly influenced by epigenetic effects, such as the forces generated by chewing (Lieberman *et al.*, 1996; Lieberman & Wood, 1999), they can be expected to mislead us more frequently than molecular evidence. Lastly, some of the techniques of molecular phylogenetics have been successfully tested on laboratory taxa of known phylogeny (Fitch & Atchley, 1987; Atchley & Fitch, 1991; Hillis *et al.*, 1992), whereas comparable analyses of morphological data have not been successful (Fitch & Atchley, 1987).

Within the primates, there are several examples of cladograms that are supported by multiple, independent, lines of biomolecular and karyological evidence. By any criteria, the molecular-based phylogeny for the living hominoids is well-established (Ruvolo, 1994, 1995, 1997), and we elected to use this as one test of the likely phylogenetic utility of fossil catarrhine craniodental data (Figure 6.1). Another group for which there is molecular data, albeit on a less comprehensive scale as those for the living hominids, is the papionins (Disotell, 1994, 1996; Disotell *et al.*, 1992; Harris & Disotell, 1998), and we used this as the other test group (Figure 6.2).

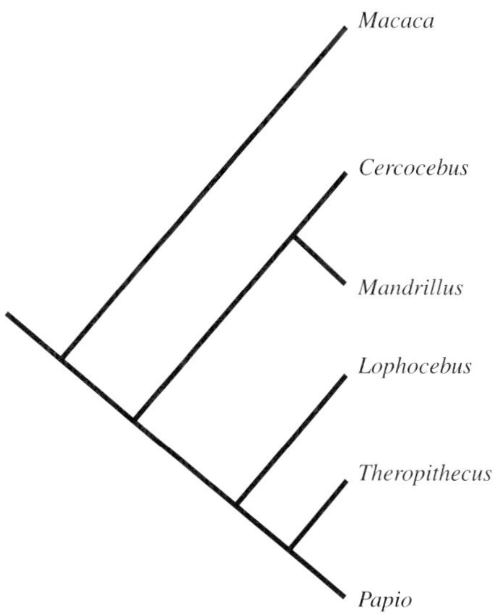

[Figure 6.2]
Papionin molecular relationships.

Materials

Morphology can be translated into character states for cladistic analysis in two main ways. The first breaks the phenotype up into anatomical components and expresses the variation within each component in terms of qualitative categories, or 'states'. Thus, an osseous prominence is 'strong', 'reduced' or 'absent', a bony contour is described as 'arched' or 'less-arched', and a feature is categorised as 'not developed' or 'developed'. To date, the majority of cladistic analyses of the catarrhines have used this approach (e.g. Delson & Andrews, 1975; Eldredge & Tattersall, 1975; Delson *et al.*, 1977; Skelton *et al.*, 1986; Skelton & McHenry, 1992; Lieberman *et al.*, 1996; Begun *et al.*, 1997; Strait *et al.*, 1997). However, we are not persuaded that it is a desirable way to express morphological variation, since it is clear that the assessment of discrete character states is often a highly subjective exercise. This is demonstrated by a recent debate concerning the Miocene hominoid *Afropithecus turkanensis*, in which some researchers scored its inferior mandibular torus as 'weakly-developed', while others considered the torus to be 'well-developed' (Leakey & Leakey, 1986; Andrews & Martin, 1987; Conroy, 1994). It is also demonstrated by the difficulty encountered by Strait *et al.* (1997) and Ahern (1998) in reproducing the scores used in previous analyses of the early hominins. Another reason for rejecting qualitative character assessment is that it is difficult to counter the confounding effects of body size differences between taxa (Kappelman, 1996). This point is exemplified by the assessment of Wood *et al.* (1998) of the likelihood of association between OH 8 and OH 35, the *Homo habilis* left talus and distal left tibia from Olduvai Gorge, Tanzania. When Wood and co-workers did not correct for body size, they obtained the same result as had been obtained in earlier discrete character assessments: the talus and the tibia appeared to have belonged to the same individual. However, when they controlled for differences in body size, they found that it was questionable whether the two bones had come from animals belonging to the same species, let alone the same individual.

The second way of expressing character state variation is to collect interlandmark distances, and then use one of a number of coding methods to break up the continuous distribution into discontinuous states. Opponents of this approach complain that measurements are unsuitable for cladistic analysis, that the coding methods break the spectrum of measurements into 'artificial' character states, and/or that cladistic analyses based on measurement data are no more than 'thinly-disguised' phenetic analyses (e.g. Pimentel & Riggins, 1987; Crisp & Weston, 1987; Cranston & Humphries, 1988; Crowe, 1994; Disotell, 1994; Moore, 1994). We contend, however, that none of these objections is valid. As Maddison *et al.* (1984), Felsenstein (1988), Swofford & Olsen (1990), Lieberman (1995) and, most especially, Rae (1998) have pointed out, there is no intrinsic difference between discrete and continuous characters as far as the cladistic methodology is concerned. The only criterion a character must fulfil for use in a cladistic analysis is that its states are homologous, and measurement-based characters can meet this criterion as well as discrete characters (Rae, 1998). This is supported by the character conflict indices obtained in cladistic analyses of the early hominins. If the metrical method of capturing information for phylogenetic analysis really is unsuitable for cladistic analysis, one would expect there to be more character conflict in studies that used measurement-based characters than in those that employed non-metrical characters. Yet, the character conflict indices obtained by Chamberlain & Wood (1987) and Wood (1991, 1992) from quantitative data are comparable with those obtained by Lieberman *et al.* (1996) and Strait *et al.* (1997) from qualitative data. The 'artificiality' argument is also easy to refute, for coding is no more 'artificial' than is the decision to break up into discontinuous states what is, with very few exceptions, such as tooth cusp and root number, continuously-distributed morphology. Moreover, a number of the methods that have been developed to convert continuously distributed characters into discrete character states are based on statistical tests, and are therefore, by convention, non-arbitrary (e.g. Thorpe, 1984; Strait *et al.*, 1996). Lastly, it is difficult to understand the argument that cladistic analyses based on measurement data are just phenetic analyses in disguise, because unlike phenetic analysis, metrical cladistics does not group taxa on the basis of overall similarity. In metrical cladistics, as in non-metrical cladistics, only those parts of the phenotype that are inferred to be shared-derived are used to group taxa into clades.

We accept that some measurements may be unsuitable because their termini span structures that have different embryonic origins, and perhaps therefore different phylogenetic histories. However, we contend that in many cases a combination of measurements can provide just as focused, but more objective, information about a structure than can an equivalent non-metrical description. It is noteworthy that few opponents complain about three other aspects of the metrical approach. First, it is quantitative, which is a desirable attribute in science. Secondly, given appropriate technical rigour, anyone can repeat the procedure and verify the observations. Thirdly, levels of intra- and interobserver error for most hominin, and presumably also other catarrhines, craniodental metrical data are low (Wood, 1991). It is for these reasons that we opted to rely principally on metrical data for our tests. In particular, we regard the requirement that the observations are replicable as paramount.

We used measurements of the cranium, mandible and dentition that have been used in hominin cladistic analyses to compile two quantitative data sets, one for the ape and human superfamily, Hominoidea, and one for the extant baboon, macaque and mangabey tribe, Papionini. The hominoid data set comprised values for 129 measurements recorded on mixed sex samples of *Gorilla, Homo, Pan, Pongo* and an outgroup. The measurements are listed in Table 6.1. Seventy-seven of the measurements were recorded on 37 *Gorilla gorilla* (20 males, 17 females), 75 *Homo sapiens* (40 males, 35 females), 35 *Pan troglodytes* (13 males, 22 females), 41 *Pongo pygmaeus* (20 males, 21 females) and 24 *Colobus guereza* (12 males, 12 females). These data were taken from Wood *et al.* (1991). The other 52 measurements were recorded on 20 *G. gorilla* (10 males, 10 females), 20 *H. sapiens* (10 males, 10 females), 20 *Pan troglodytes* (10 males, 10 females), 20 *Pongo pygmaeus* (10 males, 10 females) and 20 *C. guereza* (10 males, 10 females). These data were taken from Chamberlain (1987).

The papionin data set consisted of values for 62 measurements recorded on mixed sex samples of *Cercocebus, Lophocebus, Macaca, Mandrillus, Papio, Theropithecus* and several outgroups. The measurements are given in Table 6.2. The 62 measurements were recorded on 26 *Cercocebus galeritus/torquatus* (13 males, 13 females), 40 *Lophocebus albigena/atterimus* (20 males, 20 females), 40 *Macaca fascicularis/mulatta* (20 males, 20 females),

Table 6.1. *Hominoid metric variables*

Variable	Definition	Variable	Definition
P1	I^1 labiolingual diameter	M16	M_3 buccolingual diameter
P2	I^1 mesiodistal diameter	M17	M_3 mesiodistal diameter
P3	I^2 labiolingual diameter	M18	Maximum cusp height
P4	I^2 mesiodistal diameter	M19	Condylar height
P5	C^1 mesiodistal diameter	M20	Bicondylar breadth
P6	C^1 labiolingual diameter	M21	Coronoid height
P7	C^1 labial height	M22	Bicoronoid breadth
P8	P^3 Buccolingual diameter	M23	Right condylar head width
P9	P^3 mesiodistal diameter	M24	Right condylar head anterior-posterior breadth
P10	P^4 Buccolingual diameter		
P11	P^4 mesiodistal diameter	M25	Ramal breadth
P12	M^1 Buccolingual diameter	M26	Bigonial width
P13	M^1 mesiodistal diameter	M27	Height of mandibular body at M_1
P14	M^2 Buccolingual diameter	M28	Thickness of mandibular body of M_1
P15	M^2 mesiodistal diameter	M29	Symphyseal height
P16	M^3 Buccolingual diameter	M30	Symphyseal thickness
P17	M^3 mesiodistal diameter	M31	Inner alveolar breadth at M_3
P18	Outer alveolar breadth at M^3	M32	Maximum mandibular length
P19	Inter upper canine breadth	M33	Inter lower canine distance
P20	Palate length	M34	Mandibular corpus height at M_3
P21	Inner alveolar breadth at M^3	M35	Height of foramen spinosum
P22	Palate depth at M^1	M36	Height of mental foramen
P23	Prosthion to plane of M^3	M37	Breadth between lower second molars
P24	Maxillo-Alveolar breadth (M^2B-M^2 B)		
P25	Breadth between upper second molars (M^2L-M^2L)	M38	Lower incisor alveolar length
		M39	Lower premolar alveolar length
P26	Palate depth at incisive fossa	M40	Lower molar alveolar length
P27	Palate depth at upper second molars	F1	Right orbital breadth
		F2	Right orbital height
P28	Maxillary alveolar subtense	F3	Interorbital breadth
P29	Upper incisor alveolar length	F4	Biorbital breadth
P30	Upper premolar alveolar length	F5	Nasion-Rhinion
P31	Upper molar alveolar length	F6	Nasion-nasospinale
M1	I_1 labiolingual diameter	F7	Maximum nasal width
M2	I_1 mesiodistal diameter	F8	Nasospinale-Prosthion
M3	I_2 labiolingual diameter	F9	Bijugal breadth
M4	I_2 mesiodistal diameter	F10	Bizygomatic breadth
M5	C_1 labiolingual diameter	F11	Upper facial breadth
M6	C_1 mesiodistal diameter	F12	Lower facial breadth
M7	C_1 labial height	F13	Breadth between infraorbital foramina
M8	P_3 buccolingual diameter		
M9	P_3 mesiodistal diameter	F14	Lower nasal bone breadth
M10	P_4 buccolingual diameter	F15	Facial height
M11	P_4 mesiodistal diameter	F16	Height of infraorbital foramen
M12	M_1 buccolingual diameter	F17	Height of orbital margin
M13	M_1 mesiodistal diameter	F18	Upper malar height
M14	M_2 buccolingual diameter	F19	Lower malar height
M15	M_2 mesiodistal diameter	F20	Upper facial prognathism

Table 6.1. (*cont.*)

Variable	Definition	Variable	Definition
F21	Lower facial prognathism	C16	Breadth of mandibular fossa
F22	Malar prognathism	C17	Length of tympanic plate
F23	Naso-frontal subtense	C18	Length of petrous temporal
F24	Maxillary subtense	C19	Position of foramen ovale
C1	Glabella-Opisthocranion	C20	Position of infratemporal crest
C2	Minimum post-orbital breadth	C21	Length of foramen magnum
C3	Basion-Bregma	C22	Breadth of foramen magnum
C4	Maximum bi-parietal breadth	C23	Length of infratemporal fossa
C5	Biporionic width	C24	Breadth of infratemporal fossa
C6	Mastoid length	C25	Opisthion-infratemporal subtense
C7	Coronale-Coronale	C26	Basiooccipital length
C8	Opisthion-Inion	C27	Parietal thickness at Lambda
C9	Bimastoid width	C28	Frontal sagittal chord
C10	Posterior skull length	C29	Parietal sagittal chord
C11	Breadth across tympanic plates	C30	Parietal coronal chord
C12	Breadth between carotid canals	C31	Occipital sagittal chord
C13	Breadth between petrous apices	C32	Frontal sagittal arc
C14	Breadth between foramen ovale	C33	Occipital sagittal arc
C15	Breadth between infratemporal crests	C34	Auricular height

62 *Mandrillus leucopheus/sphinx* (42 males, 20 females), 39 *Papio anubis/cynocephalus* (20 males, 19 females), 44 *Theropithecus gelada* (22 males, 22 females), 10 *Cercopithecus aethiops* (five males, five females), 7 *Colobus badius* (three males, four females), 10 *Erythrocebus patas* (five males, five females) and 17 *Pan troglodytes* (10 males, seven females). These data were taken from Collard (1998). Fifty-five of the measurements were recorded on a further 14 *Cercocebus torquatus* (seven males, seven females), 14 *Colobus badius* (seven males, seven females) and 12 *P. troglodytes* (five males, seven females). These data were taken from Chamberlain *et al.* (unpublished data). No consistent differences were found between the data from Collard (1998) and Chamberlain *et al.* (unpublished data) using Student's two-tailed t-test.

To relate our study to as many published cladistic analyses of the fossil catarrhines as possible, we also generated a hominoid qualitative data matrix from published data. This consisted of the states of 96 cranial and dental characters recorded on specimens of *Gorilla, Homo, Hylobates, Pan, Pongo* and an outgroup. The characters were obtained from several sources. Sixty-two were characters used by Shoshani *et al.* (1996) that are wholly craniodental and which vary among the hominoids. Two characters were

Table 6.2. *Papionin metric variables*

Variable	Definition	Variable	Definition
P1	Maxillo-alveolar length	F2	Alveolar height
P2	Maxillo-alveolar breadth	F3	Superior facial breadth
P3	Incisive canal-palatomaxillary suture	F4	Bizygomatic breadth
		F5	Bimaxillary breadth
P4	Upper incisor alveolar length	F6	Anterior interorbital breadth
P5	Palatal height at M^1	F7	Orbital height
P6	Upper premolar alveolar length	F8	Minimum malar height
P7	Upper molar length	F9	Maximum nasal aperture width
P8	Canine interalveolar distance	F10	Nasal height
P9	Last premolar interalveolar distance	F11	Sagittal length of nasal bones
		F12	Superior breadth of nasal bones
P10	Second molar interalveolar distance	F13	Inferior breadth of nasal bones
		F14	Zygomaxillare – Porion
P11	I^1 mesiodistal crown diameter	F15	Upper facial prognathism
P12	I^1 labiolingual crown diameter	F16	Lower facial prognathism
P13	C^1 Mesiodistal crown diameter	C1	Glabella – opisthocranion
P14	C^1 labiolingual crown diameter	C2	Bregma – basion
P15	M^3 interalveolar distance	C3	Minimum frontal breadth
P16	Palate depth at incisive fossa	C4	Biporionic breadth
M1	Symphyseal height	C5	Glabella-Bregma
M2	Maximum symphyseal depth	C6	Postglabellar sulcus-bregma
M3	Corpus height at M_1	C7	Parietal sagittal chord
M4	Corpus width at M_1	C8	Parietal lambdoid chord
M5	Corpus height at M_3	C9	Lambda – inion
M6	Corpus width at M_3	C10	Occipital sagittal length
M7	Lower premolar alveolar length	C11	Foramen magnum maximum width
M8	Lower molar alveolar length		
M9	P_4 mesiodistal crown diameter	C12	Occipital condyle maximum length
M10	P_4 Buccolingual crown diameter		
M11	M_1 mesiodistal crown diameter	C13	Lambda thickness of parietal
M12	M_1 Buccolingual crown diameter	C14	Breadth between carotid canals
M13	M_2 mesiodistal crown diameter	C15	Breadth between petrous apices
M14	M_2 Buccolingual crown diameter	C16	Length of tympanic plate
F1	Superior facial height		

taken from Braga (1995), six from Andrews (1987), four from Schwartz (1984) and two from Delson & Andrews (1975). The other 20 characters were the craniodental characters in Groves (1986) that were neglected, without explanation, by Shoshani *et al.* (1996). The characters and states are listed in Appendix 6.1.

male + female combined? A mean value

Methods

A character state data matrix was derived from each metric data set. The confounding effects of the body-size differences between the taxa were minimised by dividing each value by the geometric mean of all the values for the appropriate specimen (Jungers *et al.*, 1995). Allometry-based size-adjustment methods were not employed as recent phylogenetic analyses have indicated that isometric and allometric methods give similar results when applied to primate craniodental data (Creel, 1986; M. Singleton, 1996, unpublished data). The size-adjusted data were then converted into discrete character states using divergence coding (Thorpe, 1984). In divergence coding, the mean values for the taxa are calculated, and the differences between them tested for statistical significance. The means are then ranked in ascending order, and a taxon-by-taxon matrix compiled. Each cell in the top row of the matrix is filled with a taxon name such that the rank of the taxa decreases from left to right. The cells of the first column of the matrix are also filled with the names of the taxa on the basis of their rank, with the highest ranked taxon being placed in the top cell and the lowest ranked taxon in the bottom cell. Thereafter, each column of the matrix is filled with −1s, +1s and 0s. A cell is filled with a −1 if the mean of the taxon in the column is greater than the mean of the taxon in the row and the difference between the means is significant. A cell is filled with a +1 if the mean of the column taxon is significantly lower than the mean of the row taxon. If the difference between the means of the column and row taxa is not significant, the cell is filled with 0. Once the matrix is completely filled, the total of 0s, −1s and +1s for each column is calculated. Lastly, an integer (in this case 10) is added to each taxon total to make them positive figures, and therefore suitable for use in computer-based phylogenetics programmes. It should be noted that divergence coding is just one of several coding methods that have been described in recent years. It should also be noted that, at the moment, there is no consensus regarding the relative effectiveness of these methods. We elected to use divergence coding because it appears to be one of the most robust of the methods that are appropriate for analysing fossil taxa. The quantitative matrices are reproduced in Appendices 6.2 and 6.3.

The quantitative and qualitative matrices were used to perform two tests of the hypothesis that conventional craniodental characters are reliable for reconstructing the phylogenetic relationships of fossil catarrhine species and genera. The first was based on parsimony analysis, which identifies the cladogram that requires the smallest number of *ad hoc* hypotheses of homoplasy to account for the observed distribution of character states. Each matrix was subjected to parsimony analysis using the branch-and-bound search routine of PAUP 3.0s (Swofford, 1991). Because the states of the metrical characters can be assumed to have evolved serially, the characters were treated as freely-reversing, linearly-ordered variables (Chamberlain & Wood, 1987; Wood, 1991, 1992; Slowinski, 1993; Rae, 1997). Some of the qualitative characters were also considered to be ordered characters, but the majority were treated as unordered variables (see Appendix 6.3 for details). Lastly, the most parsimonious cladogram or – if several equally parsimonious cladograms were favoured – the strict consensus cladogram was compared to the appropriate consensus molecular cladogram (Figures. 6.1 and 6.2). The hypothesis was considered to be supported if an analysis favoured a fully-resolved cladogram matching the molecular cladogram, or a partially-resolved cladogram comprising only molecular clades. The hypothesis was also considered supported if a strict consensus of several equally-parsimonious cladograms comprised only clades that were compatible with the molecular cladogram. These criteria were stipulated because in parsimony analysis it is not legitimate to accept some clades of a cladogram and reject others.

The second test employed the phylogenetic bootstrap, which is a resampling procedure that assigns a confidence interval to the clades that comprise the most parsimonious cladogram (Felsenstein, 1985). Using PAUP, 10 000 matrices were derived from each quantitative matrix by sampling with replacement. The bootstrap matrices were subjected to parsimony analysis, and a consensus of the most parsimonious cladograms was computed using a confidence region of 70% (Hillis & Bull, 1993). Thereafter, the clades of the consensus cladogram were compared with the appropriate molecular cladogram. The hypothesis was judged to be supported if all the clades of the consensus cladogram were compatible with the molecular cladogram.

Results

The hypothesis that catarrhine craniodental data are reliable for reconstructing the phylogenetic relationships of species and genera was not supported by the parsimony analyses. None of the matrices yielded a cladogram that was completely compatible with the group's molecular cladogram. The hominoid metric cladogram (informative characters = 118, length = 1093, consistency index [CI] = 0.77) suggested that *Homo* was the sister taxon of a (*Gorilla*, *Pan*, *Pongo*) clade, and that *Pan* was the sister taxon of a (*Gorilla*, *Pongo*) clade. The papionin metric cladogram (informative characters = 61, length = 923, CI = 0.69) suggested that *Lophocebus* is the sister of the other papionins; that *Cercocebus* is the sister of the baboons and macaques; that *Macaca* is the sister of the baboons; and that *Theropithecus* is the sister of *Mandrillus* and *Papio*. Two equally parsimonious cladograms were derived from the hominoid qualitative matrix (informative characters = 64, length = 135, CI = 0.66). The first agreed with the hominoid molecular cladogram in locating *Hylobates* as the basal hominoid. However, it differed from the molecular cladogram in positing a sister group relationship between *Pan* and *Gorilla*, and another between *Homo* and *Pongo*. The second cladogram was wholly incompatible with the molecular cladogram. It suggested that *Homo* is the sister of a clade comprising *Gorilla, Hylobates, Pan* and *Pongo*; that *Pongo* is the sister of *Gorilla, Hylobates* and *Pan*; and that *Hylobates* is the sister of *Gorilla* and *Pan*.

The bootstrap analyses also failed to uphold the hypothesis. None of the clades supported by 70% or more of the bootstrap samples was compatible with the consensus molecular cladograms. The hominoid quantitative analysis supported a (*Gorilla, Pan, Pongo*) clade at 95%, and a (*Gorilla, Pongo*) clade at 73%. The papionin quantitative analysis supported a (*Cercocebus, Macaca*, baboon) clade at 98%; a (*Macaca*, baboon) clade at 78%; a baboon clade at 97%; and a (*Mandrillus, Papio*) clade at 73%. The analysis of the hominoid qualitative data yielded one clade, which incorrectly linked *Gorilla* and *Pan* to the exclusion of the other taxa (92%).

Discussion

The results of the parsimony and bootstrap tests suggest that cladistic analyses based on catarrhine craniodental morphology cannot be relied on to recover phylogenetic relationships. Indeed, the outcomes of the tests show that the methods can generate results that are positively misleading. For example, in a number of the parsimony analyses of the quantitative data, the 'true' cladograms were less parsimonious than a substantial number of 'false' cladograms. Likewise, the bootstrap-based tests indicate that craniodental data can return impressive levels of statistical support for patterns of phylogenetic relationship that are most likely incorrect. For instance, in the hominoid analyses, the 'false' (*Gorilla, Pan, Pongo*) clade was identified in more than 70% of the bootstrap cladograms. Likewise, the 'false' (*Mandrillus, Papio*) clade was supported by more than 70% of the bootstrap cladograms in several of the papionin analyses. In other words, cladistic analyses of catarrhine gross craniodental morphology may yield not only 'false-positive' results, but 'false-positive' results that, by a substantial margin, pass the statistical test favoured by many researchers. These results are in line with those of Hartman (1988) and Harrison (1993). The former found that hominoid molar morphology was uninformative for cladistic analysis, while the latter concluded that his attempts to use cladistics to resolve the inferred relationships *among* closely related fossil primates, such as the early Miocene catarrhines from East Africa or the Eurasian pliopithecids, had been 'largely unsuccessful'. Our results are also in line with Pilbeam's (1996) conclusion that we currently know little about the phylogenetic relationships of the Miocene hominoids.

The implication of our results, and those described by Hartman (1988), Harrison (1993) and Pilbeam (1996), is that phylogenetic hypotheses for fossil hominins and other fossil catarrhines that are based solely on craniodental evidence may not be reliable. Most likely, these hypotheses reflect a mixture of the 'true' phylogeny and the phylogenetically-misleading effects of convergence, parallelism, reversal and/or behaviourally-induced morphogenesis. If anything, the results of the present study are likely to have over-estimated the reliability of fossil phylogenetic hypotheses, since our study did not account for two other factors that routinely complicate analyses of the hominin and hominid fossil records, namely contentious alpha taxonomy and intraspecific morphological change through time. In addition, as part of another study we have applied the same logic to two other groups of living primates, the platyrrhines and strepsirhines (Collard & Wood, unpublished data). These groups have less well supported molecular phylogenies than is the case for the hominoids and papionins, but the conclusions are similar. Primate craniodental data perform poorly in attempts to use them to recover the relevant phylogenetic history generated from molecular evidence.

How can the reliability of fossil catarrhine phylogenetic hypotheses be improved? One strategy is to devise techniques for characterising catarrhine craniodental morphology that are more sensitive to any phylogenetic signal than the methods presently in use (Rae, 1999). Recent studies suggest that such techniques may include the construction of metavariables using discriminant function analysis and principal component analysis (Aiello *et al.*, 1999; Collard, unpublished data). Since exogenetic stimuli can be expected to confound phylogenetic reconstruction (Lieberman, 1995, 1997, 1999), another approach is to focus on characters that are known to be minimally affected by such stimuli, for example, dental enamel and the structures of the middle and inner ear (Masali, 1968; Rak & Clarke, 1979a,b; Beynon *et al.*, 1998; Spoor & Zonneveld, 1998; Collard & Moggi-Cecchi, unpublished data). A third strategy is to develop rigorous comparative methods for discriminating between phylogenetically-informative and phylogenetically-misleading craniodental similarities. For example, the pursuit of detailed information about the ontogeny of characters may help identify convergences, parallelisms and reversals (Wood, 1988; Bromage, 1989; Lieberman *et al.*, 1996), while functional analyses may enable researchers to predict where resemblances resulting from behaviourally-induced morphogenesis are likely to occur in the hominid cranium (Lieberman, 1995, 1997, 1999; Lieberman *et al.*, 1996). A fourth approach is to develop techniques for assigning post-cranial specimens to taxa in the absence of associated skeletons, thereby overcoming the taphonomy-imposed focus on craniodental morphology and enabling hominin cladistic analyses to be based on a wider sample of the phenotype (e.g. Aiello & Wood, 1994; Wood *et al.*, 1998). We also suggest that more attention should be paid to non-morphological lines of evidence that may have a bearing on the phylogenetic relationships of fossil catarrhines, such as biogeography, stratigraphy and behavioural indicators (e.g. Turner & Wood, 1993; Augustí *et al.*, 1996; Collard *et al.*, 1999). Lastly, it is worth noting that, even if craniodental data prove to be inadequate by themselves for phylogenetic reconstruction, this does not mean that they cannot be used to recover information about evolutionary history. To adapt a phrase used in connection with the punctuated equilibrium model of evolution, homoplasies are data. The presence of homoplasies suggests that different clades responded in similar ways to biotic influences, and, providing we can eventually obtain a reliable phylogeny for the fossil catarrhines, craniodental homoplasies promise to be a rich source of information about the history of catarrhine adaptations.

Acknowledgements

We are grateful to the following individuals for providing access to the collections in their care: Paula Jenkins and Rob Kruszynski of the Natural History Museum, London; Nina Bahr and Prof. Martin of the Anthropologisches Institut und Museum, Universität Zürich-Irchel, Zurich; Dr. Roguin and Dr. Baud of the Muséum d'Histoire Naturelle, Genève; Dr. Cuisin and Prof. Denys of the Muséum d'Histoire Naturelle, Paris; and Dr. Angermann of the Museum für Naturkunde, Humboldt-Universitat zu Berlin, Berlin. This research was supported by The Wellcome Trust, The Henry Luce Foundation, The Leverhulme Trust and NERC. The comments of an anonymous reviewer are gratefully acknowledged.

Appendix 6.1. Characters for hominoid qualitative analysis

Unless otherwise indicated, the character state descriptions in the following are taken verbatim from the references for the characters.

1. Depth of subarcuate fossa

Ref.: Shoshani *et al.* (1996) #12.

States: (0) deep; (1) moderately deep to shallow; (2) very shallow to non-existent.

Dist.: *Homo* 2; *Pan* 2; *Gorilla* 2; *Pongo* 2; *Hylobates* 1; *Colobus* 0.

Notes: States as per Shoshani *et al.* (1996). Treated as ordered character in analysis – *contra* Soshani *et al.* (1996) – because states are clearly additive.

2. Morphology of the mandibular symphysis

Ref.: Shoshani *et al.* (1996) #29.

States: (1) elongated and spout-like with an angle of 150°–145°; (2) symphysis with an angle of 137°–115°; (3) angle of mandibular symphysis (excluding the simian shelf) to horizontal ramus is narrow, approaching vertical when observed dorsally and laterally, with a mandibular symphysis angle of about 100°–90° or less.

Dist.: *Homo* 3; *Pan* 2; *Gorilla* 1; *Pongo* 2; *Hylobates* 2; *Colobus* 1.

Notes: Treated as unordered because it was not clear that the states formed a straight-forward additive sequence.

3. Distinctiveness of angular process of mandible

Ref.: Shoshani *et al.* (1996) #33.

States: (0) distinct, with posterior projection; (1) not distinct.

Dist.: *Homo* 1; *Pan* 1; *Gorilla* 1; *Pongo* 1; *Hylobates* 0; *Colobus* 1.

4. Direction of incisive (anterior palatine) foramen

Ref.: Shoshani *et al.* (1996) #36.

States: (0) opening is directed dorsoventrally as in most mammals and the observer can see through the foramen; (1) foramen is directed diagonally, from anterior-ventral to posterior-dorsal, leads to a tube-like structure, and one cannot see through the foramina.

Dist.: *Homo* 1; *Pan* 1; *Gorilla* 1; *Pongo* 1; *Hylobates* 0; *Colobus* 0.

5. Carotid canal morphology when viewed from ventral side of cranium

Ref.: Shoshani *et al.* (1996) #40.

States: (0) canal perforates bulla away from basicranium and is clearly within it, opening of canal is directed medially, ventrally or ventro-medially, but the imaginary lines (one from each side) which emerge from these openings do not cross at the foramen magnum, or cross at its anterior border at the level of the occipital condyles; (1) canal perforates bulla away from basicranium and is clearly within it, opening is directed postero-medially and the imaginary lines which emerge from these openings cross the foramen magnum posterior to the occipital condyles, or caudal to the foramen magnum itself.

Dist.: *Homo* 0; *Pan* 0; *Gorilla* 0; *Pongo* 0; *Hylobates* 1; *Colobus* 0.

Notes: According to Shoshani *et al.*, to view states (1) and (2) place straight wires inside the carotid canals and note the point of intersection of the imaginary lines in continuation of these wires. In state (0), the lines cross at anterior end of the foramen magnum or in front of it, whereas in state (1) these imaginary lines cross posterior to the occipital condyles or caudal to the foramen magnum itself.

6. Size of upper first incisor relative to upper second incisor

Ref.: Shoshani *et al.* (1996) #47.

States: (0) about the same size; (1) enlarged; (2) much enlarged.

Dist.: *Homo* 1; *Pan* 1; *Gorilla* 1; *Pongo* 2; *Hylobates* 0; *Colobus* 0.

Notes: Treated as an ordered variable.

7. Honing in males (back of upper canine sharpens against third lower premolar).

Ref.: Shoshani *et al.* (1996) #48.

States: (1) present, i.e. P3 bilaterally compressed (sectorial) and modified for honing on C1, P3 is larger than P4 especially mesiodistally, also may involve honing C1 on C1; (2) honing reduced, P3 slightly buccolingually compressed, P3 is larger than P4 especially mesiodistally; (3) honing further reduced, P3 about the same size as P4 in length in occlusal view.

Dist.: *Homo* 2; *Pan* 2; *Gorilla* 1; *Pongo* 1; *Hylobates* 1; *Colobus* 0.

Notes: Treated as ordered character.

8. Interorbital pillar width.

Ref.: Shoshani *et al.* (1996) #101.

States: (0) wide; (1) narrow.

Dist.: *Homo* 0; *Pan* 0; *Gorilla* 0; *Pongo* 1; *Hylobates* 0; *Colobus* 0.

9. Depth of middle ear

Ref.: Shoshani *et al.* (1996) #102.

States: (0) shallow; (1) deepened, more than 8.5mm.

Dist.: *Homo* 1; *Pan* 1; *Gorilla* 1; *Pongo* 0; *Hylobates* 0; *Colobus* ?.

10. Axis of ear bones

Ref.: Shoshani *et al.* (1996) #103.

States: (0) acute angle; (1) right angle or more.

Dist.: *Homo* 0; *Pan* 0; *Gorilla* 1; *Pongo* 1; *Hylobates* 1; *Colobus* ?.

11. Area of inner ear

Ref.: Shoshani *et al.* (1996) #104.

States: (0) low, $< 50\text{mm}^2$; (1) increased, $> 50\text{mm}^2$.

Dist.: *Homo* 1; *Pan* 1; *Gorilla* 1; *Pongo* 0; *Hylobates* 0; *Colobus* ?.

12. Klinorhynchy (a deep foreshortened facial skeleton which bends downward with respect to the cranial base)

Ref.:	Shoshani *et al.* (1996) #106.
States:	(0) airorynch or straight; (1) more klinorhynch; (2) strongly klinorhynch.
Dist.:	*Homo* 2; *Pan* 2; *Gorilla* 2; *Pongo* 0; *Hylobates* 1; *Colobus* 0.
Notes:	Treated as ordered character.

13. Frontozygomatic suture

Ref.:	Shoshani *et al.* (1996) #107.
States:	(0) vertical; (1) medially directed.
Dist.:	*Homo* 0; *Pan* 0; *Gorilla* 0; *Pongo* 1; *Hylobates* 0; *Colobus* 0.

14. Relative height of upper face

Ref.:	Shoshani *et al.* (1996) #108.
States:	(0) high, index about 70; (1) reduced.
Dist.:	*Homo* 1; *Pan* 0; *Gorilla* 0; *Pongo* 0; *Hylobates* 1; *Colobus* 0.

15. Facial index (upper face height as a percentage of facial breadth)

Ref.:	Shoshani *et al.* (1996) #109.
States:	(0) low, index about 50; (1) increased.
Dist.:	*Homo* 0; *Pan* 1; *Gorilla* 1; *Pongo* 1; *Hylobates* 0; *Colobus* 0.

16. Height of mandibular symphysis relative to length of the lower toothrow

Ref.:	Shoshani *et al.* (1996) #110.
States:	(0) low, its height about 60% of toothrow length; (1) deepened, at least 75% of tooth row length.
Dist.:	*Homo* 0; *Pan* 1; *Gorilla* 1; *Pongo* 1; *Hylobates* 0; *Colobus* 1.

17. Presence/absence of frontal sinus

Ref.:	Shoshani *et al.* (1996) #111.
States:	(0) absent; (1) present.
Dist.:	*Homo* 1; *Pan* 1; *Gorilla* 1; *Pongo* 0; *Hylobates* 0; *Colobus* 0.

18. Pyriform aperture

Ref.:	Shoshani *et al.* (1996) #112.
States:	(0) narrow; (1) widened; (2) very wide.
Dist.:	*Homo* 2; *Pan* 2; *Gorilla* 2; *Pongo* 1; *Hylobates* 2; *Colobus* 0.
Notes:	Treated as ordered.

19. Position of infraorbital foramina relative to zygomaxillary suture

Ref.:	Shoshani *et al.* (1996) #113.
States:	(0) close to suture; (1) further from suture.
Dist.:	*Homo* 0; *Pan* 0; *Gorilla* 0; *Pongo* 1; *Hylobates* 0; *Colobus* 1.

20. Orientation of zygomatic bone

Ref.: Shoshani *et al.* (1996) #114.

States: (0) more frontally; (1) more superolaterally; (2) still further superolaterally.

Dist.: *Homo* 2; *Pan* 2; *Gorilla* 2; *Pongo* 1; *Hylobates* 0; *Colobus* 1.

Notes: Treated as ordered character in analysis.

21. Frontal bone

Ref.: Shoshani *et al.* (1996) #115

States: (0) flat; (1) more convex; (2) strongly convex.

Dist.: *Homo* 2; *Pan* 0; *Gorilla* 0; *Pongo* 2; *Hylobates* 1; *Colobus* 2.

Notes: Treated as ordered character in analysis.

22. Glabella prominence

Ref.: Shoshani *et al.* (1996) #116

States: (0) strong; (1) reduced; (2) absent.

Dist.: *Homo* 0; *Pan* 0; *Gorilla* 0; *Pongo* 1; *Hylobates* 2; *Colobus* 0.

Notes: Treated as ordered character in analysis.

23. Number of incisive foramina

Ref.: Shoshani *et al.* (1996) #117

States: (0) double, i.e. one on each side of the midline; (1) single, confluency of two foramina, at least close to the surface.

Dist.: *Homo* 1; *Pan* 0; *Gorilla* 0; *Pongo* 1; *Hylobates* 0; *Colobus* 0.

24. Maxillary sinus

Ref.: Shoshani *et al.* (1996) #118

States: (0) small; (1) expanded.

Dist.: *Homo* 0; *Pan* 0; *Gorilla* 0; *Pongo* 1; *Hylobates* 0; *Colobus* 0.

25. Supraorbital development

Ref.: Shoshani *et al.* (1996) #119

States: (0) weak; (1) more-marked; (2) torus-like.

Dist.: *Homo* 2; *Pan* 2; *Gorilla* 2; *Pongo* 0; *Hylobates* 1; *Colobus* 1.

Notes: Treated as ordered character in analysis.

26. Supraorbital contour

Ref.: Shoshani *et al.* (1996) #120

States: (0) arched; (1) less arched.

Dist.: *Homo* 0; *Pan* 0; *Gorilla* 1; *Pongo* 0; *Hylobates* 0; *Colobus* 1.

27. Orbits

Ref.: Shoshani *et al.* (1996) #121.

States: (0) as wide as high; (1) high-oval.

Dist.: *Homo* 0; *Pan* 0; *Gorilla* 0; *Pongo* 1; *Hylobates* 0; *Colobus* 0.

28. Supraorbital trigon

Ref.: Shoshani *et al.* (1996) #122.
States: (0) not developed; (1) developed.
Dist.: *Homo* 0; *Pan* 1; *Gorilla* 1; *Pongo* 1; *Hylobates* 0; *Colobus* 1.
Notes: Supraorbital trigon is the triangular area enclosed by the torus and the backwardly converging temporal lines.

29. Nasal width

Ref.: Shoshani *et al.* (1996) #123.
States: (0) broad; (1) reduced.
Dist.: *Homo* 0; *Pan* 0; *Gorilla* 0; *Pongo* 1; *Hylobates* 0; *Colobus* 0.

30. Length of nasals

Ref.: Shoshani *et al.* (1996) #124.
States: (0) long; (1) shortened.
Dist.: *Homo* 1; *Pan* 0; *Gorilla* 0; *Pongo* 0; *Hylobates* 1; *Colobus* 1.

31. Size of zygomatic foramina

Ref.: Shoshani *et al.* (1996) #126.
States: (0) very small; (1) enlarged.
Dist.: *Homo* 0; *Pan* 0; *Gorilla* 0; *Pongo* 1; *Hylobates* 0; *Colobus* 0.

32. Position of zygomatic foramina

Ref.: Shoshani *et al.* (1996) #127.
States: (0) at or below plane of orbital rim; (1) above plane of orbital rim.
Dist.: *Homo* 0; *Pan* 0; *Gorilla* 0; *Pongo* 1; *Hylobates* 0; *Colobus* 1.

33. Size of incisive foramina

Ref.: Shoshani *et al.* (1996) #128.
States: (0) large; (1) reduced in size; (2) tiny.
Dist.: *Homo* 1; *Pan* 1; *Gorilla* 1; *Pongo* 2; *Hylobates* 0; *Colobus* 0.
Notes: Treated as ordered character in analysis.

34. Size and shape of palatine foramina

Ref.: Shoshani *et al.* (1996) #129.
States: (0) large and wide; (1) small and narrow.
Dist.: *Homo* 0; *Pan* 0; *Gorilla* 0; *Pongo* 1; *Hylobates* 0; *Colobus* 0.

35. Premaxillary suture in adult

Ref.: Shoshani *et al.* (1996) #130.
States: (0) patent; (1) obliterated.
Dist.: *Homo* 1; *Pan* 1; *Gorilla* 0; *Pongo* 0; *Hylobates* 0; *Colobus* 0.

36. Foramen lacerum medium

Ref.: Shoshani *et al.* (1996) #131.

States: (0) absent; (1) present.

Dist.: *Homo* 1; *Pan* 0; *Gorilla* 0; *Pongo* 1; *Hylobates* 0; *Colobus* 1.

Notes: This is a small space, bilateral to the anterior edge of the basioccipital, just behind the suture with the basisphenoid; bordered laterally by the anterior end of the petrosal. In humans it is covered over with cartilage but pierced by the ascending pharyngeal artery. It is large in *Homo*, small in *Pongo*, and absent in *Pan* in which the medial side of the anterior petrosal fills up the gap.

37. Posterior convergence of temporal lines

Ref.: Shoshani *et al.* (1996) #132.

States: (0) converge posteriorly; (1) do not converge.

Dist.: *Homo* 1; *Pan* 0; *Gorilla* 0; *Pongo* 0; *Hylobates* 1; *Colobus* 1.

Notes: This character is apparently not redundant with #28 (supraorbital trigon) as the distribution of states is different.

38. Mesial groove on male canine

Ref.: Shoshani *et al.* (1996) #159.

States: (0) extends onto root; (1) present; (2) absent.

Dist.: *Homo* 1; *Pan* 1; *Gorilla* 1; *Pongo* 2; *Hylobates* 1; *Colobus* 0.

Notes: Shoshani *et al.*'s states (0 = present; 1 = extends onto root; 2 = absent) changed so that character can be treated as an ordered character.

39. Relative height of male canine

Ref.: Shoshani *et al.* (1996) #160.

States: (0) high relative to mesiodistal length; (1) lower relative to mesiodistal length.

Dist.: *Homo* 1; *Pan* 1; *Gorilla* 1; *Pongo* 1; *Hylobates* 0; *Colobus* 1.

40. Upper I2 occlusal edge

Ref.: Shoshani *et al.* (1996) #161.

States: (0) slopes distally; (1) does not slope distally.

Dist.: *Homo* 1; *Pan* 1; *Gorilla* 0; *Pongo* 0; *Hylobates* 0; *Colobus* 0.

41. Robusticity of canines

Ref.: Shoshani *et al.* (1996) #162.

States: (0) slender; (1) more robust.

Dist.: *Homo* 1; *Pan* 1; *Gorilla* 1; *Pongo* 1; *Hylobates* 0; *Colobus* 0.

42. Basal keel of lower canines

Ref.: Shoshani *et al.* (1996) #163.

States: (0) present; (1) absent.

Dist.: *Homo* 1; *Pan* 1; *Gorilla* 0; *Pongo* 0; *Hylobates* 0; *Colobus* 0.

43. Basal area of paracone of upper premolars

Ref.: Shoshani *et al.* (1996) #164.
States: (0) subequal to protocone; (1) smaller than protocone.
Dist.: *Homo* 0; *Pan* 0; *Gorilla* 0; *Pongo* 1; *Hylobates* 1; *Colobus* 1.

44. Molar cingulum

Ref.: Shoshani *et al.* (1996) #165.
States: (0) prominent, shelf-like; (1) reduced, incomplete, (2) fragmented or absent.
Dist.: *Homo* 2; *Pan* 2; *Gorilla* 1; *Pongo* 2; *Hylobates* 1; *Colobus* 1.
Notes: Treated as ordered character in analysis.

45. Protoconid apex on lower dP3

Ref.: Shoshani *et al.* (1996) #166.
States: (0) more lingual from the median axis; (1) truncated buccally from the median axis.
Dist.: *Homo* 1; *Pan* 0; *Gorilla* 0; *Pongo* 1; *Hylobates* 0; *Colobus* ?.

46. Metaconid of lower dP3

Ref.: Shoshani *et al.* (1996) #167.
States: (0) present; (1) absent.
Dist.: *Homo* 0; *Pan* 1; *Gorilla* 1; *Pongo* 0; *Hylobates* 0; *Colobus* ?.

47. Protocristid of lower dP3

Ref.: Shoshani *et al.* (1996) #168.
States: (0) aligned with tooth mesiodistal axis; (1) angled.
Dist.: *Homo* 1; *Pan* 0; *Gorilla* 0; *Pongo* 1; *Hylobates* 0; *Colobus* ?.

48. Talonid basin of lower dP3

Ref.: Shoshani *et al.* (1996) #169.
States: (0) open distally; (1) closed.
Dist.: *Homo* 1; *Pan* 0; *Gorilla* 0; *Pongo* 1; *Hylobates* 0; *Colobus* ?.

49. Metaconid of lower dP4

Ref.: Shoshani *et al.* (1996) #170.
States: (0) subequal to protoconid; (1) increased relative to protoconid on lower dP4.
Dist.: *Homo* 0; *Pan* 1; *Gorilla* 1; *Pongo* 0; *Hylobates* 0; *Colobus* ?.

50. Crista obliqua on lower dP4

Ref.: Shoshani *et al.* (1996) #171.
States: (0) does not reach protoconid apex; (1) reaches protoconid apex.
Dist.: *Homo* 1; *Pan* 0; *Gorilla* 0; *Pongo* 1; *Hylobates* 1; *Colobus* ?.

51. Talonid basin on lower dP4

Ref.: Shoshani *et al.* (1996) #172.
States: (0) open distally; (1) closed.
Dist.: *Homo* 1; *Pan* 0; *Gorilla* 0; *Pongo* 1; *Hylobates* 1; *Colobus* ?.

52. Protocone of upper dP3, in crown view

Ref.: Shoshani *et al.* (1996) #173.

States: (0) larger than paracone; (1) smaller than paracone.

Dist.: *Homo* 1; *Pan* 0; *Gorilla* 0; *Pongo* 1; *Hylobates* 0; *Colobus* ?.

53. Preprotocrista of upper dP4

Ref.: Shoshani *et al.* (1996) #174.

States: (0) weak; (1) more developed.

Dist.: *Homo* 1; *Pan* 1; *Gorilla* 1; *Pongo* 1; *Hylobates* 0; *Colobus* ?.

54. Postprotocrista of upper dP4

Ref.: Shoshani *et al.* (1996) #175.

States: (0) poor; (1) more developed; (2) still more developed.

Dist.: *Homo* 2; *Pan* 2; *Gorilla* 2; *Pongo* 1; *Hylobates* 1; *Colobus* 0.

Notes: Treated as ordered character in analysis.

55. Protocristid grooves of molars

Ref.: Shoshani *et al.* (1996) #176.

States: (0) prominent; (1) barely visible.

Dist.: *Homo* 0; *Pan* 1; *Gorilla* 0; *Pongo* 1; *Hylobates* 1; *Colobus* 0.

56. Lingual marginal ridges of molars

Ref.: Shoshani *et al.* (1996) #177.

States: (0) hardly appreciable; (1) more prominent; (2) very prominent.

Dist.: *Homo* 1; *Pan* 1; *Gorilla* 2; *Pongo* 1; *Hylobates* 1; *Colobus* 0.

Notes: Treated as ordered character in analysis.

57. Thickness of molar enamel

Ref.: Shoshani *et al.* (1996) #178.

States: (0) thin; (1) increased thickness; (2) very thick.

Dist.: *Homo* 2; *Pan* 0; *Gorilla* 0; *Pongo* 1; *Hylobates* 0; *Colobus* ?.

Notes: Treated as ordered character in analysis.

58. Proportion of Pattern 3 enamel

Ref.: Shoshani *et al.* (1996) #179.

States: (0) high; (1) reduced; (2) very reduced.

Dist.: *Homo* 0; *Pan* 2; *Gorilla* 2; *Pongo* 1; *Hylobates* 0; *Colobus* ?.

Notes: Treated as ordered character in analysis.

59. Insertion of genioglossus

Ref.: Shoshani *et al.* (1996) #185.

States: (0) above inferior transverse torus of internal (or posterior) of mandibular symphysis; (1) shifted to inferior transverse torus.

Dist.: *Homo* 0; *Pan* 1; *Gorilla* 1; *Pongo* 1; *Hylobates* 0; *Colobus* 0.

60. Insertion of geniohyoideus

Ref.: Shoshani *et al.* (1996) #186.

States: (0) basally on inferior transverse torus; (1) higher on inferior transverse torus; (2) above inferior transverse torus.

Dist.: *Homo* 2; *Pan* 2; *Gorilla* 1; *Pongo* 0; *Hylobates* 1; *Colobus* 0.

Notes: Treated as ordered character in analysis.

61. Insertion of digastric

Ref.: Shoshani *et al.* (1996) #187.

States: (0) posterior to inferior transverse torus; (1) inferior transverse torus; (2) not on symphysis.

Dist.: *Homo* 1; *Pan* 1; *Gorilla* 0; *Pongo* 2; *Hylobates* 0; *Colobus* 0.

Notes: Treated as unordered character in analysis.

62. Encephalization

Ref.: Shoshani *et al.* (1996) #220.

States: (0) low, <1.2; (1) increased, >1.2–1.9; (2) high >1.9.

Dist.: *Homo* 2; *Pan* 1; *Gorilla* 0; *Pongo* 1; *Hylobates* 2; *Colobus* 0.

Notes: Shoshani *et al.*'s (1996) character states ($0=$low, <10; $1=$increased, 10–11; $2=$high >11) and distributions (*Homo* 2; *Pan* 2; *Gorilla* 0; *Pongo* 0; *Hylobates* 1; *Colobus* ?) updated using Kappelman's (1996) data. Treated as ordered character in analysis.

63. Retroarticular canal

Ref.: Braga (1995).

States: (0) absent; (1) present.

Dist.: *Homo* 0; *Pan* 0; *Gorilla* 0; *Pongo* 1; *Hylobates* ?; *Colobus* ?.

64. Condylar canal

Ref.: Braga (1995).

States: (0) absent; (1) present.

Dist.: *Homo* 1; *Pan* 1; *Gorilla* 1; *Pongo* 0; *Hylobates* ?; *Colobus* ?.

65. Incisive fossa

Ref.: Andrews (1987)

States: (0) absent; (1) deep; (2) extends through palate.

Dist.: *Homo* 1; *Pan* 1; *Gorilla* 1; *Pongo* 0; *Hylobates* 2; *Colobus* ?.

Notes: Treated as an ordered character in analysis.

66. Molar dentine horns

Ref.: Andrews (1987)

States: (0) high; (1) low.

Dist.: *Homo* 0; *Pan* 0; *Gorilla* 0; *Pongo* 1; *Hylobates* 0; *Colobus* ?.

67. Molar enamel wrinkling

Ref.: Andrews (1987)

States: (0) smooth or slight wrinkling; (1) deep secondary wrinkling.

Dist.: *Homo* 0; *Pan* 0; *Gorilla* 0; *Pongo* 1; *Hylobates* 0; *Colobus* ?.

68. Postorbital sulcus

Ref.: Andrews (1987)

States: (0) absent; (1) present.

Dist.: *Homo* 1; *Pan* 1; *Gorilla* 1; *Pongo* 0; *Hylobates* 0; *Colobus* ?.

69. Ethmoid-lacrymal contact

Ref.: Andrews (1987)

States: (0) long, 100%; (1) short, 40–90%.

Dist.: *Homo* 0; *Pan* 1; *Gorilla* 1; *Pongo* 0; *Hylobates* 0; *Colobus* ?.

70. Fronto-maxillary contact in orbits

Ref.: Andrews (1987)

States: (0) no contact; (1) contact, 30–50%.

Dist.: *Homo* 0; *Pan* 1; *Gorilla* 1; *Pongo* 0; *Hylobates* 0; *Colobus* ?.

71. Nasal floor morphology

Ref.: Andrews (1987).

States: (0) nasal floor stepped; (1) nasal floor unstepped.

Dist.: *Homo* 0; *Pan* 0; *Gorilla* 0; *Pongo* 1; *Hylobates* 0; *Colobus* ?.

72. Palatine fenestrae reduced in size

Ref.: Schwartz (1984)

States: (0) no; (1) yes.

Dist.: *Homo* 1; *Pan* 1; *Gorilla* 1; *Pongo* 1; *Hylobates* 0; *Colobus* ?.

73. Cheek tooth height

Ref.: Schwartz (1984).

States: (0) low; (1) medium; (2) medium-high; (3) high.

Dist.: *Homo* 0; *Pan* 2; *Gorilla* 3; *Pongo* 0; *Hylobates* 1; *Colobus* ?.

Notes: This may be a corollary of thick enamel (Andrews, 1987). Treated as an ordered character.

74. Lower M3 smaller than lower M2

Ref.: Schwartz (1984); Andrews (1987).

States: (0) no; (1) yes.

Dist.: *Homo* 1; *Pan* 1; *Gorilla* 0; *Pongo* 1; *Hylobates* 1; *Colobus* ?.

Notes: States for *Pan* and *Hylobates* are from Andrews (1987). Others from Schwartz (1984).

75. Number of zygomatic foramina

Ref.: Schwartz (1984).

States: (0) 1–2; (1) 1–2+.

Dist.: *Homo* 0; *Pan* 0; *Gorilla* 1; *Pongo* 1; *Hylobates* 0; *Colobus* ?.

Notes: States and distribution from Schwartz (1984).

76. Post talonid basin

Ref.: Groves (1986) #201.
States: (0) absent; (1) small; (2) narrow.
Dist.: *Homo* 1; *Pan* 2; *Gorilla* 2; *Pongo* 3; *Hylobates* 0; *Colobus* ?.
Notes: Treated as unordered character because it was not clear that states form a linear transformation series.

77. Relative depth of mandible

Ref.: Delson & Andrews (1975) Table 2 #1.
States: (0) deep/moderate; (1) moderate; (2) shallow.
Dist.: *Homo* 1; *Pan* 1; *Gorilla* 1; *Pongo* 1; *Hylobates* 2; *Colobus* 0.
Notes: Treated as an ordered character in analysis.

78. Mandibular shape

Ref.: Delson & Andrews (1975) Table 2 #2.
States: (0) shallows mesially/constant; (1) constant; (2) deepens.
Dist.: *Homo* 1; *Pan* 1; *Gorilla* 1; *Pongo* 1; *Hylobates* 2; *Colobus* 0.
Notes: Treated as an ordered character in analysis.

79. Ethmo-sphenoid contact

Ref.: Groves (1986) #24.
States: (0) none/very short, 0–39%; (1) short, 40–90%; (2) long, 91–100%.
Dist.: *Homo* 2; *Pan* 1; *Gorilla* 1; *Pongo* 2; *Hylobates* 0; *Colobus* ?.
Notes: Data from Groves (1986). States adapted from Andrews (1987) states for Ethmoid-lacrymal contact (#69 in this list). Treated as an ordered character in analysis.

80. Zygomatic bone

Ref.: Groves (1986) #31.
States: (0) curved; (1) flattened.
Dist.: *Homo* 0; *Pan* 0; *Gorilla* 0; *Pongo* 1; *Hylobates* 0; *Colobus* ?.

81. Relative face height

Ref.: Groves (1986) #31.
States: (0) 19–24; (1) 27–30.
Dist.: *Homo* 0; *Pan* 1; *Gorilla* 1; *Pongo* 1; *Hylobates* 0; *Colobus* ?.

82. Canine length as percentage of upper M1 (male)

Ref.: Groves (1986) #177.
States: (0) short, 61–81%; (1) longer, 101–182%.
Dist.: *Homo* 0; *Pan* 1; *Gorilla* 1; *Pongo* 1; *Hylobates* 1; *Colobus* ?.

83. Canine length as percentage of upper M1 (female)

Ref.: Groves (1986) #178.
States: (0) short, 61–81%; (1) longer, 92–144%.
Dist.: *Homo* 0; *Pan* 1; *Gorilla* 1; *Pongo* 1; *Hylobates* 1; *Colobus* ?.

84. Canine length as percentage of upper P4 (male)

Ref.: Groves (1986) #179.
States: (0) short, 116–160%; (1) longer, 215–543%.
Dist.: *Homo* 0; *Pan* 1; *Gorilla* 1; *Pongo* 1; *Hylobates* 1; *Colobus* ?.

85. Canine length as percentage of upper P4 (female)

Ref.: Groves (1986) #179.
States: (0) short, 116–178%; (1) longer, 187–273%; (2) still longer, 307–543%.
Dist.: *Homo* 0; *Pan* 1; *Gorilla* 0; *Pongo* ?; *Hylobates* 2; *Colobus* ?.
Notes: Treated as an ordered character in analysis.

86. Angle between tooth rows

Ref.: Groves (1986) #182.
States: (0) low, −5–16° + ; (1) high, 20–40°.
Dist.: *Homo* 1; *Pan* 0; *Gorilla* 0; *Pongo* 0; *Hylobates* 0; *Colobus* ?.

87. Eruption after upper I2

Ref.: Groves (1986) #183.
States: (0) PCPM; (1) MPPC.
Dist.: *Homo* 0; *Pan* 1; *Gorilla* 1; *Pongo* 1; *Hylobates* 1; *Colobus* ?.

88. Eruption after lower I2

Ref.: Groves (1986) #184.
States: (0) CPPM; (1) MPPC.
Dist.: *Homo* 0; *Pan* 1; *Gorilla* 1; *Pongo* 1; *Hylobates* 1; *Colobus* ?.

89. Upper I1 lingual crenulations

Ref.: Groves (1986) #187.
States: (0) absent; (1) marginal; (2) whole surface.
Dist.: *Homo* 1; *Pan* 1; *Gorilla* 1; *Pongo* 2; *Hylobates* 1; *Colobus* ?.
Notes: Treated as an ordered character in analysis.

90. Upper I1 cingulum tubercle

Ref.: Groves (1986) #188.
States: (0) present; (1) absent.
Dist.: *Homo* 0; *Pan* 1; *Gorilla* 1; *Pongo* 0; *Hylobates* 1; *Colobus* ?.
Notes: Groves' (1986) three states recognises three states: (1) usually present; (2) incipient; and (3) absent. As 'incipient' is clearly encompassed by usually present, the two states were collapsed into one.

91. Number of upper I1 ridges

Ref.: Groves (1986) #189.
States: (0) one; (1) one or more than one; (2) always more than one.
Dist.: *Homo* 1; *Pan* 0; *Gorilla* 2; *Pongo* 1; *Hylobates* 0; *Colobus* ?.
Notes: Treated as an ordered character in analysis.

92. Canine sexual dimorphism

Ref.: Groves (1986) #191.
States: (0) monomorphic; (1) dimorphic.
Dist.: *Homo* 0; *Pan* 1; *Gorilla* 1; *Pongo* 1; *Hylobates* 0; *Colobus* ?.

93. Canine elongation

Ref.: Groves (1986) #193.
States: (0) buccolingual; (1) none; (2) mesiodistal.
Dist.: *Homo* 0; *Pan* 2; *Gorilla* 2; *Pongo* 2; *Hylobates* 1; *Colobus* ?.
Notes: Treated as an ordered character.

94. Lower P3 metaconid

Ref.: Groves (1986) #197.
States: (0) absent; (1) tiny; (2) small
Dist.: *Homo* 2; *Pan* 0/1; *Gorilla* 1; *Pongo* 2; *Hylobates* 0; *Colobus* ?.
Notes: Treated as an ordered character in analysis.

95. Trigonid basin

Ref.: Groves (1986) #199.
States: (0) narrow slit; (1) fair; (2) wider.
Dist.: *Homo* 1; *Pan* 2; *Gorilla* 2; *Pongo* 1; *Hylobates* 0; *Colobus* ?.
Notes: Groves' states are: *Homo* fair; *Pan* fairly wide; *Gorilla* rather wide; *Pongo* fair; *Hylobates* narrow slit; Outgroup (monkeys) varies. Treated as an ordered character in analysis.

96. Sulcus obliqus

Ref.: Groves (1986) #200.
States: (0) weak to moderate definition; (1) strong to very strong definition.
Dist.: *Homo* 0; *Pan* 0; *Gorilla* 1; *Pongo* 1; *Hylobates* 0; *Colobus* ?.
Notes: Groves' (1986) identifies five states: poor; present; fair; strong; very strong.

Appendix 6.2. Quantitative character state data matrix used in hominoid analyses

Characters P1 P2 P3 P4 P5 P6 P7 P8 P9 P10 P11 P12 P13 P14 P15 P16 P17 P18 P19 P20 P21 P22 P23 P24 P25 P26 P27 P28 P29 P30 P31 M1 M2 M3 M4 M5 M6 M7 M8 M9 M10 M11 M12 M13 M14 M15 M16 M17 M18 M19 M20 M21 M22 M23 M24 M25 M26 M27 M28 M29 M30 M31 M32 M33 M34 M35 M36 M37 M38 M39 M40 F1 F2 F3 F4 F5 F6 F7 F8 F9 F10 F11 F12 F13 F14 F15 F16 F17 F18 F19 F20 F21 F22 F23 F24 C1 C2 C3 C4 C5 C6 C7 C8 C9 C10 C11 C12 C13 C14 C15 C16 C17 C18 C19 C20 C21 C22 C23 C24 C25 C26 C27 C28 C29 C30 C31 C32 C33 C34

Colobus	E E B C B 8 6 E 8 8 6 D 6 A 6 8 6 A B 7 9 A 6 8 A B D C C 7 6 A C B E 8 C 6 E ? E 6 E 8 C 7 B 6 6 A 9 A 8 D B 8 E 6 C B 7 C 6 E 7 8 B B B 6 6 6 7 C 8 D 9 E E 8 8 8 B B 9 C D E C C 6 9 7 6 7 8 8 C A B E 9 D D E C A D 7 8 C C 7 8 8 E 7 6 9 E 8 E 8 9 B E 9 E A
Gorilla	B A B B 8 8 8 6 8 8 9 A A A 8 B 8 E B 7 E 8 8 E E B 8 9 A 9 8 C C B A 9 8 A 6 7 7 9 9 9 7 7 7 8 A 7 E 7 C 7 9 A B D 6 8 7 E B A A 9 7 E B A 8 D E 9 C 6 6 9 A B D B 8 B 6 6 8 6 6 6 A 9 9 B 7 B E E E E C E 6 D A 8 D C D E 7 8 A 8 8 B C A 7 A C B D D E A C A E
Homo	B C C 7 E E E B D C D 6 8 6 A 6 D 6 E E 6 B E A 6 B D E E E E E C E C E E E B D A D 6 7 7 B 7 B 8 E 6 E 6 D E E 7 A B E E 6 E C E 9 E 6 E E E 8 7 6 6 D E 6 C 6 7 A D 7 9 E D C C E A E E 8 E 6 6 6 6 7 6 6 8 6 6 E 6 6 7 6 E E E E E 6 7 E D A E 6 6 6 6 6 6 6 6
Pan	8 7 6 6 A C B B D E D C D E D E C A 6 B A C C 9 A B 8 8 7 C C 7 7 7 6 A 7 A B B C D C D E E E E D C C D C A A B B D A B B A 9 6 C E B 8 7 C C A B 9 A 9 A 9 7 B D 7 C 7 C 9 8 9 A 9 A 9 A B 9 B A 9 8 B 9 9 B 9 C 8 B B A B A 8 A 8 8 9 C A D 7 8 B A 9 9 A 9 A A
Pongo	6 7 A D 7 8 B 8 8 8 9 9 D A D B B A 8 B B 9 A 9 A 6 8 7 7 8 A 7 7 7 8 9 8 A 8 9 7 9 9 D A B B B D 7 9 7 C 7 6 7 7 8 B 6 B 8 9 8 7 A 7 B 7 8 A D B E E 9 B C 7 E 9 E 6 E E 9 8 9 8 9 E 9 A E 9 E C 9 C 7 9 C C 9 8 8 A 8 D B 7 8 9 C C A C A 8 9 8 8 D D A A E A A

Appendix 6.3. Quantitative character state data matrix used in papionin analyses

Characters	P1 P2 P3 P4 P5 P6 P7 P8 P9 P10 P11 P12 P13 P14 P15 P16 M1 M2 M3 M4 M5 M6 M7 M8 M9 M10 M11 M12 M13 M14 F1 F2 F3 F4 F5 F6 F7 F8 F9 F10 F11 F12 F13 F14 F15 F16 C1 C2 C3 C4 C5 C6 C7 C8 C9 C10 C11 C12 C13 C14 C15 C16
Cercocebus	C 6 F 9 C 6 B B A 9 9 6 C 9 C A C B 9 6 8 8 A C 8 5 B 6 C 6 B A 9 A 7 B 8 D D B B D 7 9 A B B A 9 9 9 7 C 7 6 5 8 B 8 8 A B
Lophocebus	C 8 9 4 B C E 6 6 7 5 6 D G 5 B C C 6 C 5 F C E D E E D E F F G 7 8 4 A 8 8 C E E 9 D 9 6 D 4 6 4 7 6 4 6 4 6 7 4 5 C 4 5 C
Macaca	C 5 F C 5 B 8 B A A A 6 B 9 8 5 B C E C C 6 B 8 8 A 6 B 8 9 B B 8 7 B G 4 9 D A A G 7 9 6 A 7 6 6 A 7 A 6 C D D 8 B B 8 A C
Mandrillus	6 E 4 A B 4 9 B E F D D 5 8 G 9 5 5 A A C 9 4 8 5 C C G 8 B 6 B D F F E D D D 5 4 8 E 9 G 6 G G E F G E G F A D F B C G F 8
Pan	G 9 9 9 C D G 4 4 4 5 E A 7 6 F D G G 7 G 8 G G G 5 C 6 G D F 4 4 8 C 4 8 G 4 G G 4 E 9 D G 7 6 9 4 D E 6 7 8 6 8 G 4 8 5 B
Papio	6 E 9 A 5 C 8 B A A C B B B A A C 9 B G C F 8 8 8 E B C 8 9 5 B D F E A D 7 7 5 6 7 8 G D 8 E E C F D E C C B D F B C D A C
Theropithecus	6 E 9 G E C 4 G G F G G 8 A D A 5 5 4 7 5 9 9 4 8 A 4 6 4 7 7 6 G 7 7 6 G 4 8 9 9 D 7 9 6 6 B C G A 6 7 C D G D C 5 B D F 4

References

Ahern, J. C. M. (1998). Underestimating intraspecific variation: the problem with excluding Sts 19 from *Australopithecus africanus*. *Am. J. Phys. Anthrop.* **105**: 461–80.

Aiello, L. C. & Wood, B. A. (1994). Cranial variables as predictors of hominine body mass. *Am. J. Phys. Anthropol.* **95**: 409–26.

Aiello, L. C., Wood, B. A., Key, C. & Lewis, M. (1999) Morphological and taxonomic affinities of the Olduvai ulna (OH 36). *Am. J. Phys. Anthrop.* **109**: 89–110.

Andrews, P. (1978). Taxonomy and relationships of fossil apes. In: Chivers, D. J. & Joysey, K. A. (eds.), *Recent Advances in Primatology*, vol. 3: *Evolution*, pp. 43–56. London: Academic Press.

Andrews, P. (1987). Aspects of hominoid phylogeny. In Patterson, C. (ed.), *Molecules and Morphology in Evolution: Conflict or Compromise?*, pp. 23–53. Cambridge: Cambridge University Press.

Andrews, P. (1992) Evolution and environment in the Hominoidea. *Nature* **360**: 641–6.

Andrews, P. & Martin, L. B. (1987). Cladistic relationships of extant and fossil hominoids. *J. Hum. Evol.* **16**: 101–18.

Arsuaga, J. L., Carretero, J. M., Martínez, I. & Gracia, A. (1991). Cranial remains and long bones from Atapuerca/Ibeas (Spain). *J. Hum. Evol.* **20**: 191–230.

Atchley, W. R. & Fitch, W. M. (1991). Gene trees and the origin of inbred strains of mice. *Science* **254**: 554–8.

Augustí, J., Köhler, M. & Moyà-Solà, S., Cabrera, L., Garcés, M. and Parés, J. M. (1996). Can Llobateres: the pattern and timing of the Vallesian hominoid radiation reconsidered. *J. Hum. Evol.* **31**: 143–55.

Begun, D. R., Ward, C. V. & Rose, M. (1997). Events in hominoid evolution. In Begun, D. R., Ward, C. V. & Rose, M. (eds.), *Function, Phylogeny, and Fossils: Miocene Hominoid Evolution and Adaptations*, pp. 389–415. New York: Plenum Press.

Benefit, B. R. (1993). The permanent dentition and phylogenetic position of *Victoriapithecus* from Maboko Island, Kenya. *J. Hum. Evol.* **25**: 83–172.

Beynon, A. D., Dean, M. C., Leakey, M. G., Reid, D. J. & Walker, A. (1998). Comparative dental development and microstructure of *Proconsul* teeth from Rusinga Island, Kenya. *J. Hum. Evol.* **35**: 163–209.

Braga, J. (1995). Emissary canals in the Hominoidea and their phylogenetic significance. *Folia Primatol.* **65**: 144–53.

Bromage, T. G. (1989). Ontogeny of the early hominid face. *J. Hum. Evol.* **18**: 751–73.

Cameron, D. W. (1997) A revised systematic scheme for the Eurasian Miocene fossil Hominidae. *J. Hum. Evol.* **33**: 449–77.

Chamberlain, A. T. (1987). *A Taxonomic Review and Phylogenetic Analysis of* Homo habilis. PhD thesis, University of Liverpool.

Chamberlain, A. T. & Wood, B. A. (1987). Early hominid phylogeny. *J. Hum. Evol.* **16**: 119–33.

Collard, M. (1998). *Morphological Evolution of the Hominoids and Papionins: Implications for Palaeoanthropological Cladistics.* Ph.D. thesis, The University of Liverpool.

Collard, M., Lockwood, C. A. & Wood, B. A. (1999). Stratigraphic consistency in hominin phylogeny. *Am. J. Phys. Anthrop.* **28** (Suppl.): 110.

Conroy, G. (1994). *Otavipithecus*: or how to build a better hominid – not. *J. Hum. Evol.* **27**: 373–83.

Corruccini, R. S. (1994). How certain are hominoid phylogenies? The role of confidence intervals in cladistics. In Corruccini, R. S. & Ciochon, R. L. (eds.), *Integrative Approaches to the Past: Paleoanthropological Advances in Honor of F. Clark Howell*, pp. 167–83. Englewood Cliffs, NJ: Prentice Hall.

Corruccini, R. S. & McHenry, H. M. (1980). Cladometric analysis of Pliocene hominids. *J. Hum. Evol.* **9**: 209–21.

Creel, N. (1986). Size and phylogeny in hominoid primates. *Syst. Zool.* **35**: 81–99.

Cranston, P. & Humphries, C. (1988). Cladistics and computers: a chironomid conundrum? *Cladistics* **4**: 72–92.

Crisp, M. & Weston, P. (1987). Cladistics and legume systematics, with an analysis of the Bossiaecae, Brongniartieae and Mirbelieae. In Stirton, C. (ed.), *Advances in Legume Systematics*, Part 3, pp. 65–130. London: Royal Botanical Gardens.

Crowe, T. (1994). Morphometrics, phylogenetic methods, and cladistics: means to an end, or much a do about nothing? *Cladistics* **10**: 77–84.

Delson, E. (1977). Catarrhine phylogeny and classification: principles, methods and comments. *J. Hum. Evol.* **6**: 433–59.

Delson, E. & Andrews, P. (1975). Evolution and interrelationships of the catarrhine primates. In Luckett, W. P. & Szalay, F. S. (eds.), *Phylogeny of the Primates: A Multidisciplinary Approach*, pp. 405–46. New York: Plenum Press.

Delson, E., Eldredge, N. & Tattersall, I. (1977). Reconstruction of hominid phylogeny: a testable framework based on cladistic analysis. *J. Hum. Evol.* **6**: 263–78.

Disotell, T. R. (1994). Generic level relationships of the Papionini (Cercopithecoidea). *Am. J. Phys. Anthrop.* **94**: 47–57.

Disotell, T. R. (1996). The phylogeny of Old World monkeys. *Evol. Anthrop.* **5**: 18–24.

Disotell, T. R., Honeycutt, R. L. & Ruvolo, M. (1992). Mitochondrial DNA phylogeny of the Old World monkey tribe Papionini. *Mol. Biol. Evol.* **9**: 1–13.

Eldredge, N. & Tattersall, I. (1975). Evolutionary models, phylogenetic reconstruction and another look at hominid phylogeny. In Szalay, F. S. (ed.), *Contributions to Primatology 5: Approaches to Primate Paleobiology*, pp. 218–42. Basel: Karger.

Felsenstein, J. (1985). Confidence limits on phylogenetics: an approach using the bootstrap. *Evolution* **39**: 783–91.

Felsenstein, J. (1988). Phylogenies and quantitative characters. *Ann. Rev. Ecol. Syst.* **19**: 445–71.

Fitch, W. M. & Atchley, W. R. (1987). Divergence in inbred strains of mice: a comparison of three different types of data. In Patterson, C. (ed.), *Molecules and Morphology in Evolution: Conflict or Compromise?*, pp. 203–16. Cambridge: Cambridge University Press.

Groves, C. P. (1986). Systematics of the great apes. In Swindler, D. R. & Erwin, J. (eds.), *Comparative Primate Biology*, Volume 1: *Systematics, Evolution and Anatomy*, pp. 187–217. New York: Alan R. Liss.

Harris, E. E. & Disotell, T. R. (1998). Nuclear gene trees and the phylogenetic relationships of the mangabeys (Primates: Papionini). *Mol. Biol. Evol.* **15**: 892–900.

Harrison, T. (1982). *Small-bodied apes from the Miocene of East Africa*. PhD thesis, University of London.

Harrison, T. (1989). New postcranial remains of *Victoriapithecus* from the Middle Miocene of Kenya. *J. Hum. Evol.* **18**: 3–54.
Harrison, T. (1993). Cladistic concepts and the species problem in hominoid evolution. In W. H. Kimbel & L. B. Martin (eds.), *Species, Species Concepts and Primate Evolution*, pp. 345–71. New York: Plenum Press.
Hartman, S. E. (1988). A cladistic analysis of hominoid molars. *J. Hum. Evol.* **17**: 489–502.
Hillis, D. M., Bull, J. J., White, M. E. Badgett, M. R. & Molineux, I. J. (1992). Experimental phylogenetics: generation of a known phylogeny. *Science* **255**: 589–92.
Hillis, D. M. & Bull, J. J. (1993). An empirical test of bootstrapping as a method for assessing confidence in phylogenetic analysis. *Syst. Biol.* **42**: 182–92.
Jungers, W. L., Falsetti, A. B. & Wall, C. E. (1995). Shape, relative size, and size-adjustments in morphometrics. *Ybk. Phys. Anthrop.* **38**: 137–61.
Kappelman, J. (1996). The evolution of body mass and relative brain size in fossil hominids. *J. Hum. Evol.* **30**: 243–76.
Kelley, J., Anwar, M., McCollum, M. A. & Ward, S. C. (1995). The anterior dentition of *Sivapithecus parvada*, with comments on the phylogenetic sigificance of incisor heteromorphy in Hominoidea. *J. Hum. Evol.* **28**: 503–17.
Kluge, A. (1998). Total evidence or taxonomic congruence: cladistics or consensus classification *Cladistics* **14**: 151–8.
Leakey, R. E. & Leakey, M. G. (1986). A new Miocene hominoid from Kenya. *Nature* **324**: 143–6.
Lieberman, D. E. (1995). Testing hypotheses about recent human evolution from skulls: integrating morphology, function, development, and phylogeny. *Curr. Anthrop.* **36**: 159–97.
Lieberman, D. E. (1997). Making behavioral and phylogenetic inferences from hominid fossils: considering the developmental influence of mechanical forces. *Ann. Rev. Anthrop.* **26**: 185–210.
Lieberman, D. E. (1999). Homology and hominid phylogeny: problems and potential solutions *Evol. Anthropol.* **7**: 142–51.
Lieberman, D. E., Wood, B. A. & Pilbeam, D. R. (1996). Homoplasy and early *Homo*: an analysis of the evolutionary relationships of *H. habilis sensu stricto* and *H. rudolfensis*. *J. Hum. Evol.* **30**: 97–120.
Lieberman, D. E. & Wood, B. A. (1999). Patterns of craniofacial variability in living primates and P. boisei. *Am. J. Phys. Anthrop.* **28** (Suppl.): 280.
McCrossin, M. L. & Benefit, B. R. (1997). On the relationships and adaptations of Kenyapithecus, a large bodied hominoid from the Middle Miocene of Eastern Africa. In Begun, D. R., Ward, C. V. & Rose, M. (eds.), *Function, Phylogeny, and Fossils: Miocene Hominoid Evolution and Adaptations*, pp. 241–67. New York: Plenum Press.
Maddison, W. P., Donoghue, M. J. & Maddison, D. R. (1984). Outgroup analysis and parsimony. *Syst. Zool.* **33**: 83–103.
Masali, M. (1968). The ear bones and vertebral column as indicators of taxonomic and postural distinction among Old World Primates. In Chiarelli, B. (ed.), *Taxonomy and Phylogeny of Old World Primates with References to the Origin of Man*, pp. 69–94. Turin: Rosenberg and Sellier.

Moore, J. H. (1994). Putting anthropology back together again: the ethnographic critique of cladistic theory. *Am. Anthrop.* **96**: 925–48.
Moyà-Solà, S. & Köhler, M. (1993). Recent discoveries of *Dryopithecus* shed new light on evolution of great apes. *Nature* **365**: 543–5.
Moyà-Solà, S. & Köhler, M. (1995). New partial cranium of *Dryopithecus* Lartet, 1863 (Hominoidea, Primates) from the Upper Miocene of Can Llobateres, Barcelona, Spain. *J. Hum. Evol.* **29**: 101–39.
Pilbeam, D. R. (1996). Genetic and morphological records of the Hominoidea and hominid origins: a synthesis. *Mol. Phyl. Evol.* **5**: 155–68.
Pimentel, R. & Riggins, R. (1987). The nature of cladistic data. *Cladistics* **3**: 201–9.
Rae, T. C. (1997). The early evolution of the hominoid face. In Begun, D. R., Ward, C. V. & Rose, M. (eds.), *Function, Phylogeny, and Fossils: Miocene Hominoid Evolution and Adaptations*, pp. 59–77 New York: Plenum Press.
Rae, T. C. (1998). The logical basis for the use of continuous characters in phylogenetic systematics. *Cladistics* **14**: 221–8.
Rae, T. C. (1999). Mosaic evolution in the origin of the Hominoidea. *Folia Primatol.* **70**, 125–35.
Rak, Y. & Clarke, R. J. (1979a). Aspects of the middle and external ear of early hominids. *Am. J. Phys. Anthrop.* **51**: 471–4.
Rak, Y. & Clarke, R. J. (1979b). Ear ossicle of *Australopithecus robustus. Nature* **279**: 62–3.
Rose, M. D., Leakey, M. G., Leakey, R. E. F. & Walker, A. (1992). Postcranial specimens of *Simiolus enjiessi* and other primitive catarrhines from the Early Miocene of Lake Turkana, Kenya. *J. Hum. Evol.* **22**: 171–237.
Ruvolo, M. (1994). Molecular evolutionary processes and conflicting gene trees: the hominoid case. *Am. J. Phys. Anthrop.* **94**: 89–113.
Ruvolo, M. (1995). Seeing the forest and the trees. *Am. J. Phys. Anthrop.* **94**: 89–114.
Ruvolo, M. (1997). Molecular phylogeny of the hominoids: inferences from multiple independent DNA data sets. *Mol. Biol. Evol.* **14**: 248–65.
Schwartz, J. H. (1984). On the evolutionary relationships of humans and orang-utans. *Nature* **308**: 501–5.
Shoshani, J., Groves, C. P., Simons, E. L., & Gunnell, G. F. (1996). Primate phylogeny: morphological vs. molecular results. *Mol. Phy. Evol.* **5**: 102–54.
Skelton, R. R, McHenry, H. M. & Drawhorn, G. M. (1986). Phylogenetic analysis of early hominids. *Curr. Anthrop.* **27**: 21–43.
Skelton, R. R. & McHenry, H. M. (1992). Evolutionary relationships among early hominids. *J. Hum. Evol.* **23**: 309–49.
Skelton, R. R. & McHenry, H. M. (1998). Trait list bias and a reappraisal of early hominid phylogeny. *J. Hum. Evol.* **34**: 109–14.
Slowinski (1993). 'Unordered' versus 'ordered' characters. *Syst. Biol.* **42**: 155–65.
Smith, A. B. (1994). *Systematics and the Fossil Record: Documenting Evolutionary Patterns*. Oxford: Blackwells.
Spoor, C. F. & Zonneveld, F. (1998). A comparative review of the human bony labyrinth. *Ybk. Phys. Anthrop.* **41**: 211–51.
Strait, D. S., Moniz, M. & Strait, P. (1996). Finite mixture coding: a new approach to coding continuous characters. *Syst. Biol.* **45**: 67–78.
Strait, D. S., Grine, F. E. & Moniz, M. A. (1997). A reappraisal of early hominid phylogeny. *J. Hum. Evol.* **32**: 17–82.

Strait, D. S. & Grine, F. E. (1998). Trait list bias? A reply to Skelton and McHenry. *J. Hum. Evol.* **34**: 115–18.

Strasser, E. & Delson, E. (1987). Cladistic analysis of cercopithecid relationships. *J. Hum. Evol.* **16**: 81–99.

Stringer, C. B. (1987). A numerical cladistic analysis for the genus *Homo*. *J. Hum. Evol.* **16**: 135–46.

Swofford, D. L. & Olson, G. J. (1990). Phylogenetic reconstruction. In Hillis, D. M. & Moritz, C. (eds.), *Molecular Systematics*, pp. 411–501 Sunderland, MA: Sinauer Associates.

Swofford, D. L. (1991). *Phylogenetic Analysis Using Parsimony, Version 3.0s.* Champaign, ILL: Illinois Natural History Survey.

Tattersall, I. & Eldredge, N. (1977). Fact, theory, and fantasy in human paleontology. *Am. Sci.* **65**: 204–11.

Thorpe, R. S. (1984). Coding morphometric characters for constructing distance Wagner networks. *Evolution* **38**: 244–55.

Turner, A. & Wood, B. A. (1993). Comparative palaeontological context for the evolution of the early hominid masticatory system. *J. Hum. Evol.* **24**: 301–18.

Wood, B. A. (1988). Are 'robust' australopithecines a monophyletic group? In Grine, F. E. (ed.), *Evolutionary History of the 'Robust' Australopithecines*, pp. 269–284. New York: Aldine de Gruyter.

Wood, B. A. (1991). *Koobi Fora Research Project*, vol. 4: *Hominid Cranial Remains.* Oxford: Clarendon Press.

Wood, B. A. (1992). Early hominid species and speciation. *J. Hum. Evol.* **22**: 351–65.

Wood, B. A. & Chamberlain, A. T. (1986). *Australopithecus*: grade or clade? In Wood, B. A., Martin, L. & Andrews, P. (eds.), *Major Topics in Primate and Human Evolution*, pp. 220–48. Cambridge: Cambridge University Press.

Wood, B. A. & Chamberlain, A. T. (1987). The nature and affinities of the 'robust' australopithecines. *J. Hum. Evol.* **16**: 625–41.

Wood, B. A., Li, Yu & Willoughby, C. (1991). Intraspecific variation and sexual dimorphism in cranial and dental variables among higher primates, and their bearing on the hominid fossil record. *J. Anat.* **174**: 185–205.

Wood, B. A., Aeillo, L. C., Wood, C. G. & Key, C. A. (1998). The use of articular surface shape to match the components of the *H. habilis* (OH8/35) talocrural joint. *J. Anat.* **193**: 61–72.

Wood, B. A. & Collard, M. (1999). The Human Genus. *Science* **284**: 65–71.

7 Cranial discrete variation in the great apes: new prospects in palaeoprimatology

José Braga

Distinguishing them from other morphological traits, the scoring methods for which are affected by major drawbacks, this chapter illustrates and defines, precisely, 19 cranial discrete traits in extant great apes and discusses their benefit for palaeoprimatological studies. An increasing number of genetic data demonstrates that the patterning of intraspecific variation among the great apes is more complex than previously thought. This study begins with an adequate sampling of the subspecies being analysed. Thereafter, using a database of 1453 individuals, the geographic variation of extant great apes is discussed and graphically illustrated for chimpanzees (Pan spp), for the five age categories distinguished, using appropriate statistical tools. The results bring to light the developmental patterns evidenced from discrete trait studies as well as the underestimate of skeletal variation due to both inadequate sampling and incomplete osteological studies. Solving these two important problems may confound the highly debated issues of early hominid, Neogene hominoid species recognition and inferences about their phylogenetic relationships and palaeobiology. In this connection, the following quote (Shea *et al.*, 1993: p. 289) exactly reflects the underlying philosophical basis for this study:

> In sum, we propose that paleoanthropology would benefit tremendously if a great deal more extensive background work were undertaken on the patterns of variation in the relevant qualitative or quantitative features, before species groupings were determined, character states delimited, polarities estimated, and phylogenies reconstructed. This would probably greatly reduce or clarify the number of equally parsimonious phylogenies, the requisite frequency of homoplasy, and the debates over basic units of analysis.

Discrete traits: definition and previous studies in the great apes

Since the pioneering work of Le Double (1903, 1906, 1912) a number of extensive anatomical and anthropological monographs have done much to synthesise data on many cranial, dental and infra-cranial discrete traits in various human populations (Saunders, 1978; Hauser & De Stefano, 1989; Scott & Turner, 1997). Moreover, Grüneberg's (1952) work in inbred strains of mice established the potential value of some of these discrete traits in genetic studies. In contrast to the amount of knowledge on discrete traits in extant human skeletal populations, as yet, little attention has been paid to geographical variation (using proper samples) and development in our closest living relatives – *Pan troglodytes, P. paniscus, Gorilla gorilla* and *Pongo pygmaeus*. Braga (1995a) attempted to fill this gap and subsequently established the value of several cranial discrete traits in fossil hominid studies (Braga & Boesch, 1997a,b; 1998; Braga *et al.*, 1998). He pointed out that variability in extant great ape taxa is often misunderstood; such a situation leads to important errors being made when interpreting the fossil record.

What do we mean by the term 'discrete traits'? As different kinds of variables may be encountered by biologists, a clear definition of these non-pathological phenotypic variants initially is crucial. The following definition is presented which takes account firstly, of the classification of variables normally used by statisticians (e.g. Stevens, 1968) and secondly, with the concept of 'fuzzy sets' developed by Zadeh (1965). *Discrete traits must be recorded as being present or absent and, when present, several clear cut expressions should be distinguished accurately and not intuitively even with a knowledge of their variability as complete as possible.* On the contrary, general morphological features are classified by some quality they possess, using a nominal scale (with various classes as, for example, 'sharp' versus 'blunt' or 'weak' versus 'moderate' versus 'strong') where absence is excluded. *Contrarily to discrete traits, general morphological features are defined by classes with uncertain borders because, intuitively, the description of a morphology as 'sharp' versus 'blunt' (for example) is partly accurate but also partly not. Such classes with uncertain borders aim to score a continuum of variable expression. There is no continuum between the putative discrete trait categories and the presence of sharp borderlines between them is of high biological meaning.* For example, the presence/absence of some emissary foramina on the skull base corresponds to the presence/absence of some emissary vein in the living. Zadeh (1965) considered arbitrary classes, aiming to describe general morphological features such as 'fussy sets'. Sangalli (1998, p. 7) wrote: 'And because fuzzy sets make room for partial membership, that is, for objects that are neither totally in nor totally out, they can accommodate better than ordinary sets the ambiguity of human language.' In other words, many general features describing shape – degree of development – are indeed non-discrete even if they are non-metrical qualitative traits, because they do not correspond to the definition given above. Such general morphological features are 'widely used in consideration of hominid phylogenetics' (Asfaw *et al.*, 1999: p. 631). Some of these features are listed for *Australopithecus garhi* by Asfaw *et al.* (1999, table 7.1). As Asfaw *et al.* (1999: p. 631) point out, because of 'arbitrary boundaries of presence or absence criteria, variability within species . . .' a 'numerical cladistic application' of such data is sometimes questionable. This is particularly due to the definition of 'arbitrary boundaries of presence or absence criteria' as already stated above by Zadeh (1965) and by Scott & Turner (1997),[1] for example. Lastly, one might say that discrete traits do not necessarily need to have the state of being absent when considering, for example, accessory or supernumerary structures (such as dental cusps, teeth or foramina). However, in such cases, the structure is regarded as accessory and/or supernumerary. This means that, intuitively, it is considered as present in comparison with a more usual phenotype (absence of the structure). The traits described herein should all be classified as present or absent, even if in some instances there is a clear identifiable cut and intermediate variation in the degree to which it expresses, when present.

Theoretical aspects: etiology and size influence

An important contribution to our understanding of the nature and formation of discrete traits is Ossenberg's classification (1969) based mainly on: (1) arrested or excessive bone formation and maturation (hypostotic and hyperostotic traits); and (2) soft tissue relationships (vascular and nervous foramina). Besides this classifying aspect, the problem of etiology is central to studies dealing with discrete traits. In different studies dealing with familial concentrations in extant humans (Saunders & Popovich, 1978) and other mammals (e.g. rhesus macaques; Cheverud & Buikstra, 1981a,b), several discrete traits were considered as highly heritable. Moreover, in human graves or cemeteries, a high association was found between non-metrical skeletal data and archaeological data, suggesting genetic links between skeletons (Crubézy *et al.*, 1998). However, the genetic basis of discrete traits is not elucidated because it is very difficult to assess the effects of environmental and genetic factors on their development and their subsequent expression in skeletal samples. This is also the case for metrical traits.

When dealing with discrete traits, another important aspect concerns

[1] Scott & Turner (1997: pp. 59–60): 'Morphological variables of the human dentition are discrete insofar as individuals within a population express or do not express a particular trait. In another respect, these variables show quantitative variation in degree of expression – a trait may be but barely visible, or very large, or somewhere between these extremes. . . . Observational difficulties are associated primarily with minimal levels of expression that fall on the boundary between trait absence and trait presence. As crown and root traits can be present or absent and variability expressed when present, different options are available to score these traits with regard to scale of measurement and recording method.'

body size influence (both within and between species) and, more specifically, the role of sexual dimorphism on trait expression (size differences between males and females being one of the main characteristics of sexual dimorphism in primates). Humans are much less dimorphic in size than gorillas and orang-utans but they have other dimorphic characters, such as, for example, metabolic rate. Often, discrete trait differences manifested through sexual dimorphism have been claimed to be minimal in human population studies (Berry, 1974). The range of opinions about discrete sexual dimorphism in extant human populations can be summarized as follows: (1) any sex differences in trait frequencies that exist are random and can be disregarded; (2) slight differences due to sexual dimorphism can be expected, but they usually do not affect distance studies; and (3) sex differences exist for some discrete traits. We reached the first conclusion in our discrete trait study on a large sample of 1453 great ape skulls because differences observed within species did not appear as consequences of differences in body size between males and females, especially within the more dimorphic species (*Gorilla gorilla* and *Pongo pygmaeus*) (Braga, 1995a). In their excellent review of 30 tooth crown and root discrete traits in extant human populations, Scott & Turner (1997, p. 129) conclude that most of them 'lack sexual dimorphism' and that 'combining data on males and females is justified in most cases'. This situation is rather different from the pattern observed for metrical traits. We therefore believe that *discrete traits offer an important advantage when dealing with morphology and fossils because their expression seems to be generally not affected by differences in body size.*

Aims of this study

By distinguishing age categories in the analysis, this study aims to test whether or not some discrete traits may represent relevant morphological discriminators of the traditional extant great ape taxa and accordingly of fossil hominoids possibly representing distinct species and/or genera. Therefore, our main question is: does skeletal discrete variation in extant great apes match, first, with the molecular evidence for relatedness among taxa and secondly, with biogeographical variation (i.e. differences among interbreeding local populations to differences between subspecies)? We consider each discrete trait as belonging to well-defined anatomical systems in terms of physiology and/or embryology and then assess its magnitude and patterning of variance.

Good illustrations 3 traits

Trait description

Every skull is classified for 19 discrete traits. Most of the following discrete traits are 'classical' features listed in studies dealing with human cranial remains (Hauser & De Stefano, 1989). The criterion for scoring these traits as present or absent through visual inspection has the major advantage to be precise and replicable.

- **Divided hypoglossal canal** (CNHY) (Figure 7.1). The hypoglossal canal (*canalis hypoglossalis*) passes in a mediolateral direction through the base of the occipital condyle. It may be totally divided by one or, rarely, two bony bridges from its inner to its outer apertures.
- **Ossified styloid process** (PRST) (Figure 7.2). The presence of the styloid process (*processus styloideus*) in the great apes is caused by the unusual ossification of two cartilaginous elements belonging embryologically to the hyoid system: a small tympanohyal and a larger stylohyal part (Braga, 1993).
- **Confluent oval and spinous foramina** (COSF) (Figure 7.3). The oval and spinous foramina may communicate by a narrow slit or a more pronounced aperture with absence of the bone separating them. This variant may be achieved through disturbances in ossification of the posterior border of the sphenoid greater wing (Braga *et al.*, 1998).
- **Infra-orbital fissure persistens** (FSIP) (Figure 7.4). This fissure corresponds to the course of the infra-orbital canal marked externally both in the orbital floor and on the anterior surface of the maxilla, respectively behind and below the infraorbital margin.
- **Accessory infra-orbital foramen** (FIAC) (Figure 7.5). An infra-orbital foramen, representing the anterior opening of the infra-orbital canal, always lies on the anterior surface of the maxilla below the infra-orbital margin. It may be accompanied by one or more accessory infra-orbital foramina corresponding to a division of the corresponding canal.
- **Craniopharyngeal canal** (CRC) (Figure 7.6). The body of the sphenoid bone may be pierced by 'a craniopharyngeal canal opening interiorly in the floor of the pituitary fossa and externally in the region where the base of the skull forms the roof of the pharynx' (Hauser & De Stefano, 1989, p.141). This inferior opening is located just behind the posterosuperior angle of the vomer.
- **Median basilar canal** (MBC) (Figure 7.7). The medial basilar canal perforates the basilar part of the occipital bone. Its inferior opening is often median and sometimes double or triple.
- **Anterior and/or palatine portion of the incisive suture completely or partially fused**. The incisive suture (*sutura incisiva*) represents the con-

tact area between the maxillary and the incisive bones. The incisive bone (*os incisivium*) is also called 'premaxillary bone' but the former name follows the International Code for Nomenclature (Braga, 1998). We consider both facial and palatal components of the incisive suture. Usually, in the great apes, when fully opened, the anterior component of the incisive suture extends, superiorly, between the nasal bone and the maxilla (Figure 7.8). The palatal component of the incisive suture (IPOB) extends, medially, from the lower aperture of the incisive canal (canalis incisivus) to, laterally, the interalveolar septum located between the lateral incisor and the canine (Figure 7.9).

- **Retroarticular foramen** (Figure 7.10). The retro-articular foramen is the inferior opening of an inconstant venous channel and is situated immediately behind the mandibular fossa (Braga, 1995b).
- **Condylar foramen** (Figure 7.11). The condylar foramen is the posterior and outer opening of the condylar canal (canalis condylaris). It lies in the condylar fossa, behind either the occipital condyle or the jugular foramen. Its inner aperture is situated at the end of the sigmoid sinus groove (sulcus sinus sigmoidei) (Braga & Boesch, 1997a,b).
- **Nasal foramen** (Figure 7.12). The nasal foramen is the outer opening of a venous and/or nervous channel piercing the nasal bone (Le Double, 1906; Braga, 1995a).
- **Parietal foramen** (Figure 7.13). The parietal foramen pierces the parietal bone near or in the sagittal suture in the obelion area.
- **Occipital foramen** (Figure 7.14). The occipital foramen pierces the occipital squama at, or slightly above the inion, but also near the posterior border of the foramen magnum (Braga & Boesch, 1997a,b).
- **Nasal spine** (SPNA) (Figure 7.15). In the great apes, a sharp and prominent bony area representing the anterior attachment of the nasal septal cartilage (cartilago septi nasi) may be observed (Braga, 1996).
- **Sphenoparietal articulation** (SPPA). The pterion is the area of the vault were the parietal, frontal, sphenoid and temporal bones meet. When the sutural connection is neither at one point nor with an accessory bone, we observe two possibilities: a frontotemporal (FTTP) or a sphenoparietal (Figure 7.16) articulation.
- **Lacrymo-ethmoidal articulation** (LAET). The medial wall of the orbit is an area were the ethmoid, frontal, lacrymal and maxillary bones meet. When the sutural connection is neither at one point nor with an accessory bone, we observe two possibilities: a frontomaxillary (FTMX) or a lacrymo-ethmoidal (Figure 7.17) articulation.

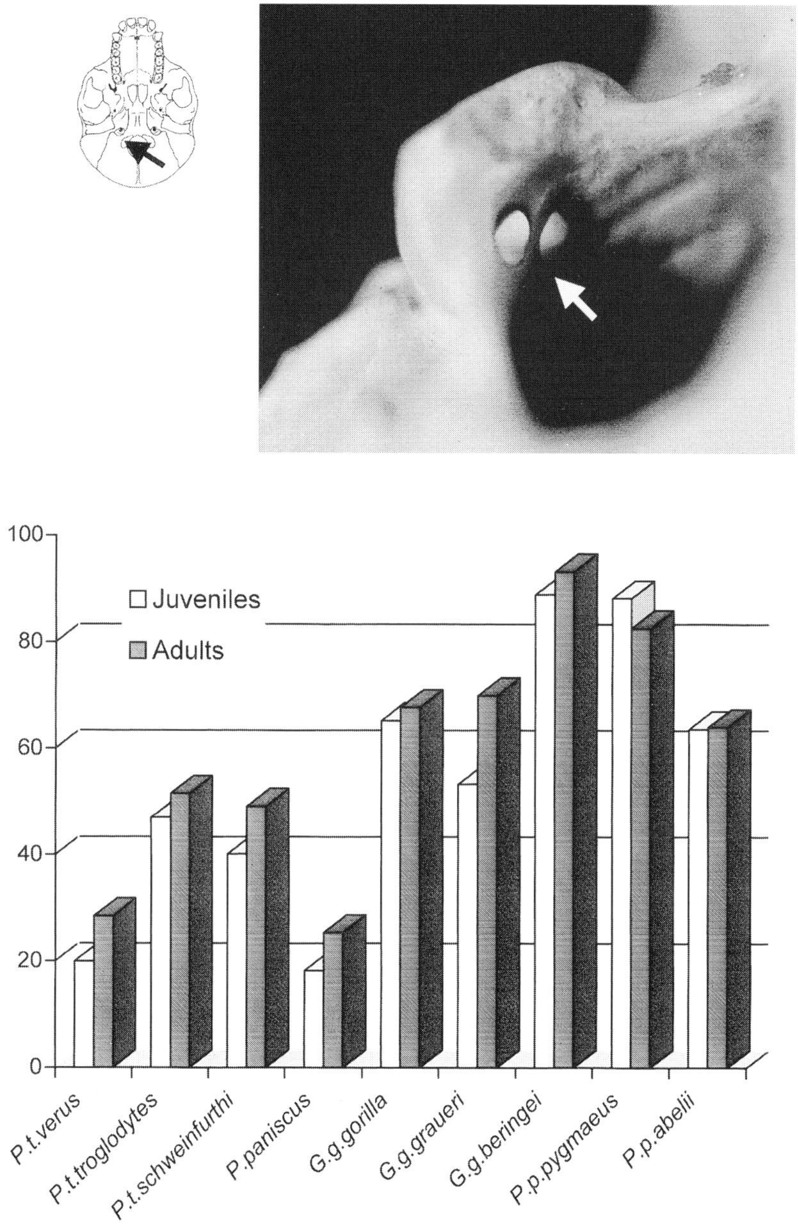

[Figure 7.1]
Calculated frequencies and posteromedial view of a right divided hypoglossal canal (*canalis hypoglossalis partitum*: CNHY) were a thin osseous septum divides symmetrically the internal aperture of the hypoglossal canal. (Species list: *Pan troglodytes verus*; *Pan t. troglodytes*; Pan t. schweinfurthi; *Pan paniscus*; *Gorilla gorilla gorilla*; *Gorilla g. graueri*; *Gorilla g. beringei*; *Pongo pygmaeus pygmaeus*; *Pongo pygmaeus abelii*.)

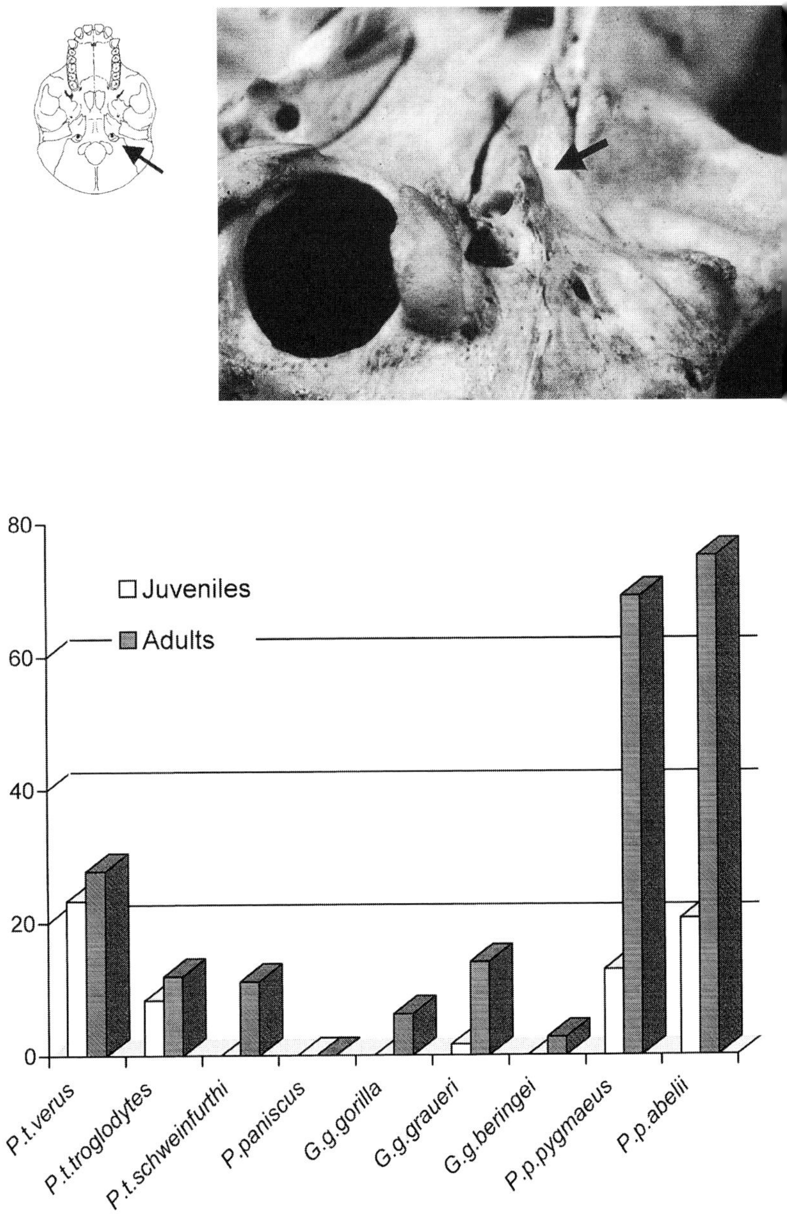

[Figure 7.2]
Calculated frequencies and view of a left ossified styloid process (*processus styloideus*: PRST) clearly projecting on the postero-lateral border of the carotid foramen. (Species list as for Figure 7.1.)

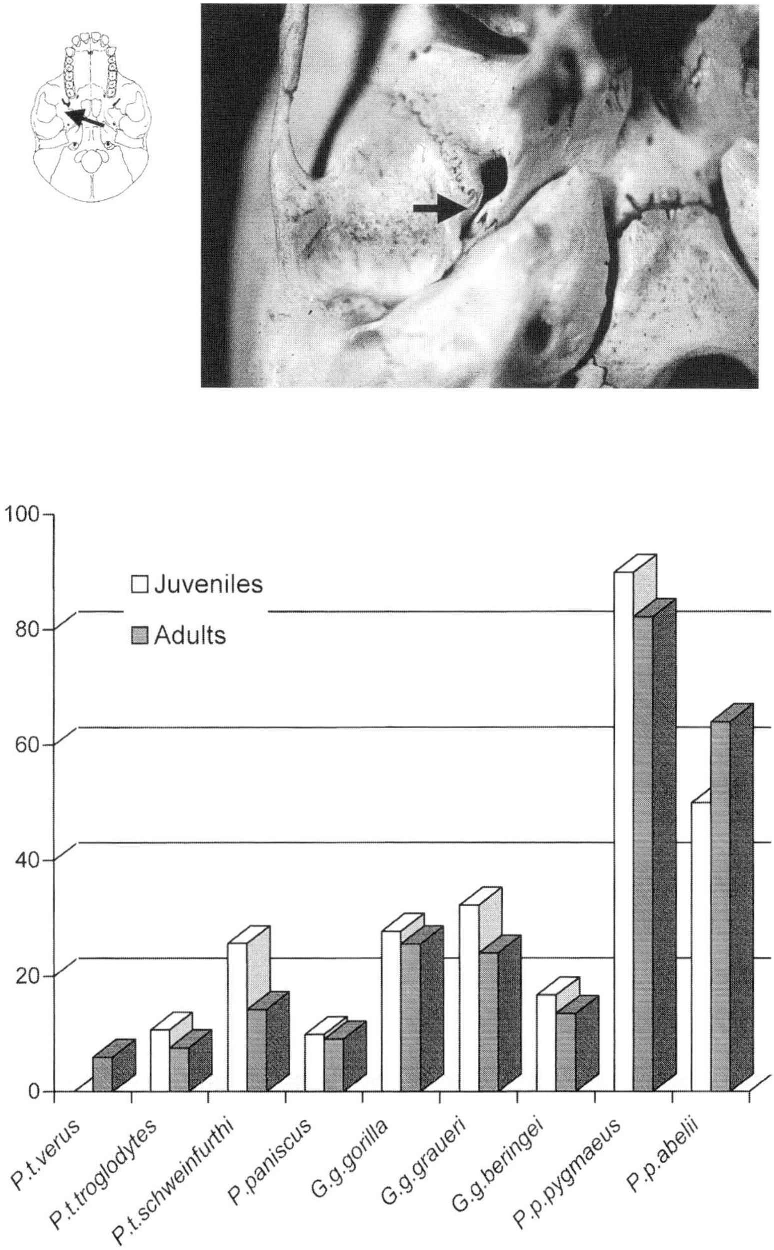

[Figure 7.3]
Calculated frequencies and view of confluent oval and spinous foramina (*foramen ovale in spinosum confluens*) (COSF) seen on the posterior border of the right sphenoid greater wing: a cleft is present in the osseous margin separating the two foramina. (Species list as for Figure 7.1.)

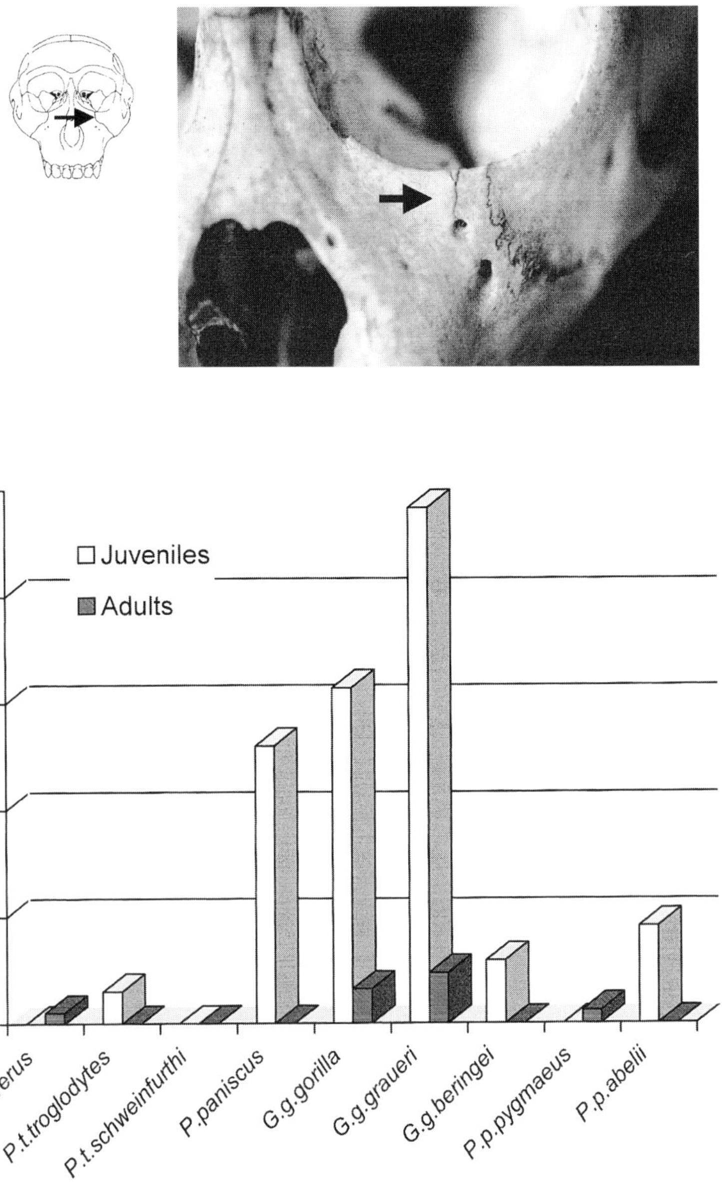

[Figure 7.4]

Calculated frequencies and view of[an infra-orbital fissure persistens (*sutura infraorbitalis persistens*: FSIP) which vertical part extends from the left infra-orbital border to an accessory infraorbital foramen. (Species list as for Figure 7.1.)

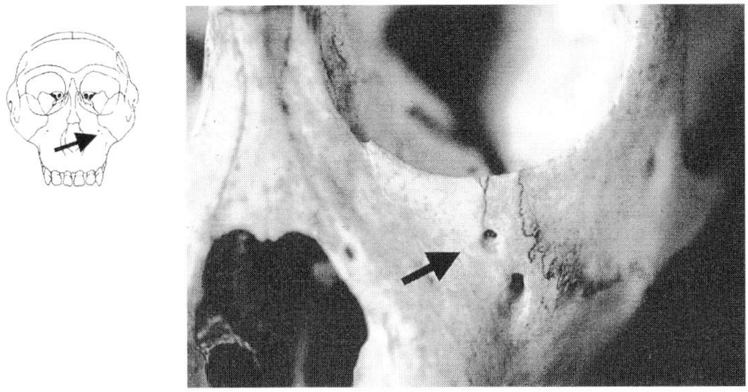

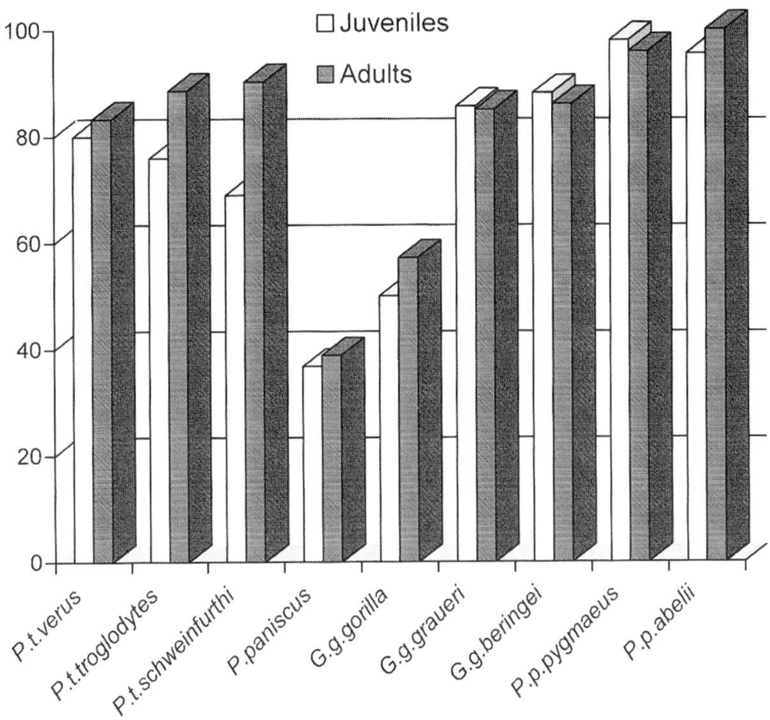

[Figure 7.5]
Calculated frequencies and view of an accessory left infra-orbital foramen (*foramen infraorbitale accessorium*) (FIAC). (Species list as for Figure 7.1.)

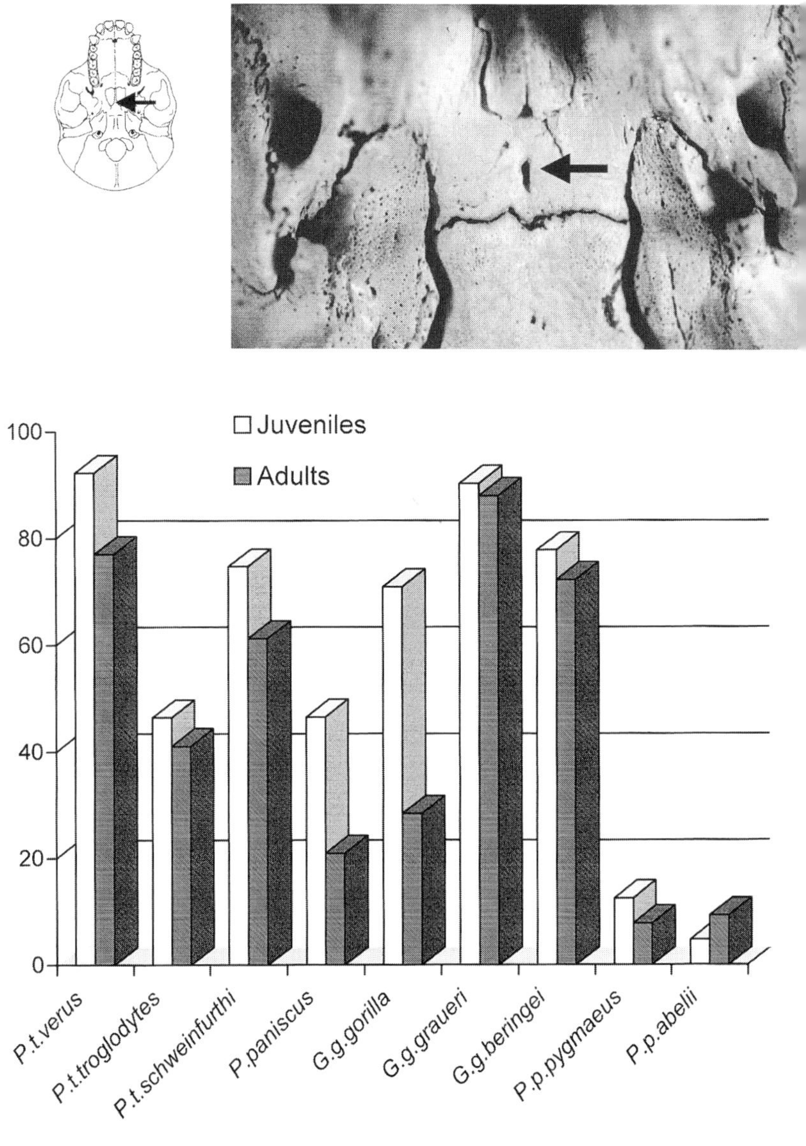

[Figure 7.6]
Calculated frequencies and view of a craniopharyngeal canal (*canalis craniopharyngeus*) (CRC) opening inferiorly on the body of the sphenoid bone, a few millimeters behind the posterosuperior angle of the vomer. (Species list as for Figure 7.1.)

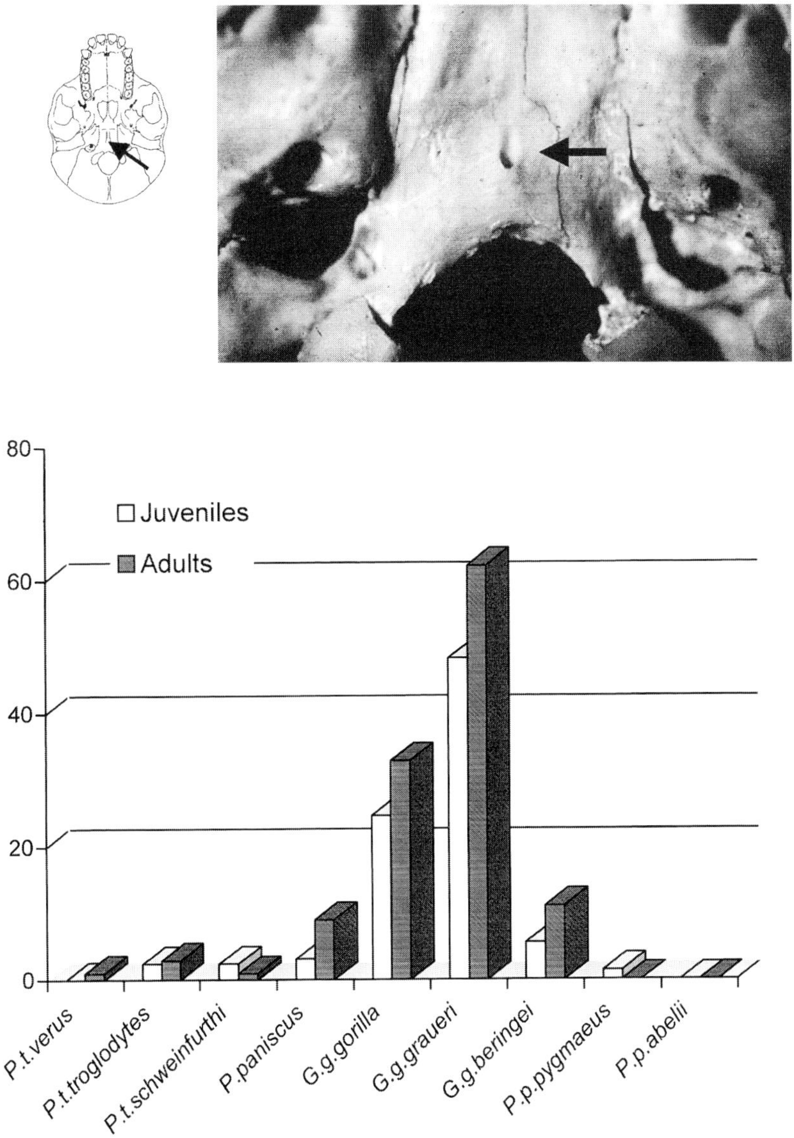

[Figure 7.7]
Calculated frequencies and view of a median basilar canal (*canalis basilaris medianus*) (MBC) opening inferiorly on the basilar part of the occipital bone in a generally median or slightly paramedian position. (Species list as for Figure 7.1.)

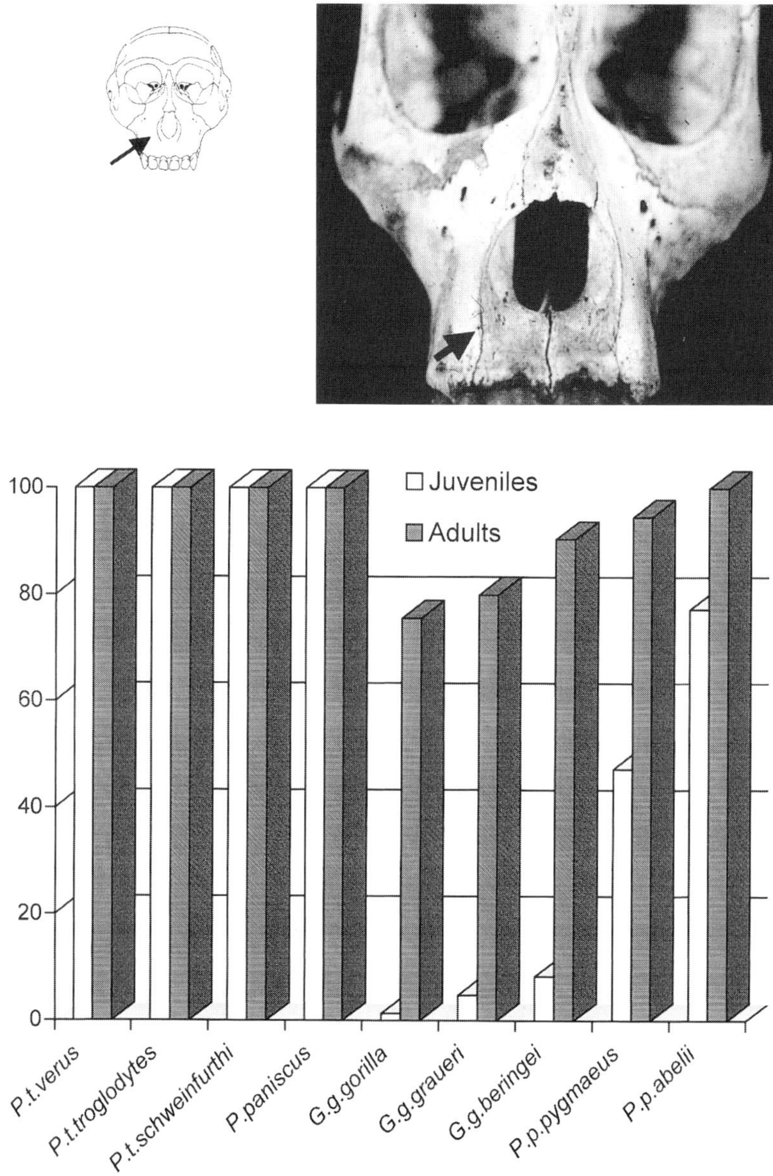

[Figure 7.8]
Calculated frequencies and view of the anterior component of the incisive suture (IAOB) (*sutura incisiva*) extending superiorly, for a small distance, between the nasal bone and the maxilla, and reaching inferiorly, the interalveolar septum located between the lateral incisor and the canine. (Species list as for Figure 7.1.)

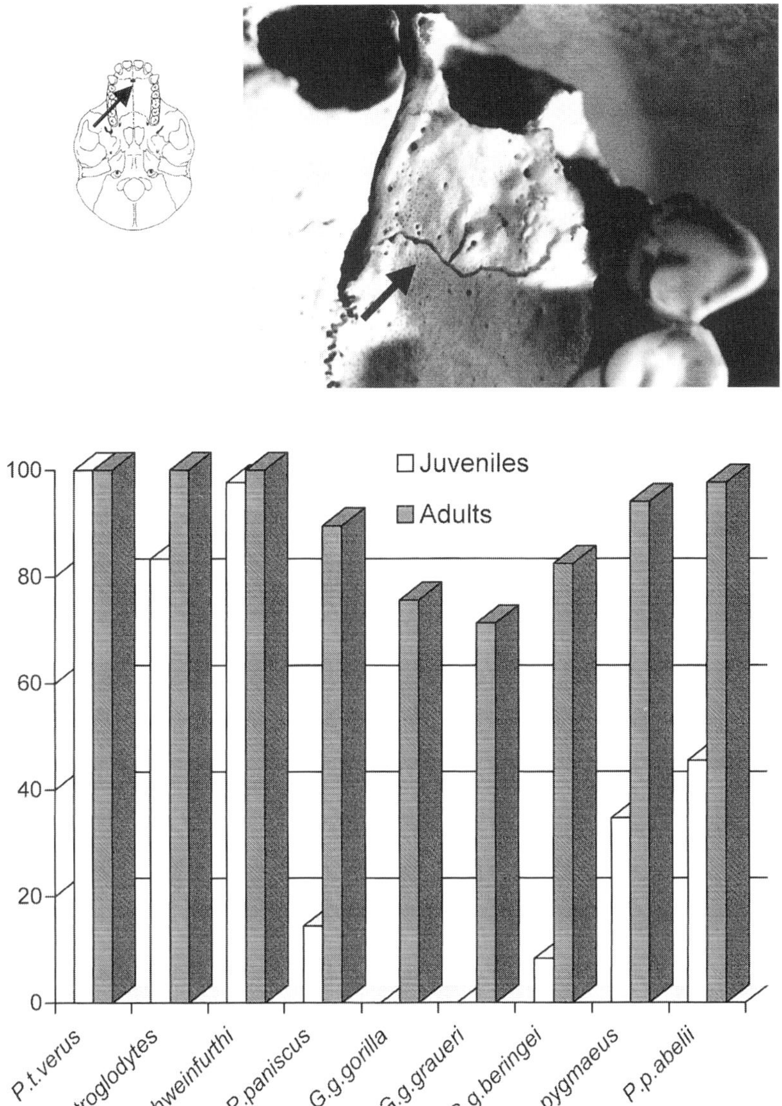

[Figure 7.9]

Calculated frequencies and view of the left palatal component of the incisive suture (IPOB) (*sutura incisiva*) extending medially, from the lower aperture of the incisive canal to, laterally, the interalveolar septum located between the lateral incisor and the canine. (Species list as for Figure 7.1.)

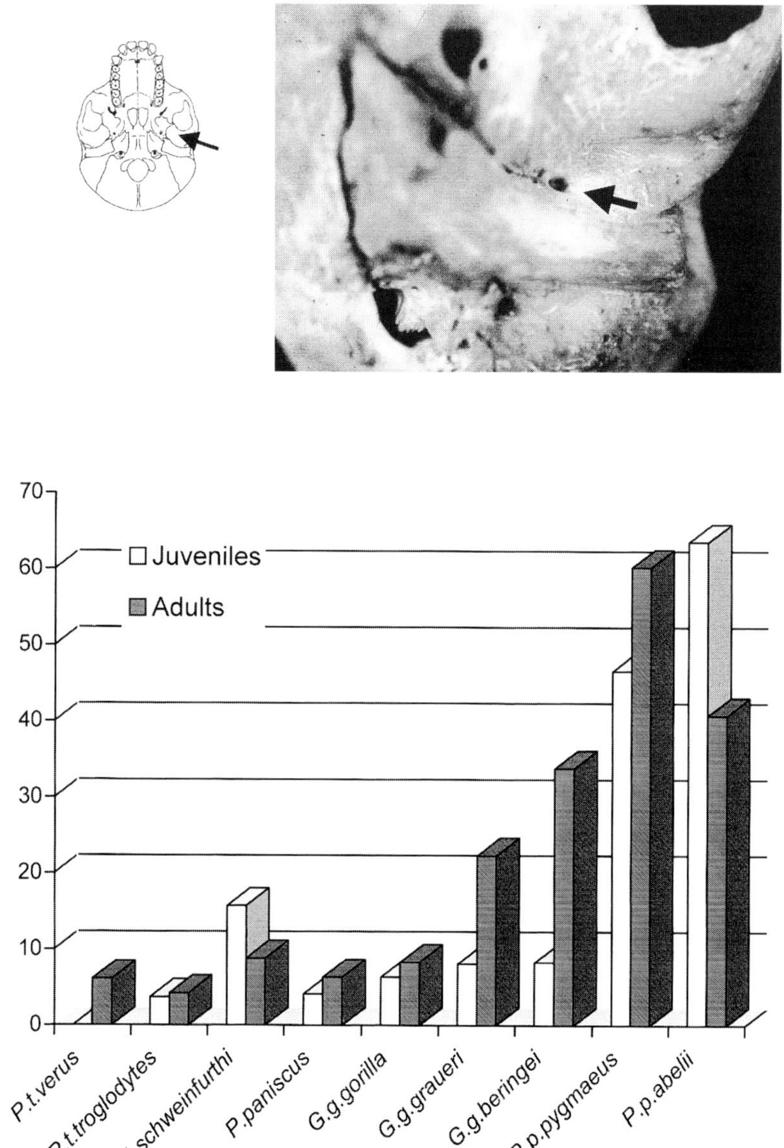

[Figure 7.10]
Calculated frequencies and view of the left retromandibular canal (*canalis retroarticularis*) opening inferiorly immediately behind the mandibular fossa. (Species list as for Figure 7.1.)

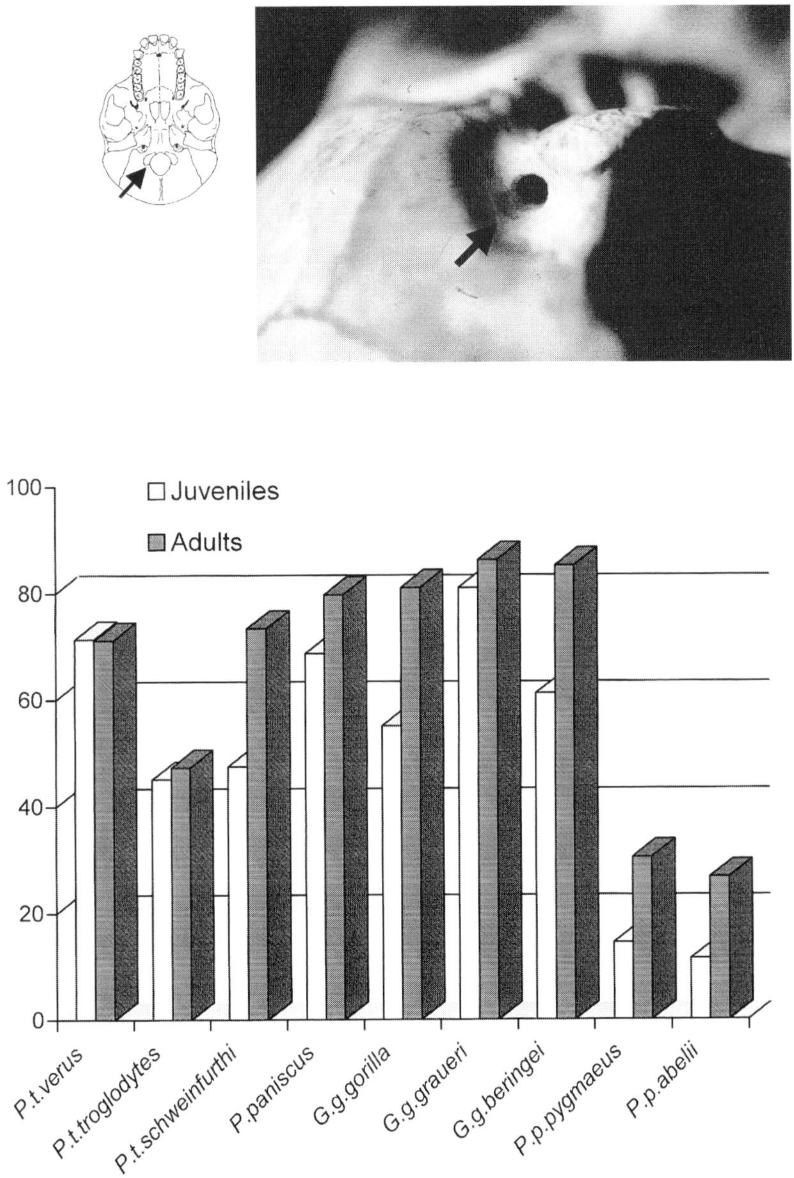

[Figure 7.11]
Calculated frequencies and view of the condylar canal (*canalis condylaris*) opening posteriorly into the right condylar fossa behind the right occipital condyle. (Species list as for Figure 7.1.)

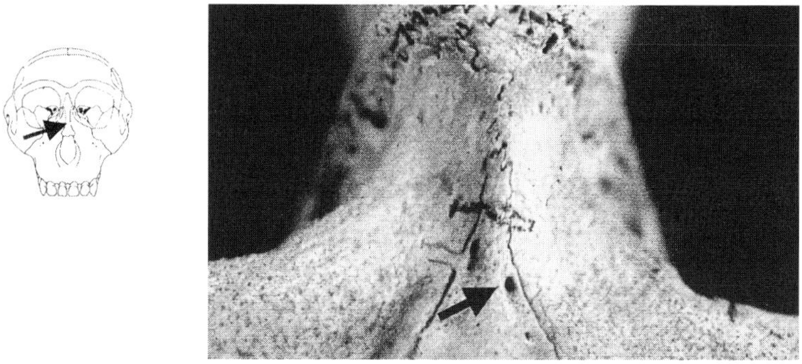

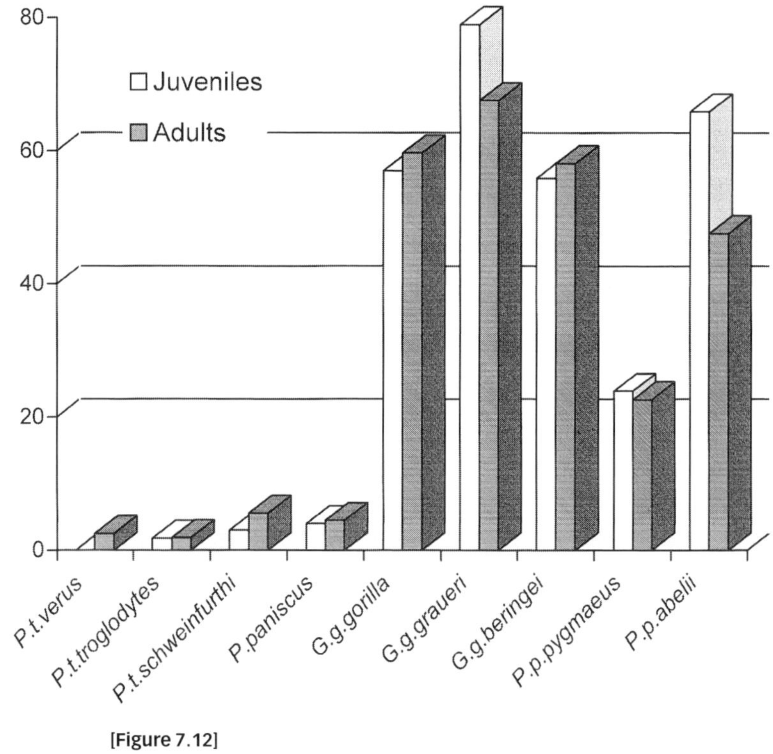

[Figure 7.12]
Calculated frequencies and view of the left nasal foramen (*foramen nasale*) piercing the nasal bone. (Species list as for Figure 7.1.)

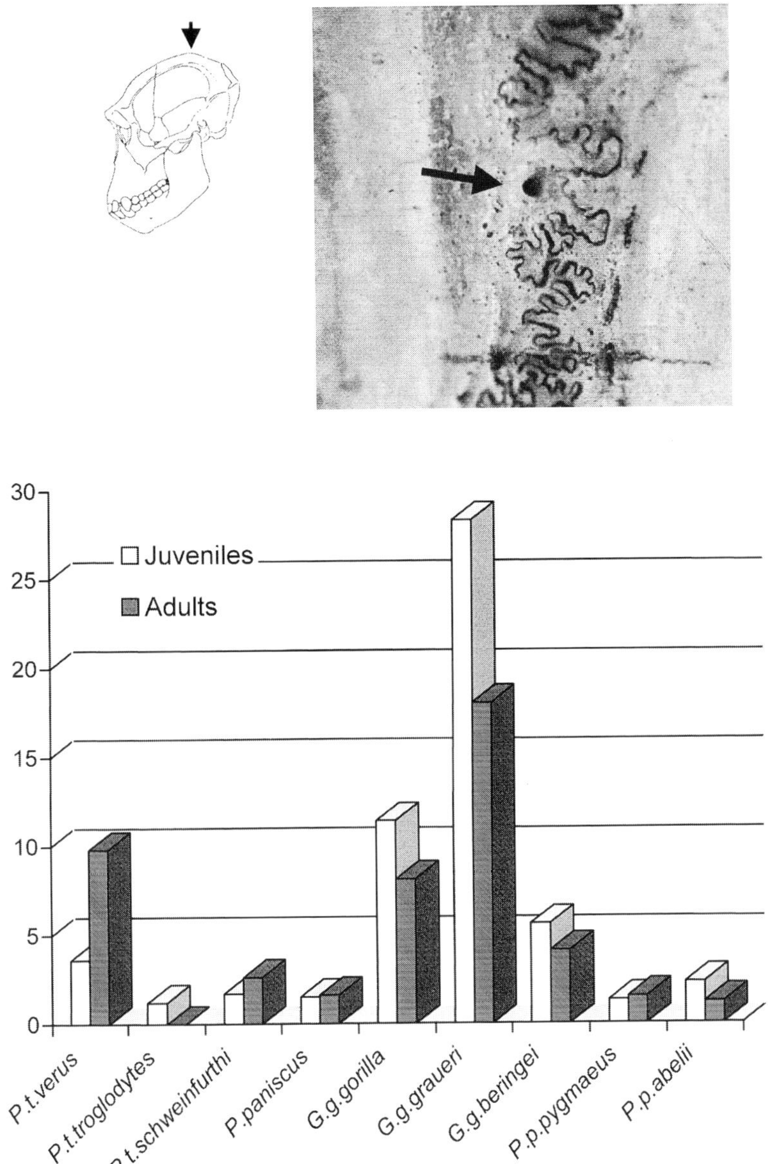

[Figure 7.13]
Calculated frequencies and view of the parietal foramen (*foramen parietale*) piercing the cranial vault in obelion area (about 4 cm in front of the lambda) on the left margin of the sagittal suture. (Species list as for Figure 7.1.)

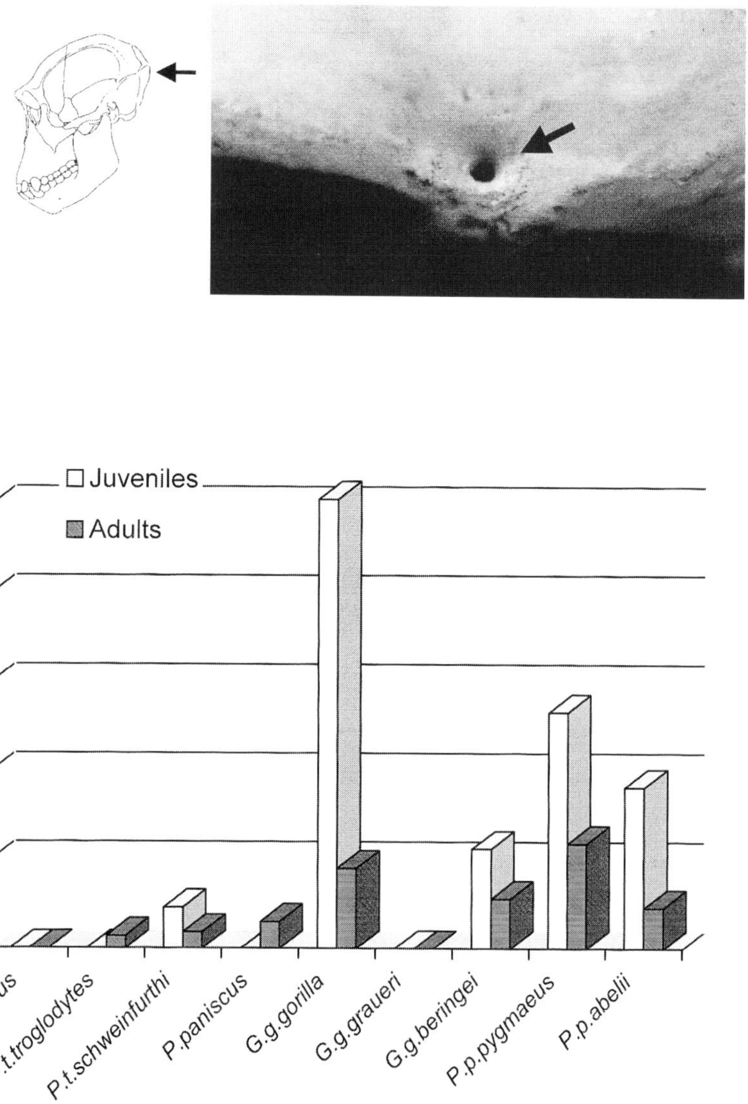

[Figure 7.14]
Calculated frequencies and view of the occipital foramen (*foramen occipitale*) piercing the occipital squama slightly above the inion. (Species list as for Figure 7.1.)

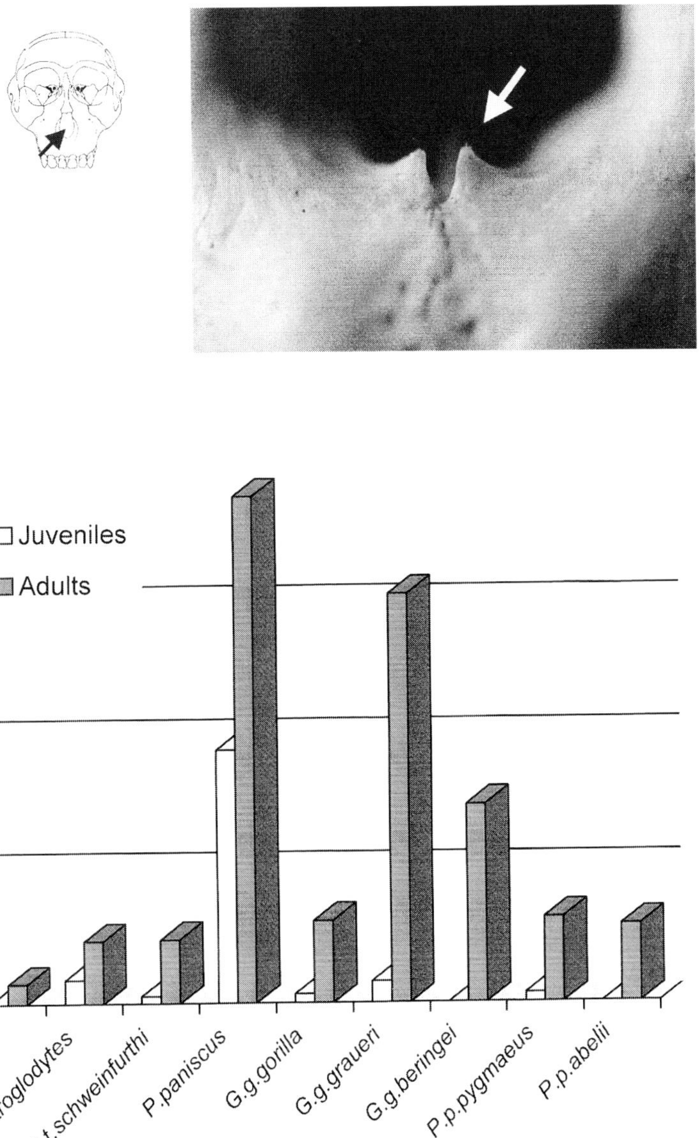

[Figure 7.15]
Calculated frequencies and view of the nasal spine (*spina nasalis* : SPNA) projecting upward and representing the anterior attachment of the nasal septal cartilage. (Species list as for Figure 7.1.)

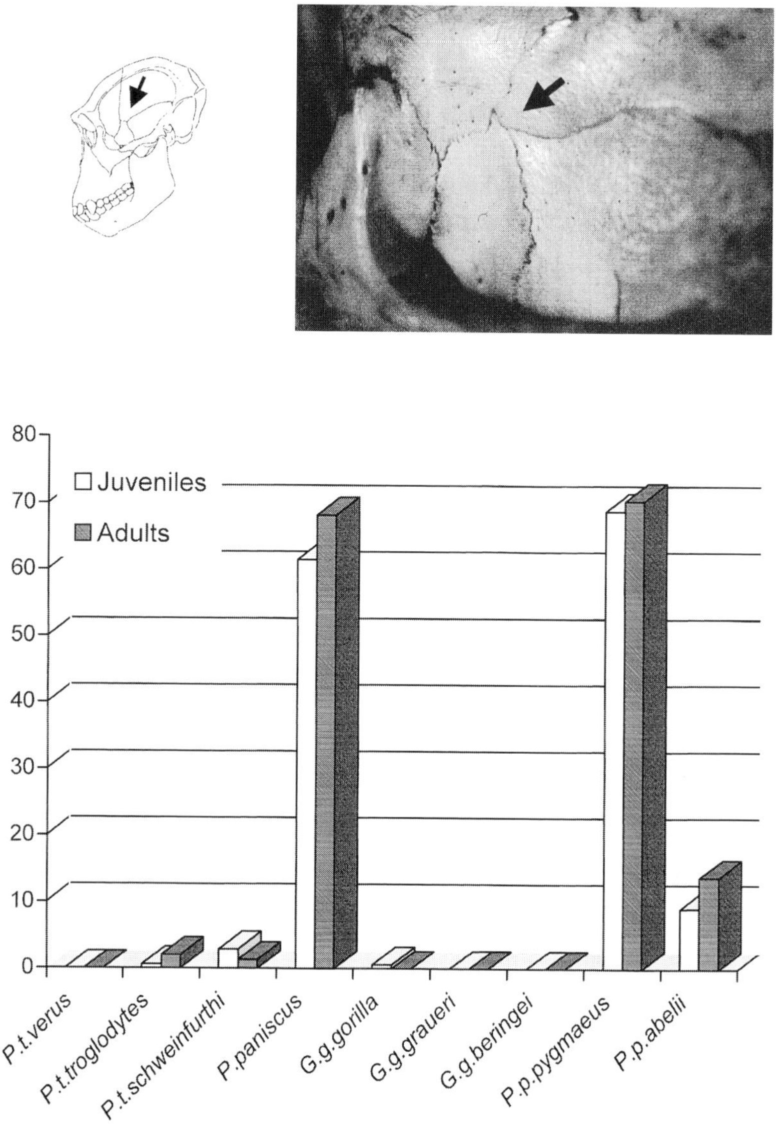

[Figure 7.16]

Calculated frequencies and view of the spheno-parietal articulation (*articulatio sphenoparietalis* : SPPA) at the left pterion where the great wing of the sphenoid bone meets the parietal bone. The frontotemporal articulation (FTTP) occurs when the frontal bone meets the squamous part of the temporal bone. (Species list as for Figure 7.1.)

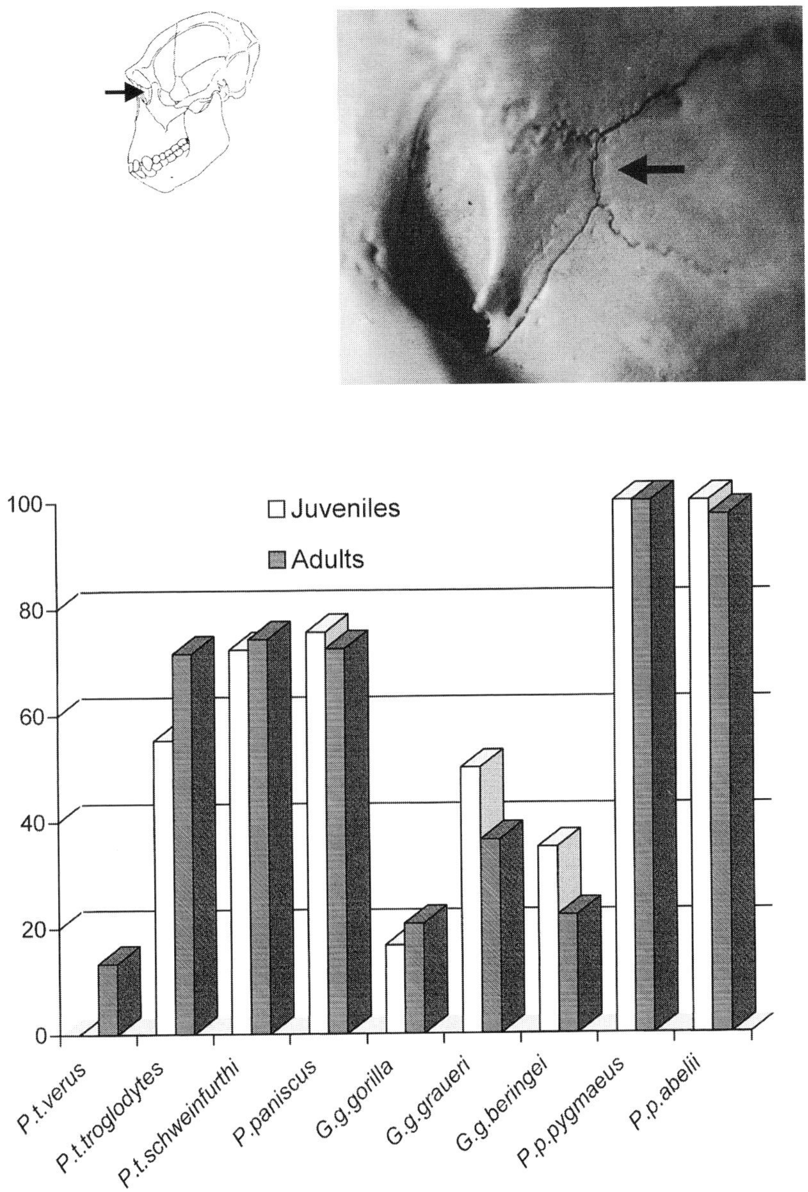

[Figure 7.17]
Calculated frequencies and view of the lacrymo-ethmoidal articulation (*articulatio lacrymo-ethmoidalis* : LAET) at the medial wall of the left orbit where the ethmoid bone meets the lacrymal bone. The frontomaxillary articulation (FTMX) occurs when frontal bone meets the maxillary bone. (Species list as for Figure 7.1.)

Discrete traits, adequate sampling and correspondence analysis in chimpanzees

In order to obtain a satisfactory picture of geographical variation (*i.e.* intra-specific variation) of each trait in all the extant great ape taxa, skulls of 1453 museum specimens of known origin were studied. As it is generally assumed that dental development is the most reliable marker of maturity because it is the least affected by environmental variability or nutritional deficiency (Eveleth & Tanner, 1990), juveniles were distinguished from adults on the basis of the eruption of the third molars (exhibiting very late emergence relative to the other teeth). Five age categories were distinguished in chimpanzee ontogeny: juvenile 1, defined as having exclusively an incomplete or complete deciduous dentition; juvenile 2, defined as not having second permanent molars erupted; juvenile 3, defined as not having third molars erupted; adult 1, defined as having a complete permanent dentition but incomplete closure of the spheno-occipital synchondrosis; and adult 2, defined as having a completely closed spheno-occipital synchondrosis. When possible, the specimens studied were combined into major geographic groups[2] (see Braga, 1995a for details by specimens), except for *Pan troglodytes verus* (the specimens coming mainly from Liberia but representing different social groups) and for *Pongo pygmaeus abelii*.

Sex and age influence, as well as differences between major geographic

[2] ***P. t. troglodytes***: **group 1**, left bank of the Sanaga river (Cameroon), between Yaoundé (03°52'N, 11°31'E), Sangmelima (02°56'N, 11°59'E), Kribi (02°56'N, 09°55'E) and Edea (03°47'N, 10°08'E), *N* = 107; **group 2**, south-east of Cameroon, between Doumé (04°15'N, 13°25'E), Batouri (04°26'N, 14°22'E) and Ouesso (01°38'N, 16°04'E), *N* = 78.

P. t. schweinfurthi: **group 3**, between the Uélé, Itimbiri and Mangala rivers (Zaïre), *N* = 49; **group 4**, between the Aruwimi and Elila rivers (Zaïre), *N* = 89.

P. paniscus: **group 5**, between the Zaïre, Lomela and Lomani rivers (Zaïre), *N* = 29; **group 6**, between the Zaïre and Lomani rivers (Zaïre), *N* = *116*.

G. g. gorilla: **group 7**, left bank of the Sanaga river; between Yaoundé (03°52'N 11°31'E), Sangmelima (02°56'N 11°59'E), Kribi (02°56'N 09°55'E) and Edea (03°47'N 10°08'E), *N* = 80; **group 8**, S.E. Cameroon, between Doumé (04°15'N 13°25'E), Batouri (04°26'N 14°22'E) and Ouesso (01°38'N 16°04'E), *N* = 132; **group 9**, South of the Dja river, *N* = *15*.

G. g. graueri: **group 10**, Massif of the Itombwe, entre Fizi et Mwenga, *N* = 29 (this group corresponds to the '*Deme* #2' defined by Albrecht & Miller (1993); **group 11**, Massif of the Tshiaberimu volcano, near Lubero (00°09'S 29°13'E), *N* = 22 (this group corresponds to the '*Deme* #1' defined by Albrecht & Miller (1993); **group 12**, Between Shabunda (02°42'S 27°20'E), Punia (01°25'S 26°25'E) and Walikale (01°25'S 28°03'E), *N* = 21 (this group corresponds to '*Deme* #3' defined by Albrecht & Miller (1993); **group 13**, between Bukavu (02°30'S 28°52'E) and Mont Kahuzi (02°15'S 28°41'E), *N* = 10.

G. g. beringei: **group 14**, Mikeno, Sabinio, Karisimbi and Muhavura volcanoes, *N* = 49; **group 15**, Kayonza forest, *N* = 6.

P. p. pygmaeus: **group 16**, N.W. of Borneo, north to the Kapuas river, *N* = 217; **group 17**, South and S.E. of Borneo; south of Kapuas and Sesayap rivers, respectively to the west and to the east, *N* = 35; **group 18**, North of Borneo, *N* = 23.

groups, subspecies and species were tested with a Chi-squared test with 2 × 2 contingency tables. In order to derive frequencies for bilateral traits, we divided the total number of times the trait occured on either side by the number of sides on which the trait could be observed if present. In order to identify effects that cannot be seen in conventional cross-tables and to show relationships between discrete traits on one chart, correspondence analysis, a multivariate analysis procedure (Greenacre, 1993), applicable only to categorical features – as, for example, discrete traits – without any other limitation, is appropriate. The important points to know about this procedure are that: (1) it shows the patterns of relationships among many categorical variables; (2) it plots the relationships between these multiple variables on a smaller number of dimensions according to variance explained; and (3) the distance between variables on each dimension is standardised (i.e. items that are close together are likely to be related). In this study, correspondence analysis is used in order to test whether the cranial discrete differences within the great apes accurately mirror their polytypism seen by the molecules and, more specifically, to identify the main discrete traits that differentiate between chimpanzee taxa in juvenile and adult specimens. We focus our attention on chimpanzees because an important amount of data on mitochondrial DNA (mtDNA) polymorphism is now available (Morin *et al.*, 1994; Gonder *et al.*, 1997; Goldberg & Ruvolo, 1997). A clear distinction between juveniles and adults has the major advantage to test whether relationships between discrete traits change during post-natal growth. The Statistica software was used for the multivariate statistical analysis which has had many other names, including 'optimal scaling', 'reciprocal averaging', 'optimal scoring', and 'appropriate scoring'. In this procedure, two categories of variables are usually selected: the 'active' and the 'supplementary' variables. The 'active' variables are discrete trait expressions. They contribute to the calculation of the main dimensions (*i.e.* principal inertias and squared cosines) of the graphical display. The supplementary variables are the species and subspecies distinguished in the analysis. They are subsequently inputted in the analysis and do not contribute to any calculation. These supplementary variables are only illustrative in order to facilitate the interpretation.

Skeletal variation among chimpanzees: how many chimpanzees for palaeoprimatology?

Some authors tend to assume that the extent of size and morphological variation among extant human populations is much greater than it would be within each great ape species. This general opinion has important consequences when interpreting the fossil record and is still vehiculated despite our increasing knowledge of variation in biological (at present, mainly genetic), behavioural and ecological features within the great ape species. For example, Ruvolo *et al.* (1994) have demonstrated, for common and pygmy chimpanzees, that within-species mtDNA variability, respectively, exceeds and approximates that within extant humans. Even if these data have to be confirmed at the nuclear level, the genus *Pan* provides us with a benchmark for comparisons, as there is now available increasing genetic evidence from wild populations. Important studies focused their attention on population history and substructure within the common chimpanzee (*Pan troglodytes*). MtDNA analyses show that variation between the two chimpanzee species exceeds, by far, that within *Pan troglodytes* (Ruvolo *et al.*, 1994). Interestingly, Morin *et al.* (1994) found an unexpectedly large genetic distance between *P. troglodytes verus* and the other two currently recognised common chimpanzee subspecies, even though their sampling distribution has been questioned (Jolly *et al.*, 1995). Additionally, Goldberg & Ruvolo (1997), conducting a geographically systematic genetic survey of the eastern-most subspecies, *P. t. schweinfurthi*, found levels of mtDNA variation which were low and similar to levels observed in extant humans implying population histories of recent demographic expansion. Moreover, sequencing mtDNA, Gonder *et al.* (1997) proposed two alternative chimpanzee taxonomies, either collapsing *P. t. troglodytes* and *P. t. schweinfurthi* into the same subspecies (the western common chimpanzee being a member of a second taxon) or creating a new West African common chimpanzee subspecies 'along with the other three named subspecies'. In most of these studies, the authors have acknowledged that anatomical and/or eco-behavioural complementary studies had to be conducted in order to confirm genetical results.[3] Therefore, in order to test whether cranial discrete traits represent relevant morphological discriminators of the traditional extant great ape taxa, a careful study of the sample including 543 *P. troglodytes* (187 juveniles and 356 adults) and 169 *P. paniscus* (102 juveniles and 67 adults) skulls is necessary (Table 7.1). This correspondence analysis is also potentially interesting for comparisons with results of previous studies investigating craniometric[4] (Shea *et al.*, 1993) and dental[5] (Uchida, 1992) variation in

[3] Morin *et al.* (1994: 1199): 'if this result is confirmed at three or more unlinked nuclear loci, and supported by eco-behavioral data, *P. t. verus* merits elevation to full species rank: *P. verus*.' Gonder *et al.* (1997: p. 337): 'These proposals need to be confirmed by more extensive sampling, sequencing of several independent nuclear loci and morphological data.'

[4] Shea *et al.* (1993): 'a significant component of the distinctveness of *P. paniscus* and the *P. troglodytes* subspecies is due to allometric factors... In the non-size-corrected analysis, *P. paniscus* most closely approximates *P. t. verus* among the common chimpanzee groups, while

samples representing populations and subspecies of common and pygmy chimpanzees. However, we should stress that the two studies used much more restricted samples for both *P. troglodytes* and *P. paniscus*. Indeed, Uchida (1992) included a total of 143 *P. troglodytes* but only 24 *P. paniscus* crania in her study while Shea *et al.* (1993) studied a total of 371 and only 44 skulls representing each species respectively. In this later study (Shea *et al.*, 1993), the three *P. troglodytes* subspecies were very disproportionately represented (only 40 *P. t. verus* versus 241 *P. t. troglodytes* versus 90 P. t. schweinfurthi). When concluding, Uchida (1992, p. 110) acknowledged this sampling problem and went on to write, 'More extensive sampling is necessary to test whether certain characters are indeed subspecies specific among *Pan troglodytes*, as opposed to being population specific without distinct separation among subspecies'.

Considering that a minimum of 90% of the total variance should be explained within each age category, Table 7.2 summarises the variance explained by each dimension. The percentage of variance explained drops below 10% on and after dimensions 5 (juvenile 1), 3 (juvenile 3) and 2 (juvenile 2, adult 1 and adult 2). The number of dimensions required to explain at least 90% of the total variance varies from dimensions

subsequent to regression correction, it is closest to *P. t. schweinfurthi*. In the analyses where only the *P. troglodytes* subspecies were included, the two most divergent groups prior to size correction are *P. t. troglodytes* and *P. t. schweinfurthi*; subsequent to regression adjustment, *P. t. verus* and *P. t. schweinfurthi* are most distinct.' (p. 275). 'The truncation of the *P. paniscus* dispersion accurately reflects its smaller overall size and the allometric underpinnings of a significant portion of its cranial shape differences relative to the *P. troglodytes* specimens.' (p. 276). 'pairwise comparisons revealed significant discrimination for both males and females in each of the *P. t. troglodytes*/*P. t. verus*, *P. t. troglodytes*/*P. t. schweinfurthi*, and *P. t. verus*/*P. t. schweinfurthi* discriminant analyses, both prior and subsequent to size correction.' (p. 277). 'adult *P. paniscus* are clearly and fairly uniformly distinguishable from adult *P. troglodytes*, particularly when subspecific and interpopulational variation within the latter is considered. In other words, the differences between the two "types" of chimpanzee are significantly greater than the considerable variation cataloged within the broadly ranging *P. troglodytes*.' (p. 278).

[5] Uchida (1992) conducted discriminant analyses on molar cusp areas, analyzing upper and lower molars separately. She went on to write (Uchida, 1992: p. 129): 'Cusp area variables discriminate *Pan paniscus* from *Pan troglodytes*, and also separate the three subspecies of *Pan troglodytes* ... Factor 1 for the upper molars and factor 2 for the lower molars are considered to reflect mainly size, discriminating *P. paniscus* from *P. troglodytes*. In both upper and lower molars, *P. t. verus* is clearly separated from the other *Pan* populations by factor 2 (large protocone) for upper molars, and factor 1 (large protoconid, small hypoconulid) for lower molars.' Analyzing further 'non-metric features', Uchida (1992: p. 131) wrote: '*Pan paniscus* has a high frequency of a well developed metaconid on LP3 ... contrasting with the quite rare occurrence of this trait in *Pan troglodytes*. There are also considerable differences in the frequency of non-metric aspects of occlusal morphology on molars among *P. troglodytes* ... *P. t. verus* is distinct in having a very high frequency of missing hypoconulids on the LM3; 24% in my samples. The hypoconulid area on the LM3, which has five cusps, is significantly reduced in *P.t.verus* compared with other two subspecies.'

Table 7.1. *Sample size for taxa, age categories and sex of great ape skulls investigated*

	Females							Males							Undetermined sex							Total						
	J1	J2	J3	Ju	A1	A2	Ad	J1	J2	J3	Ju	A1	A2	Ad	J1	J2	J3	Ju	A1	A2	Ad	J1	J2	J3	Ju	A1	A2	Ad
Pan troglodytes verus	0	1	4	5	12	23	35	0	1	5	6	12	38	50	2	1	1	4	3	12	15	2	3	10	15	27	73	100
P. t. troglodytes	8	14	12	34	16	66	82	7	11	16	34	8	26	34	5	5	6	16	3	20	23	20	30	34	84	27	112	139
P. t. schweinfurthi	5	12	8	25	3	19	22	6	10	8	24	9	24	33	9	19	11	39	16	46	62	20	41	27	88	28	89	117
P. paniscus	9	15	10	34	2	24	26	6	18	7	31	5	17	22	8	18	11	37	7	12	19	23	51	28	102	14	53	67
Gorilla gorilla gorilla	6	15	15	36	20	43	63	12	14	8	34	37	47	84	1	5	3	9	4	5	9	19	34	26	79	61	95	156
G. g. graueri	1	3	4	8	8	26	34	0	1	3	4	12	24	36	7	11	1	19	2	3	5	8	15	8	31	22	53	75
G. g. beringei	2	2	1	5	1	13	14	0	3	4	7	2	18	20	1	4	1	6	2	1	3	3	9	6	18	5	32	37
Pongo pygmaeus pygmaeus	5	10	24	39	41	72	113	10	5	17	32	58	25	83	1	1	2	4	7	2	9	16	16	43	75	106	99	205
P. p. abelii	2	4	1	7	3	9	12	3	6	2	11	8	13	21	1	2	1	4	5	5	10	6	12	4	22	16	27	43

Ju: Juveniles, J1, J2, J3, Ad: Adults, A1, A2.

Table 7.2. *Percentages of variance explained by each dimension, considering that a minimum of 90% of the total variance should be explained within each five age category*

	Juvenile 1	Juvenile 2	Juvenile 3	Adult 1	Adult 2
Dimension 1	**18.1**	**17.7**	**19.8**	**21.4**	**12.2**
Dimension 2	**12.5**	**10.1**	**12.8**	**12.6**	**11.9**
Dimension 3	**11.0**	8.8	**10.1**	9.8	9.0
Dimension 4	**10.3**	7.4	8.9	9.0	8.7
Dimension 5	8.2	7.2	7.7	7.0	7.0
Dimension 6	7.4	6.5	7.5	6.0	6.2
Dimension 7	7.1	6.0	6.6	5.9	5.7
Dimension 8	6.2	5.7	5.3	5.7	5.6
Dimension 9	4.6	5.1	4.5	5.2	5.4
Dimension 10	4.2	4.8	4.1	4.6	5.1
Dimension 11	3.2	4.5	3.5	3.5	4.8
Dimension 12	–	3.7	–	–	4.7
Dimension 13	–	3.4	–	–	3.8

Bold type: values higher than 10%.

11 (juvenile 1, juvenile 3 and adult 1) to 13 (juvenile 2 and adult 2). Table 7.3 summarizes the distances between taxa (supplementary variables) on each dimension, representing at leat 5% of variance, and within each age category. When the position for one taxon on a given dimension and within an age category, is more than 1, it is considered as markedly distant from the others. This view is very accurate, as correspondence analysis standardises the distance between variables on each dimension, items appearing close together being likely to be related and reciprocal. When the distance between taxa on a given dimension and within an age category, is equal to 0.05 at the most, these taxa are considered as very close. Interestingly, only two chimpanzee taxa can be considered as markedly distant from the others. *P. paniscus* is distant from the three currently recognised common chimpanzee subspecies within all age categories for either dimension 1 (juvenile 1 to adult 1) (representing respectively 18.1; 17.7; 19.8 and 21.4% of variance) or dimension 3 (adult 2) (representing 9% of variance).

P. t. verus is distant from both *P. paniscus* and the other two currently recognised common chimpanzee subspecies (*P. t. troglodytes* and *P. t. schweinfurthi*) within the juvenile 1 and the juvenile 2 age categories and, respectively, for four dimensions (4 and 6, 7, 8, representing a total of 31% of variance) and one dimension (2, representing 10.1% of variance). When searching for similarities rather than dissimilarities, 22 instances can be observed. Among these, six have a proximity between *P. t. troglodytes*

Table 7.3. *Distances between the taxa (or 'supplementary' variables) illustrated as coordinates in the multidimensional graphic overview. The coordinates are given for dimensions representing at least 5% of variance within each age category*

	Dim. 1	Dim. 2	Dim. 3	Dim. 4	Dim. 5	Dim. 6	Dim. 7	Dim. 8	Dim. 9	Dim. 10
Juvenile 1										
Pan paniscus	**−1.13**	0.18	0.17	−0.06	−0.06	−0.36	−0.30	*−0.11*	–	–
P. troglodytes verus	0.47	0.32	−0.74	**2.02**	*0.47*	**−1.51**	**2.57**	**1.55**	–	–
P. t. schweinfurthi	0.57	−0.29	*−0.09*	−0.15	*0.51*	0.60	−0.14	0.18	–	–
P. t. troglodytes	0.65	0.02	*−0.05*	0.03	−0.37	−0.00	0.23	*−0.14*	–	–
Juvenile 2										
P. paniscus	**−1.06**	0.13	*−0.10*	0.01	−0.00	0.12	0.10	−0.15	−0.06	–
P. t. verus	1.03	**1.15**	−0.88	*0.18*	0.30	0.98	0.56	0.18	0.28	–
P. t. schweinfurthi	*0.72*	−0.33	0.29	*0.21*	−0.26	0.03	−0.16	−0.24	0.18	–
P. t. troglodytes	*0.76*	0.01	*−0.05*	−0.25	0.24	−0.31	−0.05	0.46	−0.13	–
Juvenile 3										
P. paniscus	**−1.28**	−0.14	*−0.09*	−0.18	*−0.34*	0.24	−0.02	0.01	–	–
P. t. verus	1.10	0.21	−0.39	−0.91	*−0.30*	1.21	*−0.16*	−0.07	–	–
P. t. schweinfurthi	0.26	0.08	*−0.13*	0.54	0.01	−0.54	*−0.13*	−0.15	–	–
P. t. troglodytes	0.56	0.00	0.28	−0.05	0.34	−0.08	0.15	0.13	–	–
Adult 1										
P. paniscus	**−2.65**	0.70	−0.24	−0.21	0.57	0.32	0.25	0.03	−0.01	–
P. t. verus	0.50	0.63	0.04	*0.06*	*−0.12*	−0.20	0.42	−0.04	−0.24	–
P. t. schweinfurthi	*0.15*	−0.70	0.19	*0.03*	0.09	0.20	*−0.28*	−0.12	*0.16*	–
P. t. troglodytes	*0.14*	−0.31	−0.14	−0.03	*−0.13*	−0.05	*−0.31*	0.15	*0.14*	–
Adult 2										
P. paniscus	−0.75	0.25	**−1.28**	0.55	−0.33	*−0.06*	0.45	*0.02*	0.22	0.38
P. t. verus	*0.16*	*−0.09*	*0.23*	0.07	−0.57	0.10	*−0.03*	−0.06	−0.20	*−0.11*
P. t. schweinfurthi	*0.11*	*−0.05*	*0.19*	−0.06	*0.30*	*−0.00*	−0.14	*0.02*	0.14	*−0.15*
P. t. troglodytes	*0.11*	0.00	*0.21*	−0.22	*0.30*	*−0.04*	*−0.06*	*0.02*	−0.05	0.03

Bold type: distances higher than 1; *italic type*: distances equal to 0.05. at the most; Dim.: dimension.

and *P. t. schweinfurthi* within juvenile 1 (dimension 3), juvenile 2 (dimension 1), adult 1 (dimensions 1, 7 and 9) and adult 2 (dimension 5) age categories. A close relation is also found in six cases out of 22 between *P. t. verus* and *P. t. schweinfurthi* within juvenile 1 (dimension 5), juvenile 2 (dimension 4), juvenile 3 (dimension 7), adult 1 (dimension 4) and adult 2 (dimensions 2 and 10) age categories. It should also be stated that in five cases, *P. paniscus* was found in close association with either *P. t. troglodytes* (juvenile 1 and dimension 8; juvenile 2 and dimension 3), *P. t. schweinfurthi* (juvenile 3 and dimension 3), or both of them (adult 2 for dimensions 6 and 8). Finally, the five remaining cases of close similarities between taxa regarded the three currently recognized common chimpanzee subspecies (two instances), *P. t. verus*/*P. t. troglodytes* (two instances) and *P. paniscus*/*P. t. verus* (one case).

An important point at this stage is to search for the traits that best explain each dimension (i.e. the active variables evincing the higher contributions to inertia within each age categories and dimensions) and, consequently, some important differences between taxa. As already mentioned above, the only marked distinctions between chimpanzee taxa are for *P. paniscus* or *P. t. verus*. When considering the juvenile 1 category, the lacrymo-ethmoidal (LAET) and the spheno-parietal (SPPA) articulations play an important role for explaining dimensions 1, 4, 6 and 7. Four other traits contribute to explain the dimensions where clear distinctions appear: the accessory infra-orbital foramen (FIAC) (dimensions 4, 7 and 8); the palatine portion of the incisive suture (IPOB) (dimension 1); the nasal spine (SPNA) (dimension 7); and the divided hypoglossal canal (CNHY) (dimension 8). Juvenile 2 is the other age category where marked differences between taxa are noticeable for at least two dimensions (see above). The SPPA and the LAET articulations play an important role for explaining, respectively, dimensions 1 and 2. Moreover, the palatine portion of the incisive suture and the infra-orbital fissure persistens (FSIP) also contribute significantly to, respectively, dimensions 1 and 2. As regards the three last age categories, only one dimension evinces clear differences between *P. paniscus* and *P. troglodytes*. In juvenile 3, the spheno-parietal articulation, the nasal spine and the palatine portion of the incisive suture best explain dimension 1. In adult 1, the same three features and the accessory infra-orbital foramen best explain the first dimension. Finally, in adult 2, the nasal spine and the accessory infra-orbital foramen best explain the third dimension.

Distinguishing five age categories, two important results emerge to define the pattern of cranial discrete variation among chimpanzees, starting at the species level and pursuing the problem down to the subspecific: (1) *P. paniscus* is distant from all *P. troglodytes* subspecies; (2) *P. t. verus* is far more distinct from both *P. t. schweinfurthi* and *P. t. troglodytes* than these latter subspecies are distinct between themselves. It should also be added that no differences were found when examining the variation of discrete traits within each subspecies by studying geographic groups and comparing them (Braga, 1995a). The congruence between these results – molecular data available for chimpanzees and biogeographical data for the African tropics (i.e. spatial distribution of major endemism centres recognised by Grubb, 1982) – suggests that several cranial discrete traits could be very useful to assess macroevolutionary changes. Therefore, one should question whether skeletal variation within chimpanzees is not underestimated when forming comparative samples and attempting to interpret the fossil hominoid record. Indeed, if *P. t. verus* is both morphologically and genetically distinct from *P. t. schweinfurthi* and *P. t. troglodytes*, this has to considered in palaeoprimatological investigations.

Features distinguishing *Pan*, *Gorilla* and *Pongo* : towards a developmental prospect

Focusing our attention firstly on the African apes, important differences between gorillas and chimpanzees, in both juveniles and adults, can be established. A complete or partial closure of the anterior component of the incisive suture always occurs in the younger chimpanzees with no permanent teeth erupted yet. In gorillas of an equivalent age, the anterior component of the incisive suture is always fully opened (Table 7.4 and Figure 7.8). Significant differences between gorillas and chimpanzees also occur for the palatal component of the incisive suture (IPOB) (Table 7.5 and Figure 7.9). Moreover, Braga (1995b) already mentioned an important difference between *Pan* and *Gorilla*: the occurrence of the divided hypoglossal canal (CNHY) being significantly higher in gorillas (Table 7.6 and Figure 7.1). Braga (1995b) added that a low geographical variation was found among the different species of great apes, including *Pongo pygmaeus*. Besides these features due to modifications in the ossification of some cranial elements, a different pattern of emissary foramina is seen in gorillas, where parietal, nasal and median basilar (MBC) canals occur frequently, the differences between on the one hand, *P. paniscus* and *P. troglodytes*, and on the other hand, *Gorilla gorilla*, being always highly significant (Tables 7.7 to 7.9 and, Figures 7.13, 7.12 and 7.7). As within the African apes, some differences established between *Pan*/*Gorilla* and *Pongo* are associated with cranial maturation. Divided hypoglossal canals (CNHY) occur frequently in *Pongo* (Table 7.6, Figure 7.1). We believe these significant differences to 'represent

Table 7.4. *Chi-squared tests with 2 × 2 contingency table for differences between species concerning the complete or partial closure of the anterior component of the incisive suture*

	Pan troglodytes	*Pan paniscus*	*Gorilla gorilla*
Pan troglodytes		**No test possible**	**161.63** **$p = 0.0000$**
Pan paniscus	*No test possible*		**32.49** **$p = 0.0000$**
Gorilla gorilla	*593.48* *$p = 0.0000$*	*422.89* *$p = 0.0000$*	

Results for juveniles in italic type; results for adults in bold type.

Table 7.5. *Chi-squared tests with 2 × 2 contingency table for differences between species concerning the complete or partial closure of the palatal component of the incisive suture*

	Pan troglodytes	*Pan paniscus*	*Gorilla gorilla*
Pan troglodytes		**68.97** **$p = 0.0000$**	**192.47** **$p = 0.0000$**
Pan paniscus	*340.00* **$p = 0.0000$**		**11.83** **$p = 0.0006$**
Gorilla gorilla	*496.28* *$p = 0.0000$*	*28.21* *$p = 0.0000$*	

Results for juveniles in italic type; results for adults in bold type.

Table 7.6. *Chi-squared tests with 2 × 2 contingency table for differences between species concerning the divided hypoglossal canal*

	Pan troglodytes	*Pan paniscus*	*Gorilla gorilla*	*Pongo pygmaeus*
Pan troglodytes		**14.05** **$p = 0.0002$**	**94.29** **$p = 0.0000$**	**147.98** **$p = 0.0000$**
Pan paniscus	*31.38* *$p = 0.0000$*		**90.11** **$p = 0.0000$**	**132.32** **$p = 0.0000$**
Gorilla gorilla	*34.69* *$p = 0.0000$*	*101.75* *$p = 0.0000$*		**7.70** **$p = 0.0055$**
Pongo pygmaeus	*83.3* *$p = 0.0000$*	*159.33* *$p = 0.0000$*	*15.23* *$p = 0.0001$*	

Results for juveniles in italic type; results for adults in bold type.

some modifications in the ossification of the basioccipital and exoccipital parts of the chondrocranium' (Braga, 1995b). An ossified styloid process fused with the petrous (PRST) is also frequently seen in *Pongo* (Table 7.10, Figure 7.2). The styloid process belongs embryologically to the hyoid systems which, partly, develops from the cartilage of the second branchial arch.

Table 7.7. *Chi-squared tests with 2 × 2 contingency table for differences between species concerning the parietal foramen*

	Pan troglodytes	*Pan paniscus*	*Gorilla gorilla*
Pan troglodytes		**0.84** **$p = 0.3601$**	**21.44** **$p = 0.0000$**
Pan paniscus	*0.04* *$p = 0.8450$*		**8.99** **$p = 0.0027$**
Gorilla gorilla	*36.85* *$p = 0.0000$*	*22.73* *$p = 0.0000$*	

Results for juveniles in italic type; results for adults in bold type.

Table 7.8. *Chi-squared tests with 2 × 2 contingency table for differences between species concerning the nasal foramen*

	Pan troglodytes	*Pan paniscus*	*Gorilla gorilla*
Pan troglodytes		**0.23** **$p = 0.633$**	**502.94** **$p = 0.0000$**
Pan paniscus	*0.94* *$p = 0.3330$*		**135.22** **$p = 0.0000$**
Gorilla gorilla	*271.16* *$p = 0.0000$*	*160.93* *$p = 0.0000$*	

Results for juveniles in italic type; results for adults in bold type.

Table 7.9. *Chi-squared tests with 2 × 2 contingency table for differences between species concerning the median basilar canal*

	Pan troglodytes	*Pan paniscus*	*Gorilla gorilla*
Pan troglodytes		**8.23** **$p = 0.0041$**	**137.66** **$p = 0.0000$**
Pan paniscus	*0.00* *$p = 0.9793$*		**19.50** **$p = 0.0000$**
Gorilla gorilla	*40.35* *$p = 0.0000$*	*21.66* *$p = 0.0000$*	

Results for juveniles in italic type; results for adults in bold type.

Interestingly, the absence of the foramen spinosum (foramen ovale in spinosum confluens), as pointed out by Berry & Berry (1971), is extremely frequent in orang-utans (Table 7.11, Figure 7.3) and much more uncommon in the three African ape species. It was seen only in two (1.6%) of 123 human patients with an age range of 1–78 years by Ginsberg *et al.* (1994). We should add that the morphology of the posterior border of the sphenoid greater wing and, more precisely, the confluent oval and spinous foramina (COSF),

Table 7.10. *Chi-squared tests with 2 × 2 contingency table for differences between species concerning the styloid process*

	Pan troglodytes	*Pan paniscus*	*Gorilla gorilla*	*Pongo pygmaeus*
Pan troglodytes		**23.58**	**17.60**	**354.97**
		$p = 0.0000$	**$p = 0.0000$**	**$p = 0.0000$**
Pan paniscus	*10.34*		**10.04**	**205.48**
	$p = 0.0013$		**$p = 0.0015$**	**$p = 0.0000$**
Gorilla gorilla	*10.81*	*0.01*		**414.88**
	$p = 0.0010$	*$p = 0.9093$*		**$p = 0.0000$**
Pongo pygmaeus	*11.79*	*29.83*	*34.20*	
	$p = 0.0006$	*$p = 0.0000$*	*$p = 0.0000$*	

Results for juveniles in italic type; results for adults in bold type.

Table 7.11. *Chi-squared tests with 2 × 2 contingency table for differences between species concerning the confluent oval and spinous foramina*

	Pan troglodytes	*Pan paniscus*	*Gorilla gorilla*	*Pongo pygmaeus*
Pan troglodytes		**0.01**	**47.13**	**599.07**
		$p = 0.9291$	**$p = 0.0000$**	**$p = 0.0000$**
Pan paniscus	*4.07*		**12.54**	**217.82**
	$p = 0.0436$		**$p = 0.0004$**	**$p = 0.0000$**
Gorilla gorilla	*11.07*	*19.90*		**314.08**
	$p = 0.0009$	*$p = 0.0000$*		**$p = 0.0000$**
Pongo pygmaeus	*215.80*	*191.53*	*125.08*	
	$p = 0.0000$	*$p = 0.0000$*	*$p = 0.0000$*	

Results for juveniles in italic type; results for adults in bold type.

also depends on the composition of the vascular network and its the branching in relation to the position of bone.[6] Finally, in contrast to African apes, this study found the retroarticular canal present in at least 50% of orangutan skulls (Table 7.12, Figure 7.10). This finding has important phylogenetic implications. The retro-articular canal transmits the retroarticular vein which connects the temporal, or petrosquamous, sinus (sinus temporalis), with the retromandibular and external jugular veins. The temporal sinus is known to be present in all mamalian embryos (Butler, 1967). Although it frequently persists into postnatal life, it decreases in size with growth. We believe that 'the complete absence of a retroarticular canal can be considered as a synapomorphy supporting the human–African ape clade

[6] For example, Ginsberg *et al.* (1994: p. 289) already observed that 'congenital variants of the foramen spinosum are generally related to defects in osteogenesis or to maldevelopment of the middle meningeal artery'. The foramen spinosum transmits the middle meningeal artery, a branch of the maxillary artery that stems from the maxillary ramus of the external carotid artery. Other vascular variations may explain the absence of a foramen spinosum. For example, Curnow (1873) described hypoplasia of the *foramen spinosum* in association with origin of the middle meningeal artery from the ophthalmic artery.

Table 7.12. *Chi-squared tests with 2 × 2 contingency table for differences between species concerning the retroarticular foramen*

	Pan troglodytes	*Pan paniscus*	*Gorilla gorilla*	*Pongo pygmaeus*
Pan troglodytes		**0.04** $\mathbf{p = 0.8436}$	**26.64** $\mathbf{p = 0.0000}$	**353.74** $\mathbf{p = 0.0000}$
Pan paniscus	*3.78* $p = 0.0518$		**6.72** $\mathbf{p = 0.0096}$	**100.31** $\mathbf{p = 0.0000}$
Gorilla gorilla	*0.55* $p = 0.4596$	*1.22* *0.2691*		**187.91** $\mathbf{p = 0.0000}$
Pongo pygmaeus	*120.95* $p = 0.0000$	*102.32* $p = 0.0000$	*106.16* $p = 0.0000$	

Results for juveniles in italic type; results for adults in bold type.

and excluding *Pongo*'. Moreover, 'in orang-utans the retroarticular canal resembles the more primitive structure in prosimians and is probably homologous with it' (Braga, 1995b). As suggested by Braga (1995b) the reformulation of the biogenetic law by Nelson (1979) can be used here to validate the outgroup comparisons. Embryos of both African apes and orang-utans have a retroarticular vein (Padget, 1957; Butler, 1967) but, in African apes, with subsequent development, the emissary vein regresses and the retroarticular foramen is not formed by ossification. Given this ontogenetic character transformation, the presence of a retroarticular vein, as inferred from the presence of a retroarticular foramen, is observed to be 'more general' and its absence to be 'less general' (Nelson, 1979).

It is not surprising to find that differences between great ape taxa generally do not constitute dichotomous conditions but rather result from differing frequencies of occurrence for traits found in all species. This pattern of differences is also observed for many extant and fossil primate species. However, it is important, firstly, to note that many morphological differences between taxa are ontogenetic and secondly, that these differences are already present in the first postnatal stages of life. The data made available from the study of an important immature series indicate that the ossification of the chondrocranium does not reach its final stage before birth. Some distinctions between taxa might be due, firstly, to differences in patterns of ossification in the chondrocranium and secondly, to processes of completion occurring at different times after birth that may play an important role in the formation of typical shapes. For example, as documented by Starck (1973, p. 8), 'the lateral wall of the chondrocranium consists generally of independently originating parts which amalgamate with one another to form a longitudinally-running cartilaginous plate or ridge'. Part of this plate is ossified when osseous elements originating independently and separately from each other fuse. The varying sutural connections at the pterion (SPPA) clearly discriminating some chimpanzee taxa (see above) are probably the consequences of distinct patterns (timing, rate, incorporations of secondary cartilage, supplementary bones) of ossification. This would probably be the same for the aspect of the medial wall of the orbit (LAET) or of the posterior border of the sphenoid greater wing (COSF) (Braga *et al.*, 1998). In connection with this later example, Starck (1973; p. 17) wrote: 'At the posterior rim of the alisphenoid, the incisura ovalis is transformed later on, by supplementary bones, into a foramen ovale of the bone skull'. According to the present study, this would mean that the supplementary bone is often missing in *Pongo*. Consequently, the present study supports the argument that ontogenetic shifts are important factors in generating distinct cranial differences among the great ape species even if a further detailed study of cranial and dental maturation in living specimens, before and after birth, is needed.

timing in ontogeny

Conclusions: interpreting the fossil record

This study aimed to assess within and between group patterns of discrete trait variation in 1453 extant great ape crania in order to provide new data to interpret the fossil record. Indeed, as already stated by Shea *et al.* (1993; p. 289): 'the more information we can gather on patterns of intra- and interspecific variation in extant forms, reflecting phylogenetic and/or adaptive proximity, the better our assessments of the fossils will be'. Problems and new prospects arising from skeletal discrete variability in extant great apes are identified for the first time. It is of major interest that discrete patterns of variation in extant great apes reveal detailed and profound morphological differences within not only chimpanzees but also, to a considerably larger extent, between, *Pan*, *Gorilla* and *Pongo*, this latter evincing several primitive traits related to skeletal basicranial maturation or to cerebral vascular drainage. In this matter, *Pan* has by far the most derived skull. These differential cranial maturational or vascular patterns are very promising for investigating fossil hominoid relationships as well as their affinities to living genera. In other respects, the patterning of morphological variation among extant great apes appears more complex than is generally recognised and this issue has to be considered when using comparative samples in fossil studies. This will lead to a better reconstruction of aspects of fossil hominid palaeobiology such as, for example, development.

These findings fully contradict Patterson *et al.* (1993) when they wrote: 'Morphological evidence on hominoid relationships has probably already been pushed close to its limits, barring the possibility of extraordinary fossil discoveries'. On the contrary, future studies dealing with cranial, infracranial and dental discrete traits will certainly provide us with additional anatomical evidence on hominoid (both extant and fossil) phylogenetic relationships and will push forward palaeoprimatology. These findings are of particular interest for the study of the transition from hominoids to hominids, a widely debated topic because there is no general agreement on which Eurasian or African hominoid found in deposits aged between 11 and 7 Ma could represent the ancestor of all hominids. The assessment of skeletal and odontological discrete traits variation in samples representing all extant species and subspecies may lead to important re-evaluation of several character states defined in phylogenetic reconstructions. Indeed, the evidence from discrete trait analysis, as tested for chimpanzees in this study, agrees with the molecular evidence and defines clearly the relationships among extant taxa.

true this paper has relevance for future discoveries

Acknowledgements

I would like to thank L. de Bonis and G. Koufos for the invitation to submit a chapter in this volume. I also appreciate the helpful comments made on a previous draft of the paper by an anonymous referee.

References

Albrecht, G. H. & Miller, J. (1993). Geographic variation in primates: a review with implications for interpreting fossils. In Kimbel, W. H. & Martin, L. B. (eds.), *Species, Species Concepts, and Primate Evolution*, pp. 123–61. New York: Plenum Press.

Asfaw, B., White, T., Lovejoy, O., Latimer, B., Simpson, S. & Suwa, G. (1999). *Australopithecus garhi*: a new species of early hominid from Ethiopia. *Science* **284**: 629–35.

Berry, A. C. (1974). Non-metrical variation in the prosimian skull. In Martin, R. D., Doyle, G. A. & Walker, A. C. (eds.), *Prosimian Biology*, pp. 641–53. London: Duckworth.

Berry, A. C. & Berry, R. J. (1971). Epigenetic variation in the human cranium. *J.Anat.* **101**: 361–79.

Braga, J. (1993). Ossification du processus styloïde chez les pongidés. *Comptes Rendus de l'Académie des Sciences*, Paris, t. **317**, série II: 273–7.

Braga, J. (1995a). *Définition de certains caractères discrets crâniens chez Pongo, Gorilla et Pan. Perspectives taxonomiques et phylogénétiques*. PhD dissertation, Université Bordeaux 1.

Braga, J. (1995b). Emissary canals in the Hominoidea and their phylogenetic significance. *Folia Primatol.* **65**: 144–53.

Braga, J. (1996). Nouvelles données sur l'épine nasale chez les grands singes de grande taille. Applications en paléoanthropologie. *Comptes Rendus de l'Académie des Sciences*, Paris, t. **322**, série IIa: 1087–92.

Braga, J. (1998). Chimpanzee variation facilitates the interpretation of the incisive suture in South African Plio-Pleistocene hominids. *Am. J. Phys. Anthrop.* **105**: 121–35.

Braga, J. & Boesch, C. (1997a). Further data about venous channels in South African Plio-Pleistocene hominids. *J. Hum. Evol.* **33**: 423–47.

Braga, J. & Boesch, C. (1997b). The 'radiator bias'. Reply to Falk and Gage. *J. Hum. Evol.* **33**: 503–6.

Braga, J., Crubézy, E. & Elyatqline, M. (1998). The posterior extent of the sphenoid greater wing and its phylogenetic usefulness in human evolution. *Am. J. Phys. Anthrop.* **107**: 387–400.

Butler, H. (1967). The development of mammalian dural venous sinuses with special reference to the post-glenoid vein. *J. Anat.* **102**: 33–56.

Cheverud, J. M. & Buikstra, J. E. (1981a). Quantitative genetics of skeletal nonmetric traits in the rhesus macaques on Cayo Santiago. I. Single trait heritabilities. *Am. J. Phys. Anthrop.* **54**: 43–9.

Cheverud, J. M. & Buikstra, J. E. (1981b). Quantitative genetics of skeletal nonmetric traits in the rhesus macaques on Cayo Santiago. II. Phenotypic, genetic, and environmental correlations between traits. *Am. J. Phys. Anthrop.*, **54**: 51–8.

Crubézy, E., Telmon, N., Sevin, A., Picard, J., Rougé, D., Larrouy, G., Braga, J., Ludes, B. & Murail, P. (1998). Microévolution d'une population historique. Etude des caractères discrets de la population de Missiminia (Soudan, III°–VI° siècle). *Bull. Mém. Soc. Anthropol. Paris*, n.s. **10**: 1–212.

Curnow, J. (1873). Two instances of irregular ophthalmic and middle meningeal arteries. *J. Anat.* **8**: 155–6.

Eveleth, P. B. & Tanner, J. M. (1990). *Worldwide Variation in Human Growth.* Cambridge: Cambridge University Press.

Ginsberg, L. E., Pruett, S. W., Chen, M. Y. & Elster, A. D. (1994). Skull-base foramina of the middle cranial fossa: reassessment of normal variation with high-resolution CT. *Am. J. Neuroradiol.*, **15**: 283–91.

Goldberg, T. L. & Ruvolo, M. (1997). The geographic apportionment of mitochondrial genetic diversity in east African chimpanzees, *Pan troglodytes schweinfurthi*. *Mol. Biol. Evol.*, **14**: 976–84.

Gonder, M. K., Oates, J. F., Disotell, T. R., Forstner, M. R., Morales, J. C. & Melnick, D. J. (1997). A new west African chimpanzee subspecies? *Nature* **388**: 337.

Greenacre, M. J. (1993). *Correspondence Analysis in Practice.* London: Academic Press.

Grubb, P. (1982). Refuges and dispersal in the speciation of African forest mammals. In Prance, G. T. (ed.), *Biological Diversification in the Tropics*, pp. 537–53. New York: Columbia University Press.

Grüneberg, H. (1952). Genetical studies on the skeleton of the mouse. IV – Quasi continuous variation. *J. Genet.* **51**: 95–114.

Hauser, G. & De Stefano, G.F. (1989). *Epigenetic Variants of the Human Skull.* Stuttgart: Schweizerbart.

Jolly, C., Oates, J. F. & Disotell, T. R. (1995). Chimpanzee kinship. *Science* **268**: 185–6.

Le Double, A. F. (1903). *Traité des Variations des Os du Crâne de l'Homme. Et de leur Signification au Point de Vue de l'Anthropologie Zoologique.* Paris: Vigot Frères.

Le Double, A. F. (1906). *Traité des Variations des Os de la Face de l'Homme. Et de leur*

Signification au Point de Vue de l'Anthropologie Zoologique. Paris: Vigot Frères.
Le Double, A. F. (1912). *Traité des Variations de la Colonne Vertébrale de l'Homme. Et de leur Signification au Point de Vue de l'Anthropologie Zoologique*. Paris: Vigot Frères.
Morin, P. A., Moore, J. J., Chakraborty, R., Jin, L., Goodall, J. & Woodruff, D. S. (1994). Kin selection, social structure, gene flow, and the evolution of chimpanzees. *Nature* **265**: 1193–201.
Nelson, G. (1979). Ontogeny, phylogeny, paleontology and the biogenetic law. *Syst. Zool.*, **27**: 324–45.
Ossenberg, N. S. (1969). *Discontinuous Morphological Variation in the Human Cranium*. PhD dissertation, University of Toronto.
Padget, D. H. (1957). The development of the cranial venous system in man, from the viewpoint of comparative anatomy. *Contrib. Embryol.* **36**: 79–140.
Patterson, C., Williams, D. M. & Humphries, C. J. (1993). Congruence between molecular and morphological phylogenies. *Annu. Rev. Ecol. Syst.* **24**: 153–88.
Ruvolo, M., Pan, D., Zehr, S., Goldberg, T., Disotell, T. R. & von Dornum, M. (1994). Gene trees and hominoid phylogeny. *Proc. Natl. Acad. Sci. USA* **91**: 8900–4.
Sangalli, A. (1998). *The Importance of Being Fuzzy*. Princeton: Princeton University Press.
Saunders, S. R. (1978). *The development and distribution of discontinuous morphological variation of the human infracranial skeleton*. Archaeological Survey of Canada, No. 81. Ottawa: National Museums of Canada.
Saunders, S. R. & Popovich, F. (1978). A family study of two skeletal variants: atlas bridging and clinoid bridging. *Am. J. Phys. Anthrop.* **49**: 193–204.
Scott, G. R. & Turner, C. G. (1997). *The Anthropology of Modern Human Teeth. Dental Morphology and its Variation in Recent Human Populations*. Cambridge: Cambridge University Press.
Shea, B. T., Leigh, S. R. & Groves, C. P. (1993). Multivariate craniometric variation in chimpanzees. Implications for species identification in paleoanthropology.
In Kimbel, W. H. & Martin, L. B. (eds.), *Species, Species Concepts, and Primate Evolution*, pp. 265–96. New York: Plenum Press.
Starck, D. (1973). The skull of the fetal chimpanzee. In Bourne, G. H. (ed.), *The Chimpanzee*, vol. 6, pp. 1–33. Basel: Karger.
Stevens, S. S. (1968). Measurement , statistics, and the schemapiric view. *Science* **161**: 849–56.
Uchida, A. (1992). *Intra-species Variation Among the Great Apes. Implications for Taxonomy of Fossil Hominoids*. PhD dissertation, Harvard University.
Zadeh, L. A. (1965). Fuzzy sets. *Inf. Cont.* **8**: 338–56.

PART III

Miocene hominoids: function and phylogeny

8 Eurasian hominoid evolution in the light of recent *Dryopithecus* findings

Meike Köhler, Salvador Moyà-Solà & David M. Alba

Introduction

In the summer of 1991, remains of a facial cranium of the large fossil ape *Dryopithecus* were discovered in Can Llobateres by a team from the Institut de Paleontologia M. Crusafont in Sabadell (Barcelona, Spain). The locality of Can Llobateres (Figure 8.1) is situated near Sabadell, close to the Catalan coast in the north-west of the Iberian peninsula, and palaeomagnetic data indicate an age of 9.7 Ma for the lowest levels, and about 9.6 Ma for the highest, where the remains of *Dryopithecus* were localised (Moyà-Solà & Köhler, 1993, 1995; Agustí *et al.*, 1996). During the following years (1992–95), part of a skeleton was found, supposedly belonging to the same individual as the skull (Moyà-Solà & Köhler, 1996). Similarly, remains of two new individuals came from the same level: parts of the hindlimb of a small-sized individual (probably a female); and a milk tooth of an infant. These latest discoveries promise important findings in years to come, considering the great area that has yet to be excavated in this remarkable Catalan locality.

Without doubt, the genus *Dryopithecus*, described by Eduard Lartet in 1856 from the French locality of Saint Gaudens (Lartet, 1856), has played an important role in our understanding of extant and fossil hominoids. In the 1960s all species of large Miocene apes were included in this genus, which was considered a primitive group of species without any special relationship to living forms. However, in recent years, a growing number of specialists consider the genus *Dryopithecus* different from other forms such as *Proconsul* and *Kenyapitheus*, and its appearance is limited to Europe, where it lived from middle to upper Miocene (13 to 9 Ma). However, this is the only point of consensus. The position of this primate in the hominoid tree and its relationship to the group of extant great apes is still being discussed.

Since the 1980s, material not only from *Dryopithecus*, but also from other Eurasian Miocene hominoids, has dramatically increased. Citing as examples only, we have the extraordinary cranial and postcranial remains of *Sivapithecus* from Pakistan (Pilbeam, 1982; Pilbeam *et al.*, 1990), *Ouranopithecus* from Greece (de Bonis *et al.*, 1990), *Lufengpithecus* from China (Wu *et al.*, 1985, 1986), *Ankarapithecus* from Turkey (Andrews & Tekkaya, 1980; Alpagut *et al.*, 1996) and *Dryopithecus* from Europe (Kordos, 1997; Moyà-Solà & Köhler, 1993, 1995, 1996). One would expect the amount of information yielded by the increasing number of hominoid specimens collected in the last few years from Eurasian Miocene localities to help in

[Figure 8.1]
View of the Can Llobateres excavation.

tracing their phylogenetic relationships. Instead, the more we know about each genus, the more confused we become about their phylogenetic links. With the increasing number of recovered cranial and postcranial elements of Miocene apes, the number of characters used to unravel their phylogeny has also increased. Hence the importance of 'Jordi', the finding of a quite complete skeleton of a young *Dryopithecus* male associated with a facial cranium. This key fossil yields for the first time the means of resolving apparent contradictions between the phylogenetic information taken either from cranial or postcranial remains, to be able to reconstruct the hypothetical common ancestor of extant great apes, to discover if some of the basic skeletal features shared by extant apes and humans were acquired independently, and to find out if some of the Miocene genera could have been related to some of the extant great apes.

The cranial anatomy of *Dryopithecus* and other Upper Miocene Eurasian hominoids

Due to its well preserved state, the cranial specimen from Can Llobateres (Figure 8.2) provides new information that makes us reconsider the phylogenetic position of all the fossil European apes, and of this Miocene genus in particular. CLI-18000 (specimen abbreviation) preserves, as

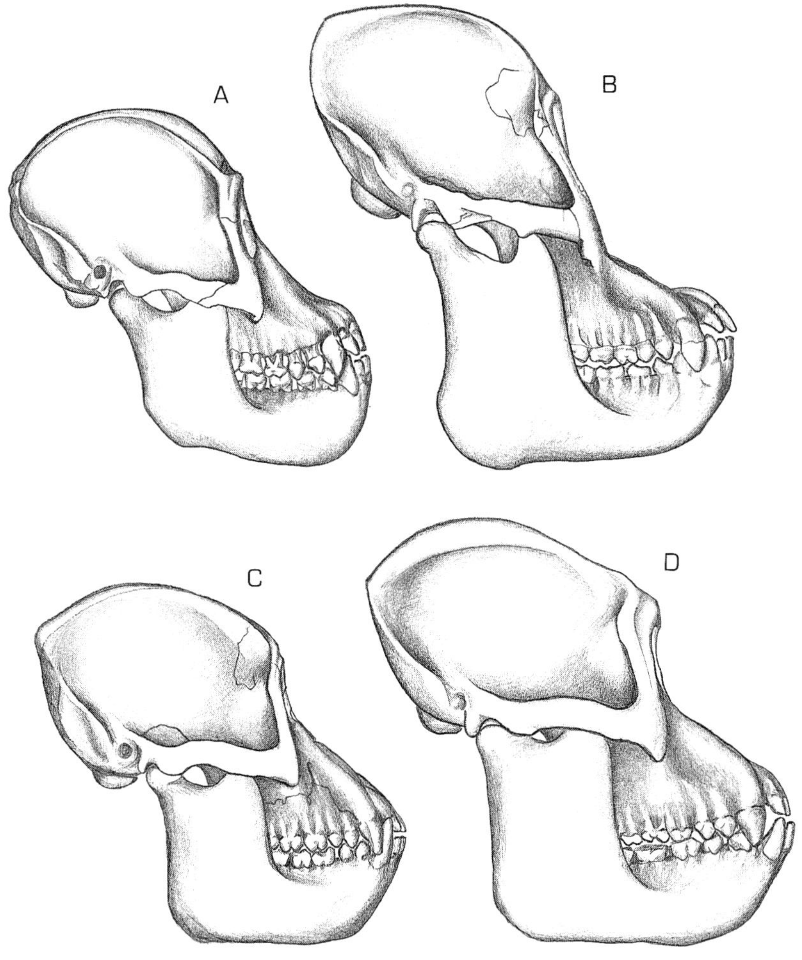

[Figure 8.2]
Cranial reconstructions. A: *Pongo pygmaeus*, female; B: *Sivapithecus indicus*, male; C: *Dryopithecus laietanus*, male; D: *Ouranopithecus macedoniensis*, male. All crania are drawn to the same scale. The fossil crania (B,C,D) are based on unpublished three-dimensional reconstructions recently made by the authors. B: *Sivapithecus*: the right half of the face was reconstructed as a mirror image of the well preserved left half of the specimen. The position of the glenoid fossa, determined by the length and the orientation of the zygomatic arch, as well as the elevated frontals permit the reconstruction of length and form of the neurocranium. C: *Dryopithecus*: the facial remains of *Dryopithecus* are much more fragmentary than those of *Sivapithecus*. Nevertheless, they are mostly broken at their sutures thus allowing us to reconstruct position and orientation of the many fragments. This reconstruction of the *Dryopithecus* cranium is our most recent interpretation and differs in two important points from

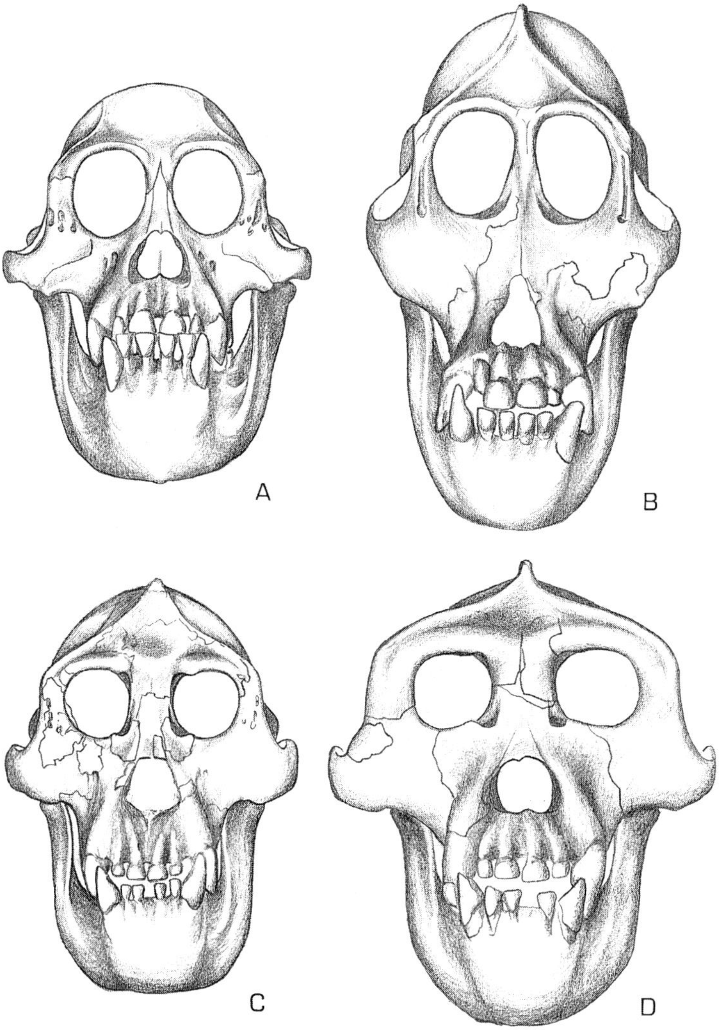

the first version (Moyà-Solà & Köhler, 1995). We reconstruct the face as being somewhat higher, leaving a reasonable space between the zygomatic bone and its maxillary root, and we now orientate the frontals more vertically, forming a less straight but more concave profile of the face, consistent with the nasofrontal angle of the Rudabànya female RU-77. Both the vertically ascending mandibular rami of several specimens and the elevated frontals indicate a short neurocranium. D: *Ouranopithecus*: in the original specimen (the male Xirochori skull), the left upper part of the face is somewhat folded diagonally over the right lower part. This deformation caused an excessive elevation of the frontals, a torsion of the interorbital pillar towards the right orbit and a compression of the right maxillary region. For our reconstruction, we cut the supraorbital region, the left part of the interorbital pillar and the maxillary tooth rows and reorientated them correspondingly.

previously mentioned, part of the temporal bone with the petrosal (ear bone). In its endocranial face, the internal acoustic aperture (*meatus acusticus*) can be perfectly observed, and what is most important for phylogenetic interpretations, the *fossa subarcuata* is completely absent. The loss of this fossa is related to the disappearance of the paraflocculus, an extension of the cerebellum into the petrosal. This feature has been given a special phylogenetic transcendence as its absence is considered a derived character of the extant great apes. However, it has been recently suggested that this character may be functionally linked to the concurrence of a large body size together with the absence of a tail (Spoor & Leakey, 1996). Yet the same authors believe that there might be even more factors involved than those mentioned above. If we admit the absence of the *fossa subarcuata* as a valid character, then it suggests, together with another set of features, such as enlarged maxillary sinus, shallow cheek region and high zygomatic root, deep glenoid fossa with prominent entoglenoid process and other dental features (Andrews, 1992; Begun, 1992), that *Dryopithecus* would be closely related to the extant great apes. There are three possible phylogenetic alternatives:

(1) If *Dryopithecus* is the common ancestor of all extant great apes, it should be morphologically plesiomorphic (primitive) in comparison to them and could not share any derived character with any of the members of this group.
(2) If *Dryopithecus* is the ancestor of the African group, it should display derived characters in common with some representatives of this clade.
(3) If *Dryopithecus* is the ancestor of the extant orang-utan, it should share derived characters with the latter.

Our analysis revealed that *Dryopithecus* shares no derived characters with the African apes, or that it displays the majority of the most evident facial features of the Asiatic orang-utan and its supposed Miocene sister taxon *Sivapithecus*. Thus, *Dryopithecus* is, in general terms, an essentially primitive form in comparison with the extant great apes.

However, a more detailed analysis exposes several intriguing characters. Some peculiar traits of the zygomatic bone are particularly interesting (Schwartz, 1990; Moyà-Solà & Köhler, 1993, 1995). On the one hand, it is very flat and frontally oriented, and also strikingly robust relative to the overall size of the cranium. On the other hand, it has three zygomatic foramina (the central one is double), which are situated above the level of the infraorbital margin. Among primates, these characters are only found in the orang-utan. The other anthropoids have a more slender zygomatic, which is more laterally orientated, noticeably more convex and has less zygomatic foramina that, moreover, are located at the same level or below the infra-orbital margin. The frontal bones of *Dryopithecus* also reveal some significant features. The different patterns of the frontal area are considered to be diagnostic in hominoids. All the known primitive hominoids, such as *Proconsul, Afropithecus,* the slightly more modern *Otavipithecus* and the extant African hominoids *Pan* and *Gorilla,* possess a frontal sinus that occupies the whole internasal and supra-orbital area of the frontals approximately half way up the lateral side of the orbits (Pickford *et al.*, 1997). The volume of this sinus has an important allometric component, that is to say, it depends, among other factors, on body mass (Blaney, 1986). Once again, the extant orang-utan, as well as *Sivapithecus, Lufengpithecus* and *Dryopithecus,* exhibit a different morphology. In the extant form as well as *Sivapithecus,* the frontal sinus is completely absent, while in *Dryopithecus* it is very reduced, only occupying the interorbital area, without penetrating the frontal squama, above and behind the supraorbital margin (Pickford *et al.*, 1997). The supra-orbital region of *Dryopithecus* and the other European Miocene apes is very distinctive, since it shows supra-orbital *costae.* This pattern is very different from that of both the Miocene African hominoids (*Afropithecus, Proconsul, Otavipithecus*) and the extant African apes, but again it is comparable to that of *Sivapithecus* and *Pongo* (see Figure 8.2).

The absence of all these characters in the primitive Miocene hominoids and, in fact, in the majority of anthropoids, reveals that these characters are derived. However, their presence both in *Sivapithecus* and *Pongo,* and their absence in the African apes, suggest that *Dryopithecus* and the Asian apes share a common ancestor. Yet, this opinion is not unanimously accepted. Two other hypotheses, however, are currently under debate: (1) *Dryopithecus* is regarded as a member of the clade of African apes and humans, based on the assumption that *Dryopithecus* has a supra-orbital torus homologous with that of African apes (Begun, 1992), which is a homology already rejected by several authors (Moyà-Solà & Köhler, 1993, 1995; Andrews & Pilbeam, 1996); and (2) *Dryopithecus* is considered to be a primitive form, plesiomorphic to all the extant great apes (Andrews, 1992; Andrews & Pilbeam, 1996), so that no value is given to the unmistakable anatomy of the zygomatic bone of this genus.

One might wonder why the selection of valuable characters for tracing phylogenetic relationships is still a matter of debate. That it depends heavily on the always-subjective assessment of the characters at issue is best illustrated by the history of the different interpretations of some Miocene hominoids, such as *Oreopithecus bambolii* (Alba *et al.*, Chapter 13) or *Ankarapithecus meteai.* This latter taxon was first described by Ozansoy (1957) on the basis of a mandible from the Sinap formation in Turkey. Later, a lower face of a male was attributed to the same species (Andrews & Tekkaya, 1980). Striking similarities with the *Sivapithecus* material from Pakistan led several authors (Andrews & Tekkaya, 1980; Andrews & Cronin, 1982) to include *Ankarapithecus* in the genus *Sivapithecus*. At the beginning of the last decade the lower face specimen played an important role in the interpretation of hominoid evolution. It was claimed to be the first palaeontological evidence of the divergence of Asian and African apes, as asserted by molecular biologists (Andrews & Cronin, 1982). Several derived characters were considered to demonstrate convincingly for the first time a close relationship between a Miocene hominoid and an extant ape (Andrews & Cronin,1982). Shortly after, this hypothesis was reinforced by the similarities with the finding of a facial specimen of *Sivapithecus indicus* (Pilbeam, 1982). In a recent revision of the Turkish lower facial specimen, Begun & Güleç (1995) revalidated the previous name *Ankarapithecus*, confirming at the same time its position within the *Sivapithecus–Pongo* clade. Thus, *Ankarapithecus* was considered as undoubtedly the most northern representative of the Asian apes, in geographical proximity to, but different from, the contemporaneous European ape *Ouranopithecus*.

Unlike *Ankarapithecus*, the discussion of the phylogenetic relationships of *Ouranopithecus* was based mainly on characters of the upper face, such as the morphology of the supraorbital torus, and it was suggested to be a member of the African ape clade (Begun, 1992) or even claimed to be an ancestor of *Australopithecus* (de Bonis *et al.*, 1990; de Bonis & Koufos, 1994). However, the most recent finding of a nearly complete face of *Ankarapithecus* (Alpagut *et al.*, 1996) has brought a dilemma. The combination of both the lower and upper face provided a surprise; once unmasked, the upper face does not fit the expected association of pongine characters, but strongly resembles the morphology of the European Miocene apes, especially that of *Ouranopithecus* (Alpagut *et al.*, 1996). Thus, above the orbits, *Ankarapithecus* has characters believed to be synapomorphies of the African apes (Begun, 1992, 1994), while the same author (Begun & Güleç, 1995) considered the features below the orbits as derived with the Asian *Pongo*-clade. Despite its appearance, a single specimen cannot combine the synapomorphies of two different clades. Therefore, we think that the problem of this apparent contradiction is rooted in the definition and the homology of characters.

Since both the morphology of the supraorbital torus and the frontal sinus are particularly controversial, we will discuss these characters in detail. The pattern of the supraorbital torus seen in African apes has been defined as a complete bar of bone that extends across the superior margins of both orbits, incorporating a prominent *glabella* region at the midline.

Ankarapithecus as well as *Ouranopithecus* and *Dryopithecus* show only a slight thickening of the supra-orbital rim, the so-called *costa supraorbitalis* (Clarke, 1977, Pickford *et al.*, 1997). Regarding both morphologies as 'supra-orbital torus' fails to differentiate between the two patterns. Consequently, in a group with 'supra-orbital torus', we should include different forms such as the African apes, *Dryopithecus* and *Ankarapithecus,* hylobatids, *Pongo, Sivapithecus,* cercopithecids, *Australopithecus, Homo erectus, H. nearderthalensis* and Australian aborigines. If we do so, may we consider this feature homologous? Disagreement also concerns the assessment of the extension of the frontal sinus. Nearly all mammals, including primates, have a sinus that simply fills dead spaces of the cranium (Sisson & Grosssman, 1975)..The lack of a sinus is therefore a derived feature, as in the case of cercopithecids. At the same time, the presence of a sinus should only be considered derived in forms where the ancestors had no sinus or in forms where the sinus is more extensive than expected from allometric scaling.

As we mentioned previously, *Pongo, Sivapithecus, Ankarapithecus, Lufengpithecus* and *Dryopithecus,* share both the peculiar morphology of the zygomatic and the supraorbital region, and the reduction of the sinus. Since these features are not shared with primitive Miocene hominoids, such as *Afropithecus, Proconsul* and *Otavipithecus,* or the clade of African apes and humans, they should be considered as shared derived characters. Consequently, *Dryopithecus* is identified as a member of the same natural group as *Pongo* – the Ponginae. The fact that the characters shared with *Pongo* are less numerous in *Dryopithecus* than in other genera of this group, such as *Sivapithecus,* is only evidence of the primitive nature of this European form (Moyà-Solà & Köhler, 1993, 1995). In fact, we may consider *Dryopithecus* as the most primitive genus of the Ponginae.

This suggests that there are basically four groups of fossil great apes in Eurasia: the first one, comprising species belonging to the genus *Dryopithecus,* would be the most primitive one; the second group would only include *Oreopithecus,* which is now considered an endemic insular form descended from *Dryopithecus* (Moyà-Solà & Köhler, 1997; Harrison & Rook, 1997) and will not be discussed here (see Alba *et al.*, Chapter 13); a third group would include *Ankarapithecus, Ouranopithecus* and *Lufengpithecus,* which are larger than *Dryopithecus* (34 kg for male *Dryopithecus laietanus* from Spain [Moyà-Solà & Köhler, 1996] vs. 72 kg for male *Ouranopithecus macedoniensis* from Greece [de Bonis & Koufos, 1994]); and finally, the *Sivapithecus* group. The group containing the Greek, Turkish and Chinese forms is probably more homogeneous than the proliferation of genera seems to suggest and, in fact, dimorphism and allometry may explain many of the morphological differences between *Ankarapithecus*, *Ouranopithecus* and *Lufengpithecus*.

The postcranial skeleton of *Dryopithecus laietanus* from Can Llobateres

After the discovery of the *Dryopithecus* skull and its publication, the fieldwork in Can Llobateres provided more important surprises. Between 1992 and 1995, skeletal remains of a young *Dryopithecus* male, attributable to the same individual whose skull had been found earlier, were unearthed. These remains comprise about 60 specimens that supply information about most parts of the body: thorax, spinal column, arms and legs (Figures 8.3 and 8.4). This is the only well-preserved hominoid skeleton from Middle to Upper Miocene, filling the gap between the two other known skeletons: the 18 million-year-old *Proconsul* skeleton and the 3.1 million-year-old 'Lucy' (*Australopithecus afarensis*) (Moyà-Solà & Köhler, 1996).

Extant hominoids, comprising the African gorillas and chimpanzees, and the Asian gibbons and orang-utans, are characterised by erect postures of the trunk (with the trunk held upright during locomotion, e.g. vertical climbing). This posture is habitual in modern hominoids, while it is only occasional in monkeys with habitually quadrupedal postures. Habitually bipedal walking is characteristic only of humans. These positional and locomotor behaviours are reflected in the morphology of the skeleton. Two well differentiated postcranial patterns exist. A primitive pattern, with a narrow thorax, a long and flexible spinal column and arms shorter than legs, is characteristic of primates that move quadrupedally (pronograde positional repertoire) both in the trees and on the ground (for example, the extant monkeys). *Proconsul*, a hominoid from the early Miocene of Africa, falls within these forms (Ward, 1993; Ward *et al.*, 1993). The modern pattern, on the other hand, is characteristic of the extant hominoids. In these forms, the thorax is wide, the lumbar spinal column shorter and more rigid, and the arms are longer than the legs. This skeletal structure is adapted for vertically climbing trees and hanging from branches. This type of positional and locomotor behaviour requires erect postures of the spinal column (orthograde positional repertoire) and, to a certain extent, can be considered a pre-adaptation to human-like bipedality. There is no known fossil evidence of this pattern dating from prior to *c.* 4.1 Ma (Leakey *et al.*, 1998).

The *Dryopithecus* skeleton from Can Llobateres provides the first clear evidence of the extant hominoid pattern. This 9.5 million-year-old ape already shows some of the most important anatomical changes for the

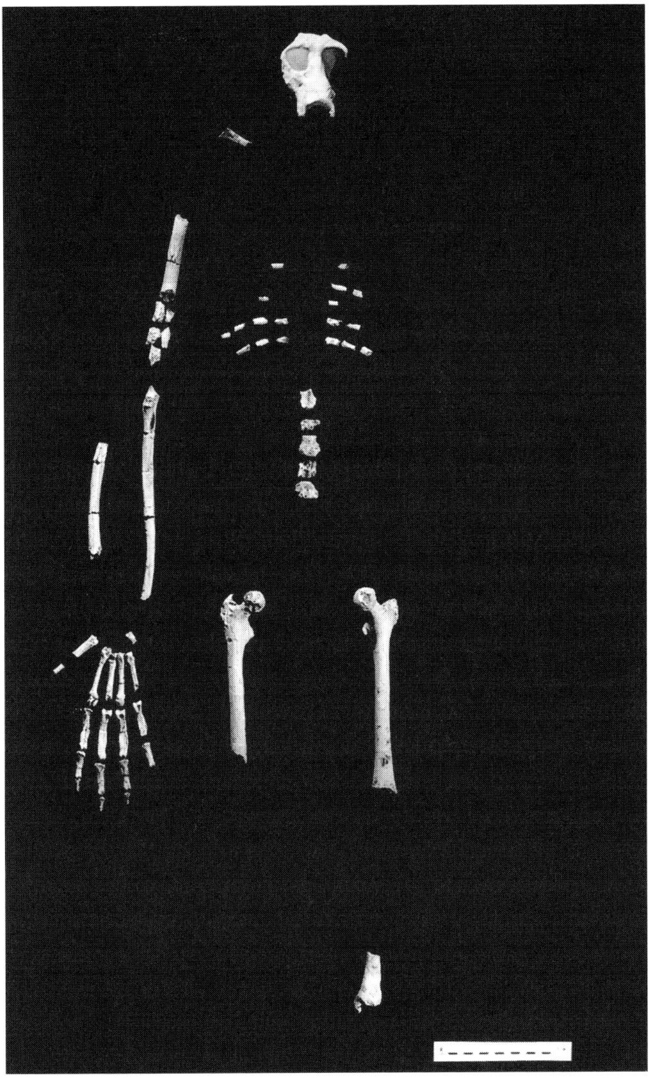

[Figure 8.3]
The skeleton CLl-18800 found in Can Llobateres.

adoption of upright postures of the trunk. Thus, several features of the vertebrae, thorax, intermembral proportions, arms and hands clearly indicate that this fossil primate possessed the basic pattern of modern apes (Moyà-Solà & Köhler, 1996). In particular, some characters of the *Dryopithecus* trunk suggest adaptations to orthograde postures. The lumbar vertebrae are proportionally shorter than those of cercopithecids and the

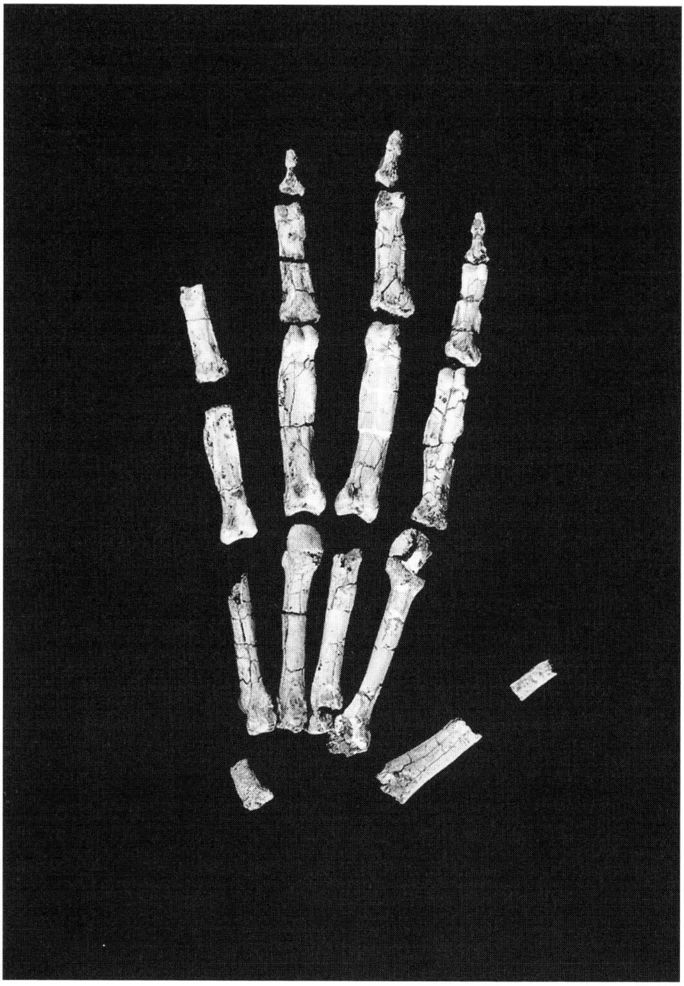

[Figure 8.4]
Detail of the hand of CLI-18800.

proconsulids of the African lower Miocene. The transverse processes of the lumbar vertebrae originate dorsolaterally, directly from the pedicle and not from the vertebral body. The more dorsal position of the articular surfaces for the ribs on the thoracic vertebrae suggests a more ventrally situated spinal column, a character related to a wider thorax. The large size of the clavicle also suggests that the scapula would have been more dorsally situated, which would also be in accordance with a wide thorax. The intermembral proportions are those of an orthograde anthropoid; thus, the intermembral index ([length of humerus + length of radius]/[length of femur + length of tibia] × 100) of 114 estimated for *Dryopithecus* (Moyà-Solà & Köhler, 1996) reveals that the arms were notably longer than the legs, opposite to the cercopithecid pattern. This value is clearly higher than that of *Pan troglodytes* (103) or *Pan paniscus* (104), and lower than that of the orang-utan (140), but very similar to that of *Oreopithecus* (119) (Moyà-Solà & Köhler, 1996).

Bearing in mind that we are comparing forms with different body masses, which as some authors have suggested, has a direct influence on the intermembral index, it is necessary to take this phenomenon into account to correctly interpret the functional significance of these morphological differences. Several authors (e.g. Aiello, 1981; Jungers, 1985) have demonstrated that gorillas and chimpanzees have the postcranial proportions that correspond to their body mass within a functional context, in which the capacity to climb trees has to be maintained (this is necessary to avoid predators and to feed from the arboreal strata, which is difficult to access for other mammals). If this statement is correct, the body proportions of *Dryopithecus*, like those of *Oreopithecus*, tend to approximate those of the extant orang-utan. We interpret the elongation of the upper extremities in these European forms as an adaptation to more frequent climbing and suspensory activities than in the African gorillas and chimpanzees. To our knowledge, currently only *Pongo* shows this locomotor behaviour. Other features, in particular the large hand, the great length of the first phalanges, the very distally inserted flexors of the fingers, and the distally strongly compressed femora, suggest once again a high capacity for climbing and suspensory activities. Even so, the differences between extant orang-utan and *Dryopithecus* are important and suggest that *Dryopithecus* was not capable of such acrobatic climbing and hanging, with such a broad spectrum of movements as the orang-utan. Possibly *Dryopithecus* still maintained a certain degree of quadrupedalism in trees, inherited from more primitive ancestors.

The skeleton of other Eurasian Miocene hominoids: the apparent contradiction of *Sivapithecus*

Despite the numerous derived characters shared with extant apes, *Dryopithecus* shows several other features that can be found in primitive hominoids of the *Proconsul* pattern or in monkeys only (e.g. somewhat elongated lumbar vertebrae, short metacarpals, small articular surfaces of metacarpals and phalanges, and other features of the hand related to palmigrady). These primitive characters suggest the retention of locomotor behaviours related to quadrupedalism and the use of the hands in palmigrade postures (Moyà-Solà & Köhler, 1996). Thus, *Dryopithecus* combines a set of characters related to vertical climbing and suspensory activities, characteristic of extant apes, while still retaining a primitive type of quadrupedalism, characteristic of primitive hominoids such as *Proconsul* or *Kenyapithecus*. There is no extant form with this association of features; only the extremes are represented. Thus, all extant apes are vertical climbers, but each of them is highly specialized either in quadrumanous climbing (*Pongo*), swinging (*Hylobates*) or secondary knuckle walking quadrupedalism (*Pan, Gorilla*), whereas monkeys are basically palmigrade quadrupeds. Yet this association of vertical climbing and palmigrade quadrupedalism is found in other Eurasian Miocene apes besides *Dryopithecus*, like for example in *Sivapithecus*. Although we do not know the intermembral proportions, the anatomy of the thorax, or even the anatomy of the lumbar region of the latter, features of the elbow joint, the carpals, and the hindlimbs suggest that it also exhibited an association of climbing and suspensory behaviour with primitive palmigrade quadrupedalism (Kelley & Pilbeam, 1986; Rose, 1994) comparable to that of *Dryopithecus*. What does this association of ape-like and monkey-like features mean in phylogenetic terms?

According to some authors, the retention of primitive characters in the European Miocene hominoids excludes all of them from the clade of the extant great apes (Pilbeam, 1996), and even perhaps from that of extant apes in general (including hylobatids), so that the cranial features shared with certain extant apes must be considered to be convergences. To other authors, however, this suggests rather that some of the derived postcranial characters shared with extant apes or great apes have been independently acquired by hylobatids, African apes and humans, and *Pongo* (Moyà-Solà & Köhler, 1995, 1996). This second hypothesis is based on the conclusions derived from the study of the cranial material, which links the Eurasian Miocene hominoids, and particularly *Sivapithecus*, to the *Pongo* clade. Furthermore, if this hypothesis is correct, it would have important implications for reconstructing the common ancestor of all the extant great apes, since this would be much more primitive than the study of the living forms suggests.

However, it is precisely this mosaic of characters that makes it difficult to know with any certainty the phylogenetic relationships of the European Miocene apes. It is evident that when the evolutionary outline drawn by the information obtained from the morphology of extant apes is applied, the European forms do not easily fit anywhere. This stems from an undeniably disturbing fact – the pervasive homoplasy in hominoid evolution (see discussion in Larson, 1998). Unfortunately, although high levels of homoplasy are easy to detect, it is much more difficult to identify which traits are truly homologous and which are homoplastic. To some authors, homoplasy can be found in certain features shared by extant apes, whereas according to others it is reflected in various characters that Miocene apes share with the extant ones.

Unfortunately, applying cladistics and the principle of parsimony is not going to solve this dilemma (see also Alba *et al.*, Chapter 13). On the one hand, the strength (tree length and correlation index) of the various hypotheses will depend on the degree of atomisation of characters chosen by the authors of each analysis. Too often it is forgotten that many characters are functionally (and even ontogenetically) related to each other, so that this interdependence often makes their segregation into independent features erroneous, thus giving a false appearance of strength to the resulting cladogram. Up to what point are the different cranial or postcranial features independent from each other? Until we can answer this question with some degree of confidence (and we certainly cannot for the moment), and until the relationship between function and morphology is more fully understood, any attempt to solve this puzzling issue by means of cladistic analysis is, in our opinion, futile. On the other hand, applying the concept of parsimony does not guarantee that we have chosen the correct hypothesis either, especially when there are several similarly parsimonious alternatives. It should not be forgotten that the way in which cladistics applies the concept of parsimony – as a methodological assumption – is not substantiated by observations from nature. Instead, it seems that convergence and parallelism have played a relevant role in ape evolution, which casts severe doubts on the applicability of cladistics and the concept of parsimony, at least when additional considerations and, especially, functional aspects, are not properly taken into account.

Sivapithecus is in the knot of this problem, given the apparently striking contradiction between its cranial and postcranial remains. Cranially, it is almost impossible not to think of *Pongo* when observing *Sivapithecus*, even if only superficially (Figure 8.2). The strongly airorhynchous cranial structure, with high orbits, narrow interorbital region, flat and anteriorly oriented zygomatic, high glenoid fossa, premaxillary–maxillary pattern, etc., is typical of *Pongo*. Therefore, from the comparison of the cranial anatomy of *Sivapithecus* with that of *Pongo*, a close phylogenetic relationship seems incontrovertible. However, the analysis of the postcranial skeleton seems to exclude it from being sister to any extant form, since it exhibits a series of primitive characters (Pilbeam *et al.*, 1990). Applying strict cladistic methodology to postcranial characters, we should place it in a clade prior to

Hylobates, because *Sivapithecus* has (according to some authors) a primitive hominoid skeleton, whereas gibbons have a modern hominoid one. What is convergent, the postcranial or the cranial skeleton? Which is misleading and where is the solution?

Evidently, if we had at our disposal a very complete fossil record, we could search the strata for the traces of the evolution of this group and more easily identify ancestor–descendant relationships. Unfortunately, this is not the case. There are enormous information gaps in the fossil record, and this, in the short-term, is not going to give us the answer. So, we have to resort to analysis of the little information we do have available – the construction of methodologically and conceptually well-founded hypotheses, verifying them with data from other disciplines such as biochronology, biogeography and molecular phylogenetics. At present, the only possibility is to analyse this additional information and see which of the two hypotheses is most convincing – *Sivapithecus* as a sister taxon of *Pongo*, or *Sivapithecus* as a primitive hominoid, sister of all living apes or great apes.

The high locomotor specialisation of the extant apes may give the answer – quite simply, the Miocene hominoids were not as specialised in climbing and suspension as are extant apes. If this is the case, the common ancestor of all of them cannot be reconstructed from information drawn from the extant forms alone because, as Harrison (1991) suggested, only the most specialised apes have survived from the Miocene hominoid radiation. If this assumption is correct, the postcranial morphology of the common ancestor of the extant great apes could have been similar in our opinion to that of *Sivapithecus indicus* and *Dryopithecus laietanus*, instead of great ape-like. The trunk and extremities would have conformed to the orthograde pattern, as *Dryopithecus* shows, but several primitive hominoid-like characters would have been also retained. In functional terms, this suggests that in spite of being specialised in climbing and hanging, palmigrade quadrupedalism was still a habitual type of locomotion for this primate. In fact, there is no evidence from the fossil record that the common ancestor of the extant apes possessed a completely modern great ape pattern.

Sivapithecus again deserves a special mention in this context. The morphology of the humerus of this genus has been described as homogeneous and fundamentally similar to that of primitive pronograde hominoids (Rose, 1994), without distinguishing *S. indicus* from *S. parvada*. However, in our opinion, there are some important differences between both species. In particular, the anatomy of the humeral shaft from *S. parvada* is so monkey-like that it really makes it difficult to imagine this genus as being phylogenetically related to *Pongo*. However, the humerus of *S. indicus* is very similar to that of *Dryopithecus fontani* from Saint Gaudens (Moyà-Solà

& Köhler, 1996) – both are slender, not retroflexed, only slightly curved laterally and, in our opinion, there is no indication that the deltoid plane was flat. This morphology, along with other postcranial characters, suggests that *S. indicus* should have had a locomotor repertoire similar to that of *Dryopithecus*, including not only climbing and suspensory behaviours, but also quadrupedalism. On the contrary, the humerus of the large-sized species *S. parvada* is robust, slightly retroflexed, medially inclined, with a flat deltoid plane, and in general close to the humeral morphology of primitive hominoids such as *Kenyapithecus* and *Proconsul* (Pilbeam *et al.*, 1990). In functional terms, the humeral morphology of *S. parvada* fits well with the quadrupedal pattern of *Proconsul* and *Kenyapithecus*. The large size of this species suggests that it moved mainly on the ground, while in the trees climbing would have been the most habitual type of locomotion.

For large primates (60–90 kg in males) life in arboreal strata is only practicable by specialising the locomotor apparatus (bones and muscles) for cautious climbing, as found in the orang-utan. The gorilla, on the other hand, has adopted a distinct strategy by spending most of the time on the ground. In *Sivapithecus parvada*, having reached this critical body mass, the bony and muscular structure probably still retained part of the architecture of monkeys. This might have favoured a habitual terrestrial quadrupedalism similar to that of the African apes, rather than the cautious climbing of the orang-utan. In these circumstances, the morphological basis from which these secondary terrestrial adaptations evolved has capital importance in understanding why these primates, which because of large body size were forced to secondarily adapt to terrestrial quadrupedalism, are so different in morphology. Most probably, in this case differences are due, among other factors, to ancestral differences in body structure that constrained the final result of the process of adaptation.

In other words, adapting to life on the ground when evolving from a cercopithecine (as in *Papio*) is not the same as descending from a chimpanzee-like creature (as in *Gorilla*). If *Sivapithecus parvada* was derived from an ancestor morphologically similar to *S. indicus* or *Dryopithecus* (morphologically, not phylogenetically), which still retained a primitive hominoid locomotor structure, the mechanical needs imposed by terrestrial locomotion on the musculoskeletal system may have led to the reappearance of certain monkey-like features such as those found in the humerus of *Sivapithecus parvada*. This hypothesis allows us to forecast that all large-bodied descendants of small Eurasian Miocene apes will tend to have more monkey-like features than large extant apes, which does not mean that these characters are primitive retentions.

In our opinion, the biogeographical data strengthen the hypothesis that

Sivapithecus and the remaining Upper Miocene Eurasian apes *Dryopithecus*, *Ankarapithecus*, *Ouranopithecus*, *Lufengpithecus* and *Oreopithecus*, are linked to the *Pongo* clade (all of them being the result of an adaptive radiation from a single colonisation event from Africa), rather than being the sister group of all extant great apes. Nevertheless, we have to admit that the common ancestor of the extant great apes was morphologically more primitive than suggested by the analysis of the extant forms.

Conclusions

Our current view of the phylogenetic relationships of the Eurasian Miocene hominoids and their links with the living forms are expressed in Figure 8.5. We propose to consider *Dryopithecus* as well as the other Upper Miocene European hominoids as primitive Ponginae. Some of them, such as *Ankarapithecus* (a genus that perhaps should include *Ouranopithecus* and *Lufengpithecus*), do possess some synapomorphies of the South Asian Ponginae, and the need for reconsideration of the whole group emerges in order to establish the true generic and specific diversity. Finally, we reconstruct the ancestral postcranial anatomy of the extant great apes as being very similar to that of *Dryopithecus laietanus*, admitting the existence of homoplasy among some of the characters of extant apes. If we accept this phylogenetic hypothesis, then we stand before the mirror image of the African ape clade. Some 13 to 8 Ma, the Eurasian landmass housed a very diversified natural group of apes, ranging from chimpanzee-sized (*Dryopithecus*) to female gorilla-sized animals (*Sivapithecus parvada*) and displaying a broad spectrum of locomotor adaptations, from climbing and suspensory behaviours to more terrestrial quadrupedalism.

In the future, the *Dryopithecus* findings will contribute to the reconstruction of what the common ancestor of the African apes and the first bipedal hominoids, such as *Australopithecus afarensis* ('Lucy'), would have looked like. Preliminary analyses suggest that the chimpanzee might not constitute a good model of this ancestor, but rather that it may have been more primitive, not specialised for quadrupedal terrestrial locomotion (as is the chimpanzee), but basically adapted to life in the trees (climbing and hanging). If this is the case, hominids would not have gone through a stage of quadrupedal terrestrial locomotion, but would have rather gone straight from arboreal to bipedal locomotion. The fact that *Oreopithecus*, an endemic descendant of *Dryopithecus*, acquired a bipedal type of terrestrial locomotion (Köhler & Moyà-Solà, 1997; Moyà-Solà *et al.*, 1999; Alba *et al.*, Chapter 13), strongly supports this possibility. Further analysis of the *Dryopithecus* skeleton may help to shed light on these and other questions.

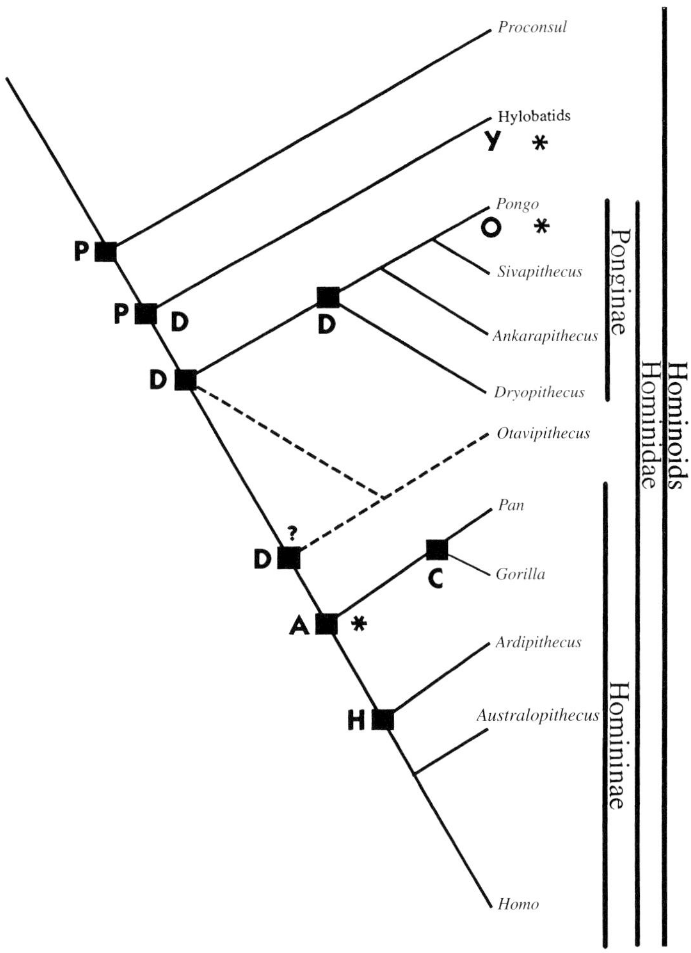

[Figure 8.5]
Cladogram showing our current view of the phylogenetic position of *Dryopithecus* and other Miocene Eurasian hominoids. On the basis of this hypothesis, we tried to reconstruct the postcranial anatomy of the different ancestors. If this phylogenetic hypothesis is correct, then the postcranial of the common ancestor of all extant great apes must have been very similar to that of *Dryopithecus laietanus*. The different nodes are: P: pronograde body plan; primitive hominoid with colobine-like postcranium and locomotion.
D: *Dryopithecus*-like postcranium; orthograde body plan with palmigrade quadrupedalism. A: specialized climbing without any quadrupedal adaptations.
C: secondary terrestrial quadrupedal adaptations (knuckle walking), maintaining a high climbing competence. H: bipedality. O: slow quadrumanous climbing.
Y: brachiation. At present, it is impossible to know whether the orthograde adaptations of hylobatids were acquired independently (ancestor with a P-type postcranium) or whether, to the contrary, they proceed from a common ancestor (D-type postcranium). *Otavipithecus* is a poorly known form from the Miocene of Namibia (Pickford *et al.*, 1997).

Acknowledgements

Financial support for this work was provided by a grant from the Wenner-Gren Foundation. D.M. Alba was supported by a predoctoral fellowship (1999FI 00765) from the Generalitat de Catalunya.

References

Agustí, J., Köhler, M., Moyà-Solà, S., Cabrera, L., Garcés, M. & Parés, J. M. (1996). Can Llobateres: the pattern and timing of the Vallesian hominoid radiation reconsidered. *J. Hum. Evol.*, **31**: 143–55.

Aiello, L. C. (1981). The allometry of primate body proportions. *Symp. Zool. Soc. Lond.*, **48**: 331–58.

Alpagut, B., Andrews, P., Fortelius, M., Kappelman, J., Temizsoy, I., Elebi, H. & Lindsay, W. (1996). A new specimen of *Ankarapithecus meteai* from the Sinap Formation of Central Anatolia. *Nature*, **382**: 349–51.

Andrews, P. (1992). Evolution and environment in the Hominoidea. *Nature*, **360**: 641–6.

Andrews, P. & Cronin, J. (1982). The relationships of *Sivapithecus* and *Ramapithecus* and the evolution of the orang utan. *Nature*, **297**: 541–6.

Andrews, P. & Pilbeam, D. (1996). The nature of evidence. *Nature*, **379**: 123–4.

Andrews, P. & Tekkaya, I. (1980). A revision of the Turkish Miocene hominoid *Sivapithecus meteai*. *Palaeontology*, **23**: 85–95.

Begun, D. & Güleç, E. (1995). Restoration and interpretation of the facial specimen attributed to *Sivapithecus meteai* from Kayincak (Yassioren), Central Anatolia, Turkey. *Am. J. Phys. Anthrop.*, **20** (Suppl.): 63–4.

Begun, D. R. (1992). Miocene fossil hominids and the chimp–human clade. *Science*, **257**: 1929–33.

Begun, D. R. (1994). Relations among the great apes and humans: new interpretations based on the fossil great ape *Dryopithecus*. *Ybk. Phys. Anthrop.*, **37**: 11–63.

Blaney, S. P. A. (1986). An allometric study of the frontal sinus in *Gorilla*, *Pan* and *Pongo*. *Folia Primatol.*, **47**: 81–96.

Clarke, R. J. (1977). *The Cranium of the Swartkrans Hominid SK 847 and its Relevance to Human Origins*. Ph.D. Thesis, University of the Witwatersrand.

de Bonis, L., Bouvrain, G., Geraad, D. & Koufos, G. (1990). New hominoid skull material from the late Miocene of Macedonia in northern Greece. *Nature*, **345**: 712–14.

de Bonis, L. & Koufos, G. (1994). Our ancestor's ancestors: *Ouranopithecus* is a Greek link in human ancestry. *Evol. Anthrop.*, **3**: 75–83.

Harrison, T. (1991). The implications of *Oreopithecus bambolii* for the origins of bipedalism. In Senut, B. & Coppens, Y. (eds.) *Origine(s) de la Bipédie Chez les Hominidés*, pp. 235–44. Paris: CNRS.

Harrison, T. & Rook, L. (1997). Enigmatic anthropoid or misunderstood ape: the phylogenetic status of *Oreopithecus bambolii* reconsidered. In Begun, D. R., Ward, C. V. & Rose, M. D. (eds.), *Function, Phylogeny and Fossils: Miocene Hominoid Origins and Adaptations*, pp. 327–62. New York: Plenum Press.

Jungers, W. L. (1985). Body size and scaling of limb proportions in primates. In Jungers, W. L. (ed.), *Size and Scaling in Primate Biology*, pp. 345–81. New York: Plenum Press.

Kelley, J. & Pilbeam, D. (1986). The dryopithecines: taxonomy, comparative anatomy, and phylogeny of Miocene large hominoids. In Swindler, D. R. & Irvin, J. (ed.), *Comparative Primate Biology*, vol. 1: *Systematics, Evolution, and Anatomy*, pp. 361–411. New York: Alan R. Liss.

Köhler, M. & Moyà-Solà, S. (1997). Ape-like or hominid-like? The positional behaviour of *Oreopithecus bambolii* reconsidered. *Proc. Natl. Acad. Sci.* **94**: 11747–50.

Kordos, L. (1997). Description and reconstruction of the skull of *Rudapithecus hungaricus* Kretzoi (Mammalia). *Ann. Hist. Nat. Mus. Natl. Hung.*, **79**: 77–88.

Larson, S. G. (1998). Parallel evolution in the hominoid trunk and forelimb. *Evol. Anthrop.*, **6**: 87–99.

Lartet, E. (1856). Note sur un grand singe fossile qui se rettache au grouppe des singes superieurs. *C.R. Acad. Sci.*, **43**: 219–23.

Leakey, M. G., Feibel, C. S., McDougall, I., Ward, C. & Walker, A. (1998). New specimens and confirmation of an early age for *Australopithecus anamensis. Nature*, **393**: 62–6.

Moyà-Solà, S. & Köhler, M. (1993). Recent discoveries of *Dryopithecus* shed new light on evolution of great apes. *Nature*, **365**: 543–5.

Moyà-Solà, S. & Köhler, M. (1995). New partial cranium of *Dryopithecus* Lartet, 1863 (Hominoidea, Primates) from the upper Miocene of Can Llobateres, Barcelona, Spain. *J. Hum. Evol.*, **29**: 101–39.

Moyà-Solà, S. & Köhler, M. (1996). A *Dryopithecus* skeleton and the origins of great-ape locomotion. *Nature*, **379**: 156–9.

Moyà-Solà, S. & Köhler, M. (1997). The phylogenetic relationships of *Oreopithecus bambolii* Gervais 1872. *Com. Red. Acad. Sci. Paris* **324**: 141–8.

Moyà-Solà, S., Quintana, J., Alcover, J. A. & Köhler, M. (1999). Endemic island faunas of the Mediterranean Miocene. In Rössner, G. E. & Heissig, K. (eds.) *The Miocene Land Mammals of Europe*, pp. 435–42. München: Verlag Fritz Pfeil.

Ozansoy, F. (1957). Faunes de mamiferes du Tertiarire de Turquie et leur révisions stratigraphiques. *Bull. Min. Resch. Explor. Inst. Turkey*, **49**: 29–48.

Pickford, M., Moyà-Solà, S. & Köhler, M. (1997). Phylogenetic implications of the first African Middle Miocene hominoid frontal bone from Otavi, Namibia. *C.R. Acad. Sci. Paris*, **325** (IIa): 459–66.

Pilbeam, D. (1996). Genetic and morphological record of the Hominoidea and hominid origins: A synthesis. *Molec. Phylogen. Evol.*, **5**: 155–68.

Pilbeam, D., Rose, M. D., Barry, J. C. & Ibrahim Shah, S. M. (1990). New *Sivapithecus* humeri from Pakistan and the relationship of *Sivapithecus* and *Pongo*. *Nature*, **348**: 237–8.

Pilbeam, D. R. (1982). New hominoid skull material from the Miocene of Pakistan. *Nature*, **295**: 232–4.

Rose, M. D. (1994). Quadrupedalism in some Miocene catarrhines. *J. Hum. Evol.*, **26**: 387–411.

Schwartz, J. H. (1990). *Lufengpithecus* and its potential relationship to an orang-utan clade. *J. Hum. Evol.*, **19**: 591–605.

Sisson, R. & Grossman, B. (1975). *Anatomy of Domestic Animals*. Philadelphia: W. B. Saunders.

Spoor, F. & Leakey, M. (1996). Absence of the subarcuate fossa in cercopithecids. *J. Hum. Evol.*, **31**: 569–75.

Ward, C. V. (1993). Torso morphology and locomotion in *Proconsul nyanzae*. *Am. J. Phys. Anthrop.*, **92**: 291–328.

Ward, C. V., Walker, A., Teaford, M. F. & Odhiambo, I. (1993). Partial skeleton of *Proconsul nyanzae* from Mfangano Island. *Am. J. Phys. Anthrop.*, **90**: 77–111.

Wu, R., Xu, Q. & Lu, Q. (1985). Morphological features of *Ramapithecus* and *Sivapithecus* and their phylogenetic relationships – morphology and comparison of the teeth. *Acta Anthrop. Sinica*, **4**: 197–204.

Wu, R., Xu, Q. & Lu, Q. (1986). Relationship between Lufeng *Sivapithecus* and *Ramapithecus* and their phylogenetic position. *Acta Anthrop. Sinica*, **5**: 1–30.

9 Functional morphology of *Ankarapithecus meteai*

Peter Andrews and Berna Alpagut

Introduction

During the early part of this century, when fossil apes were first discovered it was common practice to try to find links between them and the living apes. This tendency was effectively halted by the 1965 review of Simons and Pilbeam, who grouped almost all known fossil apes into a single clade which they called *Dryopithecus*, sinking numerous genera previously recognised into just three subgenera of *Dryopithecus*. On the other hand, these authors did support the link between what was then called *Ramapithecus* and the line leading to modern humans (Simons & Pilbeam, 1965).

More recently, the discovery of new fossil ape specimens has led to renewed suggestions of a direct relationship between fossil and recent apes. The first of these proposed a link between *Sivapithecus* and the orang-utan based on characters of the face (Andrews & Cronin, 1982) and skull (Ward & Pilbeam, 1983). One decade later, new material of *Dryopithecus* led to the suggestion that this genus was more closely related to the African apes than were other fossil apes (Begun, 1992), and an even closer relationship was suggested between *Graecopithecus* (referred to as *Ouranopithecus*) and hominines by de Bonis & Koufos (1993). Meanwhile the supposed relationship between *Ramapithecus* (now *Sivapithecus*) and humans had effectively been denied by its inclusion in the genus *Sivapithecus*. The only one of these proposed sets of relationships to find full support in a recent review of hominoid evolution (Andrews, 1992) was that linking *Sivapithecus* with the orang-utan, but even for this there is contrary evidence provided by recent discoveries of postcranial bones (Pilbeam *et al.*, 1990). For the rest, I proposed a set of relationships very similar to that of Simons & Pilbeam (1965), but with the early Miocene apes distinguised at the family level (Proconsulidae) and the middle to late Miocene apes combined at the subfamily level (Dryopithecinae) and distinguised as three distinct tribes: Afropithecinae; Kenyapithecinae; and Dryopithecinae.

The most recent discovery of Miocene ape is a face and partial skull from late middle Miocene deposits in the Sinap formation of Turkey (Alpagut *et al.*, 1996). This provides support for the view first proposed by Simons & Pilbeam (1965) that there was a middle Miocene radiation of fossil apes that left no clear survivors at the present day, a radiation that may include all of the above-mentioned fossil ape species in addition to others such as *Kenyapithecus* from Kenya and *Griphopithecus* from Turkey (Andrews,

1992). The significance of this fossil ape radiation to ape and human evolution will be discussed in the following pages.

Much of the uncertainty about hominoid evolution that still exists despite continuing research efforts is the result of the inadequate fossil record, which is both patchy and fragmentary. The occasional discovery of more complete fossils adds greatly to our knowledge, and with each new skull, new interpretations of hominoid evolution have ensued – for example the skull of *Sivapithecus* (Pilbeam, 1982), the skull of *Graecopithecus* (de Bonis & Koufos, 1993) and the skull of *Dryopithecus* (Moyà-Solà & Köhler 1994). At the same time, however, they contribute additional questions about hominoid evolution that require still further material to answer. One such specimen that has both provided new evidence and raised new questions is the recent discovery of a partial skull and associated postcranial remains from the Sinap Formation in Turkey (Alpagut *et al.*, 1996).

The skull was found in 1995 during excavation of locality 12 on Delikayincaktepe, in late Miocene deposits of the Sinap Formation (Alpagut *et al.*, 1996). The sediments consist of overbank flood plain deposits cut by channel-fill conglomerates, and the fossils occur in rich pockets in the flood plain silts (Kappelman *et al.*, 1996). There is a rich fauna of hipparionine equids, aardvarks, rodents, suids, giraffids, rhinos, carnivores, proboscideans, bovids, tortoises and turtles. Little is known of the taphonomy and palaeoecology of the site.

Description of the skull

The hominoid skull from Sinap (number AS 500) preserves much of the facial skeleton of an adult female with fully erupted M3s (Figure 9.1). Its body size has been estimated based on the orbital dimensions (Aiello & Wood, 1994) to be between 23 and 29 kg, with the lower estimate being based on orbital height and the higher estimate based on orbital breadth. This is slightly less than the size of a female bonobo. The mandible was found preserved upside down and located about 20 cm from the skull. Although the mandible was not articulated with the face, it was found at the same stratigraphic level, and its size and the identical state of wear on the upper and lower dentitions, including more advanced wear on the left side, provides strong evidence that these two specimens represent the same individual. The face includes most of the premaxilla, maxilla, zygomatics, lacrimals and nasals, and portions of the sphenoids, palatines and frontal, including a portion of the temporal line on the right. Minor breakage and distortion along the upper lateral margin of the frontal gives the left eye orbit

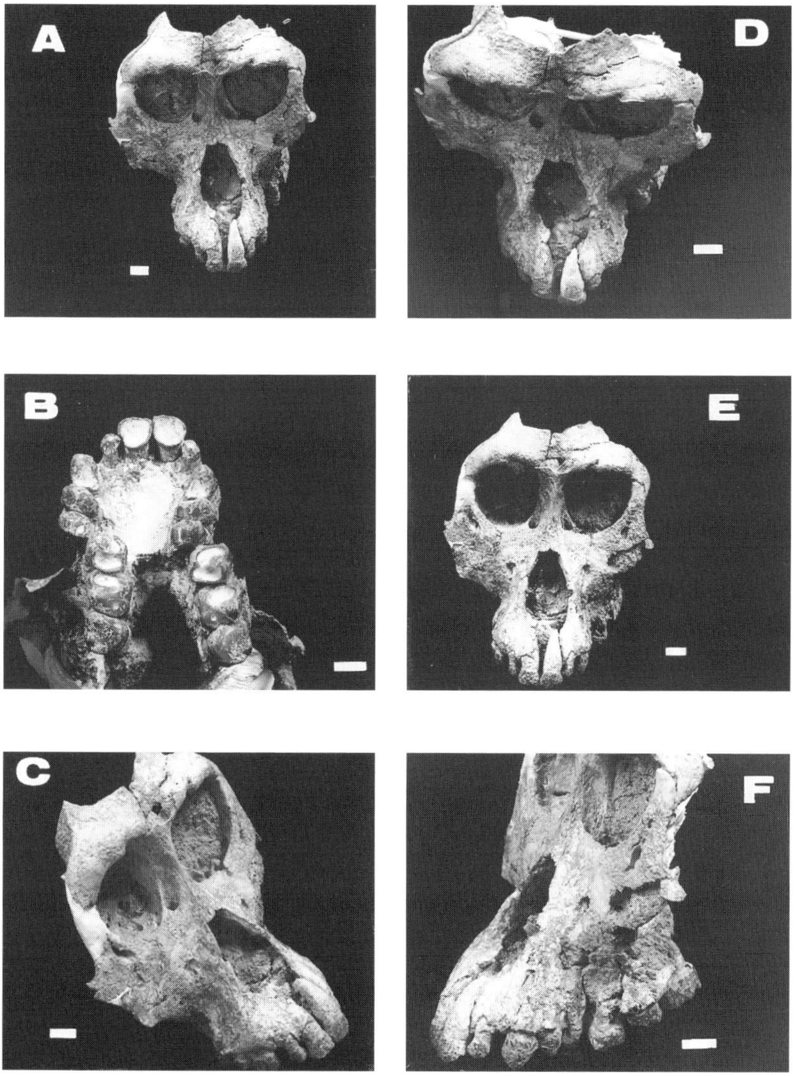

[Figure 9.1]
The skull of *Ankarapithecus meteai*. A: view of the skull from the top frontal; B: occlusal view showing the distorted maxillary toothrows; C: half-side view of the right side of the skull; D: view from the top; E: frontal view of the face; F: half-side view of the left side. (Scale bars = 1 cm.)

an angular appearance, and the parietals, occipitals, temporals and auditory ossicles are missing. The back part of the palate is displaced sideways so that the maxillary tooth rows are also displaced.

The mandible is preserved almost in its entirety without distortion (Figure 9.2). The symphysis is complete and the superior and inferior transverse tori

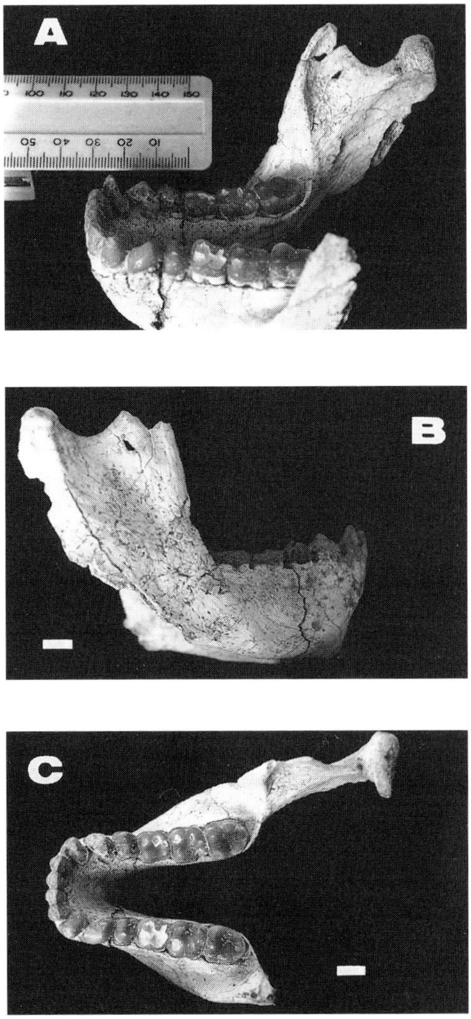

[Figure 9.2]
The mandible of *Ankarapithecus meteai*. A: side view showing the internal surface of the right mandibular body and ascending ramus; B: lateral view of the right side; C: occlusal view of the mandibular toothrows and the body of the mandible. (Scale bars = 1 cm.)

are well-defined. The body is robust and vertically fractured but not broken near the mental foramen. It is complete posteriorly past the oblique line on both the right and left sides with the internal surface showing a distinct mylohyoid line. The ascending ramus is nearly complete on the right side and includes the complete anterior vertical margin, coronoid process, mandibular notch and condyle. Only the inferior and posterior margins of the gonial angle of the right ramus is missing.

Both the upper and lower dentitions are complete. The enamel on the molars appears to be thick, and it is heavily marked by tiny root traces. The mandibular arcade is undistorted and has tooth rows that diverge slightly posteriorly. The arcade of the upper incisors through premolars is also perfectly preserved but all of the upper molars on both the right and left sides are displaced somewhat superiorly with about 1 cm left lateral offset from the anterior tooth row. The offset involves no plastic deformation and it will be possible to accurately reconstruct the complete upper arcade.

Taxonomic designation

The Sinap skull is assigned to the species *Ankarapithecus meteai* on the basis of its similarity with a larger but less complete partial face described in 1980 (MTA2125: Andrews & Tekkaya, 1980). Differences in canine size suggest that the new skull is from a female individual whereas the older face is from a male, but canine size in both is relatively small compared with molar size. The original description of the 1980 specimen recognised that it and the type mandible of A. *meteai* (Figure 9.3) were also conspecific, and on the basis of the similarities shared between these specimens and *Sivapithecus* material from the Indian subcontinent, Andrews & Tekkaya (1980) synonymised all of this material with *Sivapithecus* but retained *meteai* for the species name. The recovery of this new material has made it necessary to resurrect *Ankarapithecus* as the valid genus for this material (Begun & Güleç 1995, 1998). The history of naming of these specimens is shown in Table 9.1.

Functional morphology

Two functional complexes will be described for *A. meteai*, and their influence on the characters of the skull and postcranial skeleton will be assessed.

Powerful masticatory apparatus for chewing food

There are a number of characters of the teeth and skull that all indicate that *A. meteai* had a powerful masticatory apparatus. These relate to the cheek teeth, the dimensions of the mandible and maxilla and the development of the muscles of mastication. These will be described briefly both in terms of their functional aspects and their phylogenetic implications when degree of association between characters is taken into account (Table 9.2).

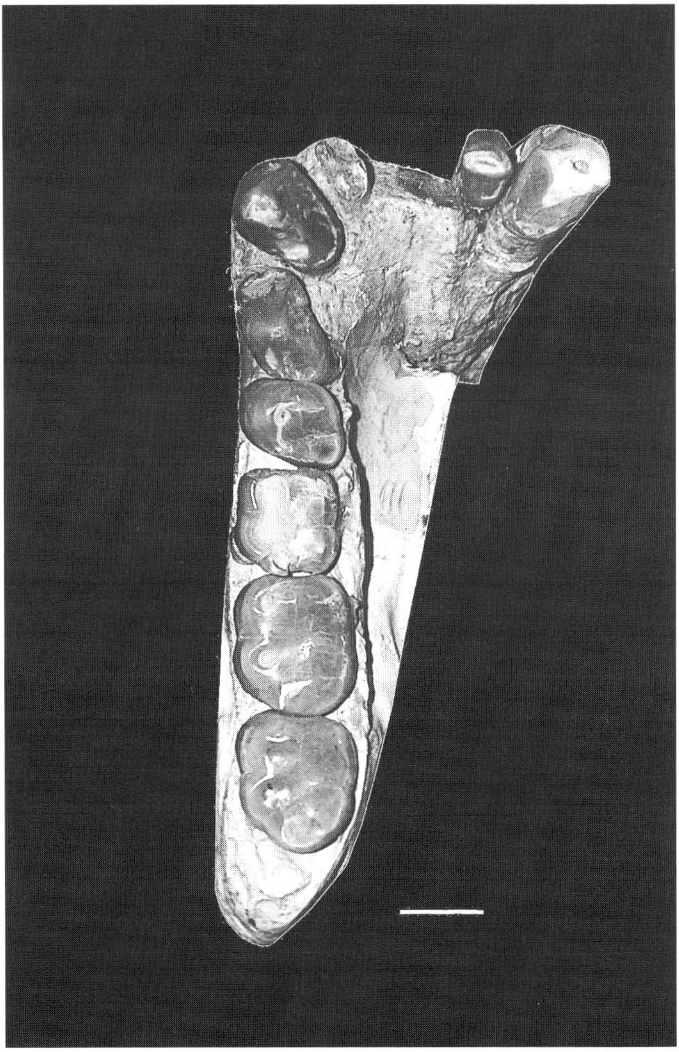

[Figure 9.3]
The type specimen of *Ankarapithecus meteai*, with symphysis and left mandibular tooth row. (Scale bar = 1 cm.)

- Teeth with thick enamel – this has not been measured directly on the teeth of *Ankarapithecus* as there are no broken edges and none of the teeth have been sectioned or radiographed (Martin, 1985). Enamel edges at the borders of dentine pits are exposed through wear, and it appears from these that the enamel was of an equivalent thickness to *Sivapithecus*, which places it in the thick category as defined by Martin (1985; Andrews

Table 9.1. *Taxonomy of* Ankarapithecus meteai.

- First mentioned – mandible of a large pongid (Ozansoy, 1955)
- Named *Ankarapithecus meteai* – type specimen MTA 2253, mandible of a large pongid (Ozansoy, 1957)
- Diagnosis of *A. meteai* based on the mandible MTA 2253 (Ozansoy, 1965)
- 1st revision: *Dryopithecus (Sivapithecus) indicus* – Dryopithecinae (Simons & Pilbeam, 1965)
- 2nd revision, 1980: *Sivapithecus meteai* – orang-utan clade based on the description of MTA 2125 maxilla (Andrews & Tekkaya, 1980)
- Resurrection of *Ankarapithecus meteai* and description as a stem great ape based on the discovery in 1995 of the skull AS 500 (Alpagut *et al.*, 1996)

Table 9.2. *Functional morphology of Ankarapithecus meteai*

A. Powerful masticatory apparatus for chewing food

teeth with thick enamel
large molars and premolars
flat occlusal planes on molars
poorly developed shearing crests
reduced canines
anteriorly positioned zygoma
strong lateral flare of zygoma
high ascending ramus of mandible
enlarged temporal foramen
strong zygomatic process
anterior origin of masseter
robust mandibular body
high floor of maxillary sinus
robust alveolar process of maxilla

B. Adaptations of anterior teeth for food preparation

broad and low crowned
heavy wear on incisors
robust premaxilla
buttressing of face

& Martin, 1991). The function of thick enamel is considered to be related to mastication of hard food items (Kay, 1981), strengthening the molar teeth against the stresses imposed by breaking down hard resistant objects such as nuts and woody fruits and seeds, but recent work has shown thick enamel to be more widespread in Miocene hominoids than previously recognised (Beynon *et al.*, 1998).

- Large molars and premolars – megadonty is frequently found associated with thick enamel, and the increased occlusal area resulting from increased overall size of the cheek teeth increases the effective area of the teeth and is associated with processing large quantities of plant food. No indication is necessarily given as to the type of food in this instance. There is a clear trend towards megadonty in australopithecines (Wood 1981), and megadonty has also been observed for some fossil apes, for example *Kenyapithecus, Afropithecus, Graecopithecus* and *Sivapithecus* (Pilbeam *et al.*, 1990; Andrews & Martin, 1991; Andrews & Wood, unpublished data). When crown size is compared with cranial dimensions in *Ankarapithecus*, a similar degree of megadonty is also observed in this fossil ape (Andrews & Wood, unpublished data).

- Flat occlusal planes on molars – this is generally associated with thick enamel mainly because the thick layer of enamel covering the dentine base of the crown tends to cover over much of the dentine relief. Also, as wear progresses on thick enamelled crowns, dentine is not exposed until late so that enamel–dentine interfaces, which produce cutting edges on occlusal surfaces, are not exposed for much of the duration of tooth wear. On both the mandibular and maxillary toothrows, *Ankarapithecus* has flat occlusal planes with little cusp relief in contrast to species of *Proconsul* that combine thicker enamel with high cusp relief. In the former, advanced wear fails to expose dentine at the cusp tips, whereas in the latter, dentine is exposed even at early stages of wear despite the thick enamel (Beynon *et al.*, 1998).
- Poorly developed shearing crests – shearing crests may be present on teeth with high enamel relief or on thin-enamelled teeth where the enamel–dentine interface produced by occlusal wear produces sharp edges for cutting tough food items (Kay, 1977; Ungar, 1996). Again, this feature is usually associated with thick enamel and flat occlusal planes, and although the shearing surfaces have not yet been measured on *Ankarapithecus*, inspection of the occlusal surfaces suggest poor development of shearing crests on this fossil ape. Work in the future on microwear should provide more evidence on the degree of shear (see next section).
- Reduced canines – projecting interlocking canine teeth may restrict lateral and anteroposterior movement of the jaws during chewing, and this may in turn restrict the grinding capacity of species with this feature (Jolly, 1970). *Ankarapithecus* has relatively small canines, both in the presumed male (MTA2125) and female (AS500) individuals, with the C/M1 ratio ranging from 73 to 109 for the upper teeth and from 78 to 96 in the lower teeth (comparing the type mandible with the mandible associated with the skull – AS500). This ratio, however, may be due to the megadont molars rather than the small canines, and when canine size is compared with body weight, for example derived from orbit dimensions, the canine size is seen not to be reduced (Kelley, see Chapter 12). In functional terms, however, the canines are small relative to the molars, which affects the grinding capacity and degree of movement of the jaws relative to one

another.

- Robust mandibular body – the body of the mandible is extremely robust in the mandible associated with the skull of *Ankarapithecus* (the type mandible MTA2253 lacks a mandibular body). The thickness/depth proportion at the level of M_1 is 51%, but at M_3 this value is 102%, so that the body is actually broader than it is tall. This is exceptional even for the sivapithecines and kenyapithecines, which are thick-enamelled apes having extremely robust mandibles, and the combination of these characters in these three groups, as well as in others such as the australopithecines, indicates a strong correlation between thick enamel, megadonty and robust mandibles. It should be noted, however, that *Graecopithecus*, which has extremely thick enamel on its molars (Andrews & Martin, 1991) has relatively deep mandibular bodies. Contrary to some recent speculation, experimental studies have shown that strains associated with mastication are present both in alveolar bone and in regions well removed from the loaded alveolus (Daegling & Hylander, 1997). Although the differential effects of a deep narrow mandibular body compared with a shallow robust body have not been analysed, it seems likely that both may be as effective as the other in countering chewing stresses.
- High floor of maxillary sinus, producing a robust alveolar process of maxilla – this character was described for *Kenyapithecus* (Andrews, 1971), and it parallels the robust mandibular body for exactly the same reasons. The alveolar process in *Ankarapithecus* is not as robust as that of *Kenyapithecus* or *Sivapithecus*.
- Anteriorly directed zygoma with strong lateral flare – one of the striking features of both the original 1980 face (Andrews & Tekkaya, 1980) and the new skull of *Ankarapithecus* (Alpagut *et al.*, 1996) is the anterior and lateral expansion of the zygomatic region of the skull. The root of the zygomatic is not anteriorly placed, being situated above M^2 as commonly seen on fossil apes, but instead of sloping posteriorly it flares out laterally in the plane of the second molar or even anterior to it in MTA 2125, so that the lower face is both broad and relatively flat. This has the effect of moving the insertion of the masseter muscles anteriorly and laterally, so that they produce both a more anteriorly directed pull on the mandible and is also laterally directed.
- Strong zygomatic process – this also relates to the action of the masseter, which inserts along the length of the zygomatic process. The great robusticity of the process in *Ankarapithecus* indicates the powerful development of this muscle, and this in turn indicates powerful chewing in *Ankarapithecus*. The lateral displacement of the masseter origin resulting from the lateral flare of the zygomatic suggests that it would have had a

lateral component in its action which would have produced lateral movement during chewing, also indicated (above) by the robusticity of the mandibular body.

- Flat frontal process of the zygomatic bone – in *Ankarapithecus* the frontal process of the zygomatic is flattened as well as being anteriorly directed. This development appears to be related to the lateral flare and anterior displacement of the zygomatic region, so that as the zygomatic region expands anterolaterally, the frontal process of the zygomatic is similarly rotated to face anteriorly. As a result, it is not an independent character. This same combination of characters is seen in *Sivapithecus* and *Pongo*, and it is also characteristic of the Madagascar lemur *Hadropithecus*, which has extreme development of the zygomatic region comparable with the robust australopithecines.
- Enlarged temporal fossa – the lateral expansion of the zygoma displaces the anterior portion of the zygomatic process laterally in *Ankarapithecus*, leaving a large space between it and the temporal bone of the cranial vault. This is the temporal fossa, through which the temporalis muscle passes from its origin on the side of the cranium to its insertion on the coronoid process of the mandible. The large size of the temporal fossa indicates the powerful development of the temporalis muscle, and in parallel with the development of the masseter this indicates the presence in this fossil ape of powerful chewing muscles. These in turn relate to the robust mandible and maxilla, the thick enamel on the teeth with flat occlusal surfaces and the large size of the teeth to indicate an extremely powerful masticatory apparatus as an adaptation to a diet of hard objects.
- Stresses from the cheek teeth pass up through posterior part of the face through the zygomatic region, passing on either side of the orbits to the supra-orbital tori. The size and direction of the zygoma in *Ankarapithecus* may therefore be related in part to dissipating chewing stresses. Similarly, the presence of well developed brow ridges in *Ankarapithecus* may also be related to the same thing and also to the requirements of forces generated during food preparation.

Adaptations of anterior teeth for food preparation

Linked with the adaptations relating to mastication, a more or less independent set of characters relate to food preparation before and during ingestion. These concern the anterior teeth, particularly the incisors, and the anterior parts of the mandible and premaxilla and their relationships with the rest of the skull.

- I^1 broad and low crowned – large central incisors are generally linked with dietary adaptations for processing relatively large items of food, particularly fruits (Kay & Hylander, 1978). All living apes have relatively large incisors for fruit processing, as do also the more frugivorous cercopithecoid monkeys, and one of the characteristics of all fossil apes throughout the Miocene is that they also have relatively large incisors. Two fossil groups stand out from this general scheme of things: one is the early Miocene genus *Rangwapithecus*, and its relatively narrow incisors in combination with greater development of molar shearing crests linked with its more folivorous diet (Kay, 1977); and the other group includes *Ankarapithecus* and *Graecopithecus*, which both have extremely broad (mesiodistally) upper central incisors relative to smaller and more caniniform lateral incisors. The upper central incisors of the latter two genera are also relatively low crowned, and in this they differ from the more usual frugivore pattern as seen, for instance, in the orang-utan and other recent and fossil apes, and this suggests use of these teeth for an exceptional form of food preparation.

- Heavy wear on incisors – in addition to the unusual size of the upper central incisors, these teeth are also very heavily worn on almost all specimens of both *Ankarapithecus* and *Graecopithecus*. This heavy wear appears early in comparison with molar tooth wear, and it suggests a behaviour involving heavy use of incisors in the preparation of hard or resistant foods.
- Robust premaxilla – the dimensions of the premaxilla appear to be closely related to the size of the incisors, particularly the central incisors. Both *Ankarapithecus* and *Graecopithecus* have elongated premaxillae, approaching the size of the premaxillae in the orang-utan and chimpanzee, both of which also have large incisors. The gorilla, by contrast, has smaller incisors and a shorter and less robust premaxilla (Ward & Pilbeam, 1983), and the gorilla proportions are generally seen in most fossil apes. To a certain extent, therefore, the premaxillary proportion is an adaptation more related to incisor size and food preparation, so that the morphological consequences of enlarged robust premaxillae, such as reduction in size of the incisive fossa and narrowing of the incisive canal, are not independent phylogenetic characters (see below).
- In parallel with the elongation and robusticity of the premaxilla, the symphysis of the mandible is relatively deep, with an inferior transverse torus in *Ankarapithecus*. The depth of the symphysis is 40 mm compared with posterior depths of the mandibular body at M2–3 of 24–25 mm, and the breadth/depth ratio at the symphysis is 0.39. This compares with the symphyseal depth on the type specimen of *Ankarapithecus meteai* (probably a male individual) of 49.2 mm. The greater depth gives greater strength to the symphysis during both longitudinal torsion and lateral movement, particularly with the internal torus at the base of the symphysis, and

this character state is similar to the condition in *Graecopithecus* and *Sivapithecus*.

Comparative morphology and phylogenetic relationships

It now remains to consider some of the phylogenetic implications of the new skull from Turkey, but in doing this it is necessary to look closely at the characters being used to determine relationships of *Ankarapithecus meteai* with other hominoid primates.

Body size estimates based on orbital breadth (not as accurate as orbital height) range from 23.2 kg for the 1995 skull (AS 500) to 35.7 kg for MTA 2125 (orbital height could not be measured on this specimen), and these estimates compare with values based on M2 size of 35.2 kg for the former and 48.1 kg for the latter. The tooth predictors of body size give values from 9 to 13 kg higher than the orbital predictors. This can be interpreted as indicating that the teeth in *Ankarapithecus* are relatively large compared with cranial dimensions.

On the basis of the teeth Simons & Pilbeam (1965) and Andrews & Tekkaya (1980) assigned the earlier Sinap specimens to *Sivapithecus*. Except for the incisors, there is close similarity with the Asian genus, but what this similarity amounts to, in fact, is the presence in both of molars with thick enamel, low cusp relief, reduced shearing, small canines but megadont molars. As shown above, these are not independent characters but are part of a single functional complex related to powerful chewing. Most of the same characters are also present in *Kenyapithecus* (Benefit & McCrossin, 1995), almost certainly denoting similar adaptation; and except for the canine dimensions the same characters are also known for *Afropithecus* and *Graecopithecus*. This is therefore a common set of characters that may be evidence of relationship, as was suggested by Martin (1985) when he postulated that thick enamel and its correlates were ancestral for the great ape and human clade (Hominidae). It is equally likely, however, that these characters evolved independently in some of these fossil taxa as a response to similar environmental conditions and in particular similar dietary specialisations. This supposes that these characters change readily, through simple genetic mechanisms, in response to similar adaptive conditions (Begun & Güleç, 1997).

Work on microwear and shearing crest analysis (Teaford, 1994; Ungar, 1996) supports the similarity in function of the thick-enamelled fossil taxa with living primates with hard fruit diets, but it tends to distinguish some of the taxa that apparently have thinner enamel – some species of the genera

Proconsul and *Dryopithecus* for example. These have microwear and shearing crest patterns indicating soft fruit diets (Ungar, 1996), and the example of *Rangwapithecus* was mentioned above, with its shearing crest pattern indicating a more folivorous diet (Kay, 1977). This range of fossil morphologies appears to represent at least three grades of dietary adaptation (Kay & Ungar, 1998), and if additional evidence is considered, for example the more strongly folivorous adaptations of *Oreopithecus* (Ungar, 1996), additional grades of evolution could be added, but it is doubtful if these grades reflect any meaningful phylogenetic distinction.

Regarding the incisors, it has been noted that all fossil and recent hominoids have relatively large and spatulate upper incisors. The living apes are variable in this regard, with gorillas and hylobatids having incisor proportions similar to some fossil apes, but chimpanzees have greatly enlarged upper incisors, both central and lateral, while orang-utans have enlarged central incisors but small laterals. No fossil ape matches the chimpanzee condition, but there is some similarity between the fossil genera *Ankarapithecus, Graecopithecus* and *Sivapithecus* and the orang-utan. The first two genera have the additional incisive features of low-crowned upper incisors with extremely heavy early wear, and it is possible that the enlarged incisors in the fossil genera are not homologous with the enlargement in *Sivapithecus* and *Pongo*. In this case, the apparent similarity could be due to functional convergence resulting from similarity in food preparation and similar diets.

When the naso-alveolar region of the lower face is considered, similar conclusions can be drawn, in this case possibly affecting *Sivapithecus* as well. The extended premaxilla in chimpanzees and orang-utans has been related to increased incisor size and heavier use of the incisors in food preparation, and it is possible that the apparent synapomorphies of this region linking *Sivapithecus* with the orang-utan could be due to functional convergence resulting from similar diets. The possibility of homoplasy is even stronger for *Ankarapithecus* and *Graecopithecus*, for not only is their premaxillary morphology less specialised in the direction of the orang-utan, but their incisor morphology has also been seen to differ in some respects. In terms of evolutionary grade, it must be concluded that *Sivapithecus* and *Pongo* should be grouped together, but *Ankarapithecus* and *Graecopithecus* constitute a separate grade, with the heavily worn low-crowned incisors of the latter two fossil apes indicating a slightly different diet from the former two. *Dryopithecus*, with its *Proconsul*-like incisors, forms a separate grade, but again it is highly questionable that this similarity indicates a phylogenetic link between it and *Proconsul*. It is interesting that the two incisor morphologies of *Griphopithecus*, one similar to *Proconsul* and

Equatorius (Ward *et al.*, 1999), with a strong lingual pillar, and the other more similar to *Kenyapithecus wickeri*, are to some extent intermediate between these grades, so that there is some indication of a phylogenetic link, but it is difficult to separate this off from the clear functional link.

Considering finally the face, a number of characters of *Ankarapithecus* can be seen to be related to the development of the masticatory apparatus rather than being phylogenetically distinct characters. The expansion of the zygomatic region, for instance, has been shown to be highly distinct in *Ankarapithecus* (Andrews & Tekkaya, 1980; Alpagut *et al.*, 1996), but if it is related solely to the expansion and direction of the muscles of mastication, in turn related to the adaptation for processing a diet of hard food objects, then the zygomatic morphology cannot be used as an independent character. Similarly, the development of the supra-orbital tori may also be related to the stresses set up during heavy mastication and cannot be used phylogenetically as an independent character. The narrow nose may be related to the same functional system, since the stresses derived from heavy use of the anterior teeth pass up through the face on either side of the nose, although in this case there is good evidence that this is a primitive retention in *Ankarapithecus*, since all earlier fossil apes retain this character in contrast to their less developed zygomatic regions and no brow ridges. The narrow interorbital distance appears not to be related to these other characters, since chewing stresses also pass up past the orbits to the brow ridges, and where the zygomatic region lateral to the orbits is robust, the narrow pillar of bone internal to the orbits is relatively gracile. This may relate to the conformation of the brow ridges themselves, since they are strongly built laterally and only cover the lateral two-thirds of the tops of the orbits.

The interorbital region has sharply keeled nasals and anteriorly oriented lacrimals. Its minimum interorbital breadth of 11.9 mm is moderately narrow like that of *Pongo* and *Sivapithecus*, but unlike these hominoids, a small indentation to the left of centre near glabella reveals the presence of an invasive frontal sinus that extends laterally to the point of contact with the temporal line. This is surprising given the moderately narrow interorbital breadth that leaves limited room for ethmoidal air cells, and future preparation will be needed to reveal the extent of the sinus in this specimen. In its present condition it is not clear if the sinus is continuous across glabella. The maxillary sinus is also invasive and extends into the root of the zygoma.

The conclusion from this brief summary is that the complex of cranial characters found individually in various configurations in other roughly contemporaneous hominoid species are found together in *Ankarapithecus*. One set of characters links it with forms found to the east, *Sivapithecus* and *Pongo*, as recognized originally (Andrews & Cronin, 1982). A second set of characters links it with forms found to the west, *Graecopithecus* and *Dryopithecus*. A full consideration of the exact relationships among what is now a diverse and better represented sample of late Miocene hominoids will require a clear and detailed definition of characters and, perhaps more importantly, character complexes that are used in phylogenetic analysis.

Many of these characters in *Ankarapithecus* are present also in *Graecopithecus*, although as presently reconstructed, the nose appears to be broader on this fossil ape (de Bonis & Koufos, 1993). The broad interorbital distance in this fossil, like that of *Afropithecus* and *Dryopithecus*, would appear to be structurally stronger to counter-act chewing forces, and this condition contrasts with that of *Ankarapithecus* and *Sivapithecus*. As in the latter genera, the brow ridges are more strongly developed laterally than medially in *Graecopithecus* and *Dryopithecus*. Microwear analysis of *Graecopithecus* showed high numbers of pits on the occlusal surfaces of its molars, and this supports the morphological evidence for it being adapted for a diet of hard objects (Ungar, 1996). Its incisor microwear was also found to be interesting, in that the abundance of laterally oriented scratches indicates that *Graecopithecus* may have stripped food laterally across its mouth during food preparation (Ungar, 1996).

The case of *Dryopithecus* is interesting, since that fossil genus has many of the same characters as *Ankarapithecus* in the zygomatic, orbital and supraorbital regions, including the flat anteriorly directed frontal region of the zygomatic bone (Moyà-Solà & Köhler, 1994, 1996), but it appears to lack the megadont thick enamelled teeth and robust jaws of *Ankarapithecus* and *Graecopithecus*. To this extent, *Dryopithecus* appears intermediate between the early Miocene apes like *Proconsul*, with relatively lightly built jaws, and lacking the megadont molars and enlarged incisors present in the later Miocene thick-enamelled forms, but sharing some of the facial characters with the latter. This could indicate that the apparent functional link between the facial and masticatory characters is not as strong as suggested here, that the facial characters were acquired independently in *Dryopithecus* from *Ankarapithecus*, and that they relate to a different functional adaptation in *Dryopithecus*. It is not considered likely that these facial characters were primitive for Hominoidea any more than the thick-enamelled, robust-jawed morphology was primitive.

Conclusions

The new skull from the Sinap Formation is assigned to *Ankarapithecus meteai* following Begun & Güleç (1995) and Alpagut *et al.* (1996). In this very brief analysis of dental and facial characters, it has been seen that most characters are inter-related, so that their separate use in phylogenetic analysis is highly questionable. Comparisons with other fossil apes based on the descriptions in the earlier publication show that *Ankarapithecus* shares some similarities with the orang-utan, but it also shares a number of characters with European Miocene hominoids, notably *Graecopithecus* and to a lesser extent *Dryopithecus*. The polarity of change of many of the characters used for assessing the phylogenetic relationships of fossil hominoids is not clear at present.

Two aspects of functional morphology of *Ankarapithecus meteai* provide an indication of its dietary behaviour. Characters of teeth and face indicate its adaptation for crushing and grinding hard food objects. Large thick-enamelled teeth with strong lateral buttressing of the jaws provide the basis for this conclusion, and this is supported by the great development and lateral expansion of the zygomatic region and the powerful development of the masticatory muscles (inferred from the robusticity of the zygomatic process and the size of the temporal fossa). Strong compressive forces would have been generated by the large masticatory muscles, and a lateral component to chewing is also indicated by the great breadth of the mandibular body.

In the second aspect of functional morphology, the large size of the incisors, the massive premaxilla and the deep symphysis of the mandible indicate an adaptation to extensive preparation of large food items before and during ingestion. Again this would have involved both resistance to compressive forces at the symphysis and premaxilla and to lateral forces across the anterior teeth.

Many of the characters found to be inter-related in the functional analysis may be considered to be inappropriate for phylogenetic analysis. A detailed phylogenetic analysis is needed to clarify the relationships of *Ankarapithecus*, and until that is done it remains uncertain if it belongs with the pongine clade, as suggested previously (Andrews & Cronin, 1982; Martin & Andrews, 1984), or if it is a stem hominid, as suggested more recently (Alpagut *et al.*, 1996).

Acknowledgements

We are grateful to J. Agustí for comments on an earlier version of this paper and to D. Begun for helpful comments on the present version.

References

Aiello, L. & Wood, B. (1994). Cranial variables as predictors of hominine body mass. *Am. J. Phys. Anthrop.* **95**: 409–26.

Alpagut, B., Andrews, P. Fortelius, M., Kappelman, J., Temizsoy, I., Celebi, H. & Lindsay, W. (1996). A new specimen of *Ankarapithecus meteai* from the Sinap Formation of central Anatolia. *Nature* **382**: 349–51.

Andrews, P. (1971). *Ramapithecus wickeri* mandible from Fort Ternan, Kenya. *Nature* **231**: 192–4.

Andrews, P. (1992). Evolution and environment in the Hominoidea. *Nature* **360**: 641–6.

Andrews, P. & Cronin, J. (1982). The relationships of *Sivapithecus* and *Ramapithecus* and the evolution of the orang utan. *Nature* **297**: 541–6.

Andrews, P. & Martin, L. (1991). Hominoid dietary evolution. *Phil. Trans. R. Soc. Lond.* **334**: 199–209.

Andrews, P. & Tekkaya, I. (1980). A revision of the Miocene hominoid *Sivapithecus meteai*. *Palaeontology* **23**: 85–95.

Begun, D. (1992). Phyletic diversity and locomotion in primitive European hominids. *Am. J. Phys. Anthrop.* **87**: 311–40.

Begun, D. & Güleç, E. 1995. Restoration and reinterpretation of the facial specimen attributed to *Sivapithecus meteai* from Kayincak (Yassioren), central Anatolia, Turkey. *Am. J. Phys. Anthrop.* **20** (Suppl.): 26.

Begun, D. & Güleç, E. (1998). Restoration of the type and palate of *Ankarapithecus meteai*: taxonomic and phylogenetic implications. *Am. J. Phys. Anthrop.* **105**: 279–314.

Benefit, B. & McCrossin, M. (1995). Miocene hominoids and hominid origins. *Ann. Rev. Anthrop.* **24**: 237–56.

Beynon, A. D., Dean, M. C., Leakey, M. G., Reid, D. J. & Walker, A. (1998). Comparative dental development of *Proconsul* teeth from Rusinga Island, Kenya. *J. Hum. Evol.* **35**: 163–209.

Daegling, D. J. & Hylander, W. L. (1997). Occlusal forces and mandibular bone strain: is the primate jaw 'overdesigned'? *J. Hum. Evol.* **33**: 705–17.

de Bonis, L. & Koufos, G. (1993). The face and mandible of *Ouranopithecus macedoniensis*: description of new specimens and comparisons. *J. Hum. Evol.* **24**: 469–91.

Jolly, C. J. (1970). The seed eaters: a new model of hominid differentiation based on a baboon analogy. *Man* **5**: 5–28.

Kappelman, J., Sen, S., Fortelius, M., Duncan, A., Alpagut, B., Crabaugh, J., Gentry, A., McDowell, F. & Viranta, S. (1996). Paleomagnetic reversal stratigraphy and chronology of the Miocene Sinap Formation of Central Turkey. In Bernor, R., Fahlbusch, V. & Rietschel, S. (ed.), *Evolution of Neogene Continental Biotopes in Central Europe and the Eastern Mediterranean*, pp. 78–95. New York: Columbia University Press.

Kay, R. F. (1977). Diet of early Miocene African hominoids. *Nature* **268**: 628–30.

Kay, R. F. (1981). The nutcrackers – a new theory of the adaptations of the Ramapithecinae. *Am J. Phys. Anthrop.* **55**: 141–51.

Kay, R. F. & Hylander, W. L. (1978). The dental structure of mammalian folivores with special reference to Primates and Phalangeroidea (Marsupalia). In *The Ecology of Arboreal Folivores*, ed. G. G. Montgomery, pp. 173–91. Washington DC: Smithsonian Institution Press.

Martin, L. (1985). Significance of enamel thickness in hominoid evolution. *Nature* **314**: 260–3.

Martin, L. & Andrews, P. (1984). The phyletic position of *Graecopithecus freybergi* Koenigswald. *Courier Forschung. Senckenberg* **69**: 25–40.

Moyà-Solà, S. & Köhler, M. (1994). New partial cranium of *Dryopithecus* Lartet 1863 (Hominoidea, Primates) from the upper Miocene of Can Llobateres, Barcelona, Spain. *J. Hum. Evol.* **29**: 101–39.

Moyà-Solà, S. & Köhler, M. (1996). A *Dryopithecus* skeleton and the origins of great ape locomotion. *Nature* **379**, 156–9.

Ozansoy, F. (1955). Sur les gisements continenteaux et les mammiferes du Neogen et du Villafranchian d'Ankara (Turquie). *C. R. Acad. Sci. Paris* **240**: 992–4.

Ozansoy, F. (1957). Faunes des mammiferes des Tertiare de Turquie et leurs revision stratigraphique. *Bull. Miner. Res. Explor., Ankara* **49**: 29–48.

Ozansoy, F. (1965). Etudes des gisements continenteaux et des mammiferes du Censozoique de Turquie. *Mem. Geol. Soc. France* **44**: 1–92.

Pilbeam, D. (1982). New hominoid skull material from the Miocene of Pakistan. *Nature* **295**: 232–4.

Pilbeam, D., Rose, M., Barry, J. & Shah, I. (1990). New *Sivapithecus* humeri in Pakistan and the relationship of *Sivapithecus* and *Pongo*. *Nature* **348**: 237–9.

Simons, E. L. & Pilbeam, D. (1965). Preliminary revision of the Dryopithecinae. *Folia Primatol.* **3**: 81–152.

Teaford, M. (1994). Dental microwear and dental function. *Evol. Anthrop.* **3**: 17–30.

Ungar, P. (1996). Dental microwear of European Miocene catarrhines: evidence for diets and tooth use. *J. Hum. Evol.* **31**: 335–66.

Ward, S. C., Brown, B., Hill, A., Kelley, J. & Downs, A. (1999). *Equatorius*: a new hominoid genus from the middle Miocene of Kenya. *Science* **285**: 1382–6.

Ward, S. & Pilbeam, D. (1983). Maxillofacial morphology of Miocene Hominoidea from Africa and IndoPakistan. In *New Interpretations of Ape and Human Ancestry*, ed. R. Ciochon & R. Corruccini, pp. 211–38. New York: Plenum Press.

Wood, B. A. (1981). Tooth size and shape and their relevance to studies of hominid evolution. *Phil. Trans. R. Soc. Lond.* **292**: 65–76.

10 African and Eurasian Miocene hominoids and the origins of the Hominidae

D. R. Begun

Introduction

Among the more controversial hypotheses regarding hominoid evolutionary relationships is a Eurasian origin of the euhominoidea[1] (Begun, 1997; Begun *et al.*, 1997). This hypothesis, while supported by or at least consistent with other analyses based on independent lines of evidence (Gebo *et al.*, 1997; McCrossin & Benefit, 1997; Stewart & Disotell, 1998), runs counter to most other recent proposals of hominoid evolutionary history (Andrews, 1992; Moyà-Solà & Köhler, 1995; Moyà-Solà *et al.*, 1999; Harrison & Rook, 1997; Ishida & Pickford, 1997; McCrossin & Benefit, 1997; Andrews & Bernor, 1999). A key aspect of the Eurasian origin hypothesis is the phyletic position of early and middle Miocene hominoids relative to the euhominoidea (see Figure 10.1). Begun, *et al.* (1997) investigated hominoid relationships using a data base consisting of 240 characters in 13 taxa. The results of this research indicate that *Kenyapithecus*[2] is the sister taxon to the clade that includes all living hominoids and all Eurasian Miocene hominids[3] (Begun *et al.*, 1997, p. 404, figure 1). *Proconsul* and *Afropithecus* are even further removed from this clade.

As suggested elsewhere (Begun, 1996; Begun *et al.*, 1997; Stewart & Disotell, 1998; Heizmann & Begun, unpublished data) there are several palaeobiogeographic implications of this hypothesis. Euhominoids first appear, and may have originated in Eurasia following a single dispersal event from Africa towards the end of the early Miocene (Heizmann & Begun, unpublished data). All currently known fossil Eurasian hominoids are cladistically hominid, being more closely related to great apes and humans than to hylobatids (Figure 10.1 and Table 10.1). The Eurasian euhominoid radiation is extensive, and lasts from about 16.5 Ma to 7 Ma (Andrews, *et al.*,

[1] Euhominoids include all living hominoids and fossils hominoids more closely related to living great apes than *Hylobates* (Figure 10.1) (Begun *et al.*, 1997). Because there is no currently accepted taxonomic level between superfamily (e.g. Hominoidea) and family (e.g. Hominidae) euhominoids remains an informal taxonomic designation, despite its justification on cladistic grounds.

[2] '*Kenyapithecus*' in Begun *et al.* (1997) is a paraphyletic amalgam of samples that share only primitive characters. More than one taxon is represented (see below, and Abel, 1902; Andrews, 1992; Harrison, 1992; Ward *et al.*, 1999; Begun, 2000).

[3] Hominids in this paper refers to the great apes and humans and their fossil relatives – i.e. all taxa more closely related to humans than to hylobatids.

Table 10.1. *A classification of the Hominoidea*

'Eohominoids'		
	Proconsul[a]	
	Xenopithecus	
	Rangwapithecus	
	Nyanzapithecus	
	Turkanapithecus	
	Heliopithecus	
	Mabokopithecus	
	Otavipithecus	
	Samburupithecus	
	Nacholapithecus	
Griphopithecidae		
	Griphopithecus	
	Afropithecus	
'Euhominoids'		
Hylobatidae		
	Hylobates	
Hominidae		
	Oreopithecus[b]	
	Lufengpithecus	
	Gigantopithecus	
	Kenyapithecus	
Homininae		
Dryopithecini		
	Dryopithecus	
	Ouranopithecus	
Hominini[c]		
	Gorilla	
	Unnamed *Pan–Homo* clade:	
		Ardipithecus
		Praeanthropus
		Australopithecus
		Paraustralopithecus
		Paranthropus
		Homo
		Pan
Pongini		
	Pongo	
	Ankarapithecus	
	Sivapithecus	

[a] The relations among the genera of primitive hominoids included here are unclear. For this reason no attempt has been made to classify them in suprageneric nomena below the superfamily. The small fossil catarrhines *Limnopithecus, Dendropithecus, Micropithecus, Kalepithecus* and *Simiolus* cannot be definitively assigned to the Hominoidea (Harrison, 1988; Rose *et al.*, 1992). In addition, *Proconsul*, including fossils known primarily from two temporally distinct site clusters (Tinderet and Kisingiri) is probably paraphyletic and includes at least two genera (Begun, unpublished data).

[b] These four genera of early hominids cannot be definitively placed among the tribes of hominines recognised here, and so are left for the moment unsorted beyond the family level.

[c] Because of uncertainties concerning relations among the Hominini they are not sorted beyond the tribal level. However, since the weight of the morphological and molecular evidence strongly favours a *Pan–Homo* clade (including this work), this level in the hierarchy of the Hominoidea is depicted graphically here.

1996). A pattern of hominid extinction events trends temporally and geographically from late Vallesian extinctions in the northwestern range of the euhominoids (Europe) to the late Turolian or Messinian extinctions in the southeast (South Asia, China), with one interesting exception (the late Turolian Italian hominid *Oreopithecus*). By the late Turolian, with the exception of *Oreopithecus*, all hominoids living north of the Tropic of Cancer are extinct. During this extinction phase, one lineage, most closely related to European fossil great apes (*Dryopithecus* and *Ouranopithecus*), appears and disperses into Africa, to give rise to the clade of the African apes and humans (Begun, 1996; Begun, *et al.*, 1997). These palaeobiogeographic implications emerge simply as a consideration of the smallest number of dispersal events necessary to explain the observed distribution of hominoid taxa and their hypothesised cladistic relations (Begun *et al.*, 1997; Stewart & Disotell, 1998). They also account for the dearth of fossil evidence of hominoids of modern aspect in Africa prior to about 5 Ma (Leakey *et al.*, 1996). While other biogeographic hypotheses for Eurasian Miocene hominoids have been proposed, they are not considered here because they are not consistent with the phylogenetic hypothesis supported here (e.g. Agustí *et al.*, 1996; Andrews *et al.*, 1996; Andrews & Bernor, 1999; Ishida & Pickford, 1997; Moyà-Solà *et al.*, 1999; Pickford, 1991; Senut, 1991).

While there are many controversial aspects to these proposals, the logic of these palaeobiogeographic reconstructions, a parsimonious perspective on hominoid dispersal events, requires that hylobatids are more closely related to Eurasian fossil hominoids and to living hominids than are early and middle Miocene African hominoids. If this were not the case, as suggested by a number of hypotheses (Andrews, 1992, and Chapter 9; Moyà-Solà & Köhler, 1995; Moyà-Solà *et al.*, 1999; Harrison & Rook, 1997; Ishida & Pickford, 1997; Rae, 1997; McCrossin *et al.*, 1998) then a separate, earlier dispersal event would have to be posited to account for the Asian distribution of hylobatids. If *Kenyapithecus* and/or *Griphopithecus* are considered to be in a clade that also includes *Dryopithecus*, among other taxa (Andrews, 1992), then four separate dispersal events are required. One each for hylobatids and *Dryopithecus* and, since Andrews (1992) suggests that *Graecopithecus* (or *Ouranopithecus*) is ancestral to the African apes and humans, and to *Samburupithecus* (Ishida & Pickford, 1997; see below), a dispersal into Europe, a Eurasian origin for this clade, and a dispersal into Africa from Europe, are all implied as well.

In this chapter I examine in more detail the evidence of the phyletic position of a number of African early and middle Miocene hominoids, particularly with regard to hylobatids and Eurasian fossil hominids. Recent suggestions of alternative placements of both early and middle Miocene hominoids and hylobatids are tested for their relative consistency with the largest published data base of fossil hominoid morphology (Begun *et al.*, 1997), which has been enlarged here following recent discoveries (Ward *et al.*, 1999, and unpublished data). A brief consideration of the phyletic position of *Samburupithecus* (Ishida & Pickford, 1997) is also given.

New data

Since Begun *et al.* (1997), new discoveries and analysis require some modification to the database used to generate the hypothesis described above. The most significant change is the recognition of two genera within the hypodigm of *Kenyapithecus*. *Kenyapithecus* in this chapter refers to *Kenyapithecus wickeri*, known only from Fort Ternan, Kenya, though possibly also from Paşalar, Turkey (Ward *et al.*, 1999; see below). *Griphopithecus* here refers to all other samples usually attributed to *Kenyapithecus africanus*, as well as the middle Miocene thickly enamelled hominoids from Germany, Slovakia and Turkey. Most African samples have been placed in the new taxa *Equatorius africanus* and *Nacholapithecus kerioi* (Ward *et al.*, 1999; Nakatsukasa *et al.*, 2000). *Equatorius africanus* is a junior synonym of *Griphopithecus africanus* (Begun, 2000), while *Nacholapithecus kerioi* is distinctive enough for the new genus. It is not discussed here. Main changes to the data base include the addition of seven new postcranial characters, six from the proximal ulna and one humeral character, and the coding of many character states from Begun *et al.* (1997) previously not known for many taxa (Table 10.1). In total, 161 character states codes have been added to the data matrix. Using the methods described in Begun *et al.* (1997) six most parsimonious cladograms were recovered from these data, and these are represented by the two cladograms in Figure 10.1.

Afropithecus and *Proconsul*

In two comparatively recent analyses, *Afropithecus* has been placed among the great apes to the exclusion of *Hylobates*, in contrast to Figure 10.1 (Andrews, 1992; Rae, 1997). Most other works in which the morphology of *Afropithecus turkanensis* is considered in some detail either fail to place the taxon in an unambiguous phyletic position (e.g. Leakey & Walker, 1997) or consider it to be more primitive than hylobatids (Gebo, *et al.*, 1997; Rose, 1997; C. Ward, 1997; Begun *et al.*, 1997). Unlike most of the other phylogenetic analyses considered here, Rae (1997) provides an explicit

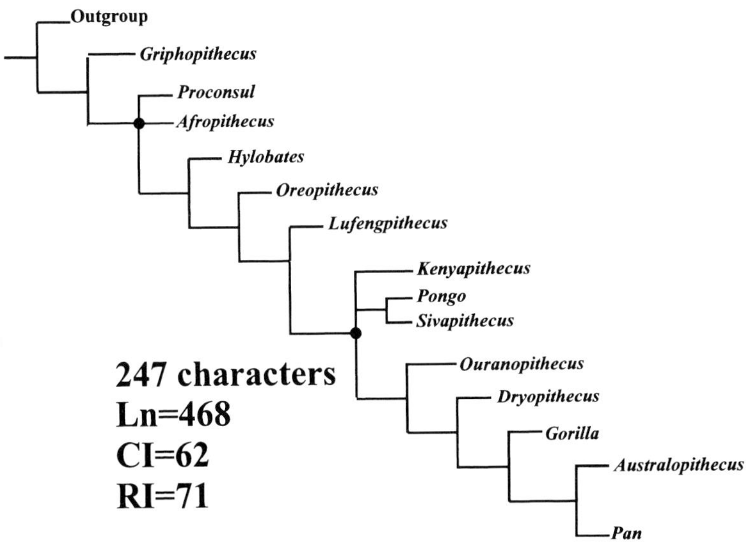

[Figure 10.1]
A cladogram representing a consensus among the six most parsimonious cladograms generated from data modified from Begun *et al.* (1997). Unresolved clades originate at the marked nodes. *Afropithecus* and *Proconsul* alternate between being sister clades or each others immediate outgroup (three cladograms). *Kenyapithecus* alternates between being the sister clade to *Sivapithecus* and the sister to all hominids after *Lufengpithecus* (two cladograms). New characters added to this matrix appear in Table 10.2. Note that the common ancestors of all clades between *Hylobates* and the African ape-human clade are Eurasian. (See text for discussion.)

definition of each character and a data matrix of character states, which facilitates critical review. In his cladogram (Rae, 1997, p. 67, figure 2), colobines are paraphyletic, suggesting some problems in resolving correct relationships with these data. Nevertheless, according to Rae (1997), *Afropithecus* shares with a few other early Miocene hominoids, *Pan* and *Pongo*, a tall naso-alveolar process compared to *Hylobates*, most other early Miocene hominoids and cercopithecoids. *Afropithecus* and *Proconsul nyanzae* also share with living great apes to the exclusion of *Hylobates* a wider anterior palate and vertical zygoma.

Because Rae (1997) treats character states as ordered, his proposed clades are linked by synapomorphies that are actually different character states in his data matrix. For example, naso-alveolar height, said to be a synapomorphy of *Afropithecus* and great apes, is actually character state 3 (tall) in *Afropithecus* and character state 4 (taller) in *Pan* and *Pongo* (Rae, 1997: p. 68, figure 3). The assumption here is that it is more likely for a naso-alveolar process to go from tall to taller than from short or intermediate to directly to tall, or from tall to short. One problem is that naso-alveolar length is a continuous character, and no specific criteria are given to determine the difference between character states such as intermediate (2), tall (3) and taller (4). Even if these characters were truly distinct, the assumption of a necessary transition from short to tall naso-alveolar processes is violated by the results of Rae's own cladogram. For example, the outgroup in Rae (1997) has a naso-alveolar process of intermediate length. Three separate clades forming an obviously paraphyletic polychotomy each have different character states for this character. From an intermediate ancestral morphotype, *Colobus* has evolved a tall process, *Presbytis* a short process and the paraphyletic group, including *Nasalis* and three cercopithecine genera, has three different character states, the ancestral morphotype most likely being a short process. In this example, naso-alveolar processes can go from intermediate to tall or short without preference. The same phenomenon has been noted for other taxa elsewhere (Begun, 1994). If a morphocline polarity apriority is falsified in one part of the cladogram, there is no reason to believe that it will hold any more validity in another part of the cladogram. Therefore, there is no evidence in Rae's data that naso-alveolar height links *Afropithecus* to living great apes. Both the other two character states said to link *Proconsul* and *Afropithecus* with living great apes are plagued by similar difficulties.

In the final analysis, Rae (1997) really presents a character tree rather than a phylogeny, given the very small number of characters (fewer than there are taxa) and restricted anatomical coverage. It is more interesting in showing the complexity of the evolution of the face in hominoids than it is useful in resolving relationships among hominoids. A phyletic link between early Miocene hominoids and living great apes to the exclusion of hylobatids is not convincingly demonstrated.

In Andrews (1992), *Afropithecus*, along with a 'heterogeneous association of taxa' (Andrews, 1992: p. 642) including *Dryopithecus* and '*Kenyapithecus*', are distinguished from *Proconsul* and hylobatids by premolar enlargement, squarish molars with low cusps, flat surfaces and reduced cingulum, reduced superior transverse torus and robust, low crowned canines. However, as noted by Andrews (1992), the premolars of *Afropithecus* are further enlarged, so that unless a morphocline polarity is assumed this feature cannot, a priori, be considered a synapomorphy with the great apes (see above). In addition, there is extensive overlap and no clear distinction in shape between the molars of *Afropithecus* and *Proconsul*. The cingula are also well developed and similar to those of *Proconsul*, particularly the later occurring Kisingeri *Proconsul*, and the same can be said for molar occlusal topology, which is not particularly flat, even in worn specimens of *Afropithecus* (e.g. the type cranium). However, it is worth noting that in all of these molar features the other taxa of Andrews' 'Dryopithecinae' share the same character state seen in living great apes. This offers strong support for the suggestion made by Andrews (1992) that this 'clade' is indeed paraphyletic.

The superior transverse torus is very variable in hominoids, and there are certainly specimens of *Proconsul* (e.g. KNM RU 2087 and 7290) with a torus of similar development to that of *Afropithecus* (Brown, 1989). Hylobatids are also variable in superior transverse torus development, and, like other living apes, they tend to have shelf-like inferior transverse tori (simian shelves). With regard to canine morphology, *Afropithecus* has very robust and low crowned canines compared to *Dryopithecus* and *Kenyapithecus sensu stricto* (Fort Ternan). Robust, low crowned canines then do not distinguish this clade from hylobatids and *Proconsul*, nor do they associate *Afropithecus* with *Dryopithecus* and '*Kenyapithecus*'. In fact, aside from their low crowns, *Afropithecus* canine morphology is more similar to that seen in other early Miocene hominoids (*Morotopithecus*, *Proconsul*) than it is to later occurring genera (*Dryopithecus*, *Lufengpithecus*, *Ouranopithecus*, *Sivapithecus*, *Oreopithecus*, *Kenyapithecus wickeri* and living hominoids), which all tend to have more labiolingually compressed canines.

The unusual dental characters of *Afropithecus* (very enlarged and flared premolars and low crowned, robust canines) are explained by Leakey & Walker (1997) as aspects of an adaptation to sclerocarp feeding. In other features of its facial and postcranial anatomy *Afropithecus* is quite primitive, according to Leakey & Walker (1997), an observation consistent with the position of *Afropithecus* in Figure 10.1.

When all the available data are considered together, 41 (10%) more steps are required to place *Proconsul* in the great ape clade to the exclusion of *Hylobates*. This is the minimum length for a tree with a *Proconsul*–great ape clade, and is essentially the same tree presented by Rae (1997). The minimum length for a tree consistent with Andrews (1992), with a polychotomy including *Dryopithecus*, '*Kenyapithecus*', and *Afropithecus* as the sister clade to the great apes, is 31 steps longer than the most parsimonious tree. Some of the reduction in homoplasy in this tree compared to the tree consistent with Rae (1997) is related to the placement of *Proconsul* out of the euhominoid clade, in contrast to Rae (1997). Part of the reduction in homoplasy, however, is also due to an artefact of failing to resolve relations at the polychotomous node. Placing *Afropithecus* in the great ape clade to the exclusion of hylobatids requires additional steps, or homoplasy (parallel evolution of character states in hylobatids and great apes) in the following areas and amounts: five carpal; three metacarpal; four phalangeal; one tibial; two talar; one cuboid; one metatarsal; three humeral; one ulnar; and nine craniodental. Although weighted toward postcranial characters, numerous homoplasies are required in the cranium and dentition of hylobatids as well, and in such functionally apparently unrelated characters as infra-orbital position, inclination of the zygomatic arch, greater palatine foramen position, molar cingula and molar size ratio. The distribution of these apparent homoplasies across several functional and structural anatomical complexes suggests that they can not reasonably be explained by a single complex functional or developmental model. In the absence of an obvious cause, it must be considered unlikely that hylobatids and great apes (including fossil great apes) would have acquired all of these characters independently. Including *Proconsul* in the great ape clade only increases the amount of homoplasy between hylobatids and great apes.

Kenyapithecus and *Griphopithecus*

More than is the case for *Afropithecus* and *Proconsul*, the disputed phyletic positions of *Kenyapithecus* and *Griphopithecus* present the greatest challenge to the Eurasian euhominoid origin hypothesis. Most recent analyses that have included hylobatids and *Kenyapithecus* and/or *Griphopithecus* have concluded that the fossil apes are more closely related to great apes than are hylobatids (Andrews, 1992, and Chapter 9; Moyà-Solà & Köhler, 1995; Harrison & Rook, 1997; Ishida & Pickford, 1997; Rae, 1997; McCrossin *et al.*, 1998). Several recent analyses, however, conclude that hylobatids are the sister clade to the great apes to the exclusion of *Kenyapithecus* and *Griphopithecus* (Begun *et al.*, 1997; Gebo *et al.*, 1997; McCrossin & Benefit, 1997; Rose, 1997; Ward, 1997), and this is essentially the pattern of relations represented in Figure 10.1.

As noted above, the data matrix used to generate the cladograms in Figure 10.1 divides '*Kenyapithecus*' into *Kenyapithecus* and *Griphopithecus*. *Griphopithecus* retains its primitive position in the cladogram, and in fact appears to be less closely related to great apes than is *Proconsul*. However, *Kenyapithecus* is found to be considerably more closely related to great apes than *Griphopithecus*, falling not only within the Euhominoidea but also in an unresolved polychotomy with the *Sivapithecus*/*Pongo* clade and Afro-European hominids (Figure 10.1 and Table 10.1). This finding appears to run counter to the conclusions presented in Begun *et al.* (1997) on which most of the conclusions presented here and in other works are based (Begun, 1996; Stewart & Disotell, 1998). However, new results from the recent revision of *Kenyapithecus* serve to explain this discrepancy (Ward *et al.*, 1999; see below).

The greatest difficulty in interpreting *Kenyapithecus sensu stricto* (*K. wickeri*) is the very small size of the hypodigm. Only a few fragmentary gnathic specimens and isolated teeth are known, as well as a distal portion of a humerus and one partial phalanx. *Griphopithecus* was until recently not much better known or published, but now two partial skeletons and a well preserved mandible add considerably to the previously published craniodental material (e.g. Le Gros Clark & Leakey, 1951; Pickford, 1982; McCrossin & Benefit, 1993, 1997; Rose *et al.*, 1996; Ward *et al.*, 1999). The maxillary morphology remains obscure, with only a few fragmentary published specimens preserving portions of the postcanine tooth rows. While the postcranial collection is much improved, with the exception of Rose *et al.* (1996) it has been described only briefly (McCrossin, 1997; McCrossin & Benefit, 1997; McCrossin *et al.*, 1998; Ward *et al.*, 1999) and sometimes in internally contradictory fashion (e.g. McCrossin 1997; McCrossin & Benefit, 1997; McCrossin, *et al.*, 1998; see below). However, because many other fossil hominoids preserve few postcrania (*Ouranopithecus, Lufengpithecus*) many comparisons to *Kenyapithecus* and *Griphopithecus* are mostly limited to gnathodental remains, and this explains some of the topology of the cladograms in Figure 10.2. Because *Kenyapithecus* is not known from anatomical regions that have proven to be the most phyletically informative, such as the anterior palate, and also because no non-gnathic cranial fossils of *Kenyapithecus* have been described, attempts to place *Kenyapithecus* phyletically relative to other fossil hominoids have proven difficult. One way to illustrate this problem is represented in Figure 10.2.

This cladogram is based on the data used to generate Figure 10.1, with all characters not known for *Kenyapithecus* removed from the data matrix – 53 characters are known for *Kenyapithecus sensu stricto*, mostly from the jaws and teeth. The cladogram in Figure 10.2 clusters taxa along what are most probably functional rather than phyletic lines, illustrating the danger of using data from limited anatomical regions to generate cladograms. While some of the relations represented here are intriguing, such as the placement of *Ouranopithecus* as sister to *Australopithecus*, with *Pan* as the outgroup, other relations are very unlikely. There is no evidence for a sister group relationship between *Pongo* and Afro- European hominids that excludes *Sivapithecus*. Although the preponderance of the evidence indicates that *Sivapithecus* and *Pongo* share a sister taxon relationship (Andrews & Cronin, 1982; Pilbeam, 1982; Ward & Pilbeam, 1983; Brown & Ward, 1988; Andrews, 1992; Ward, 1997), even among those who hold this proposed relationship in doubt (e.g. Pilbeam, 1996; McCrossin & Benefit, 1997), none would suggest that *Sivapithecus* is the outgroup to the group that includes Pongo, *Oreopithecus, Dryopithecus, Lufengpithecus* and African hominids.

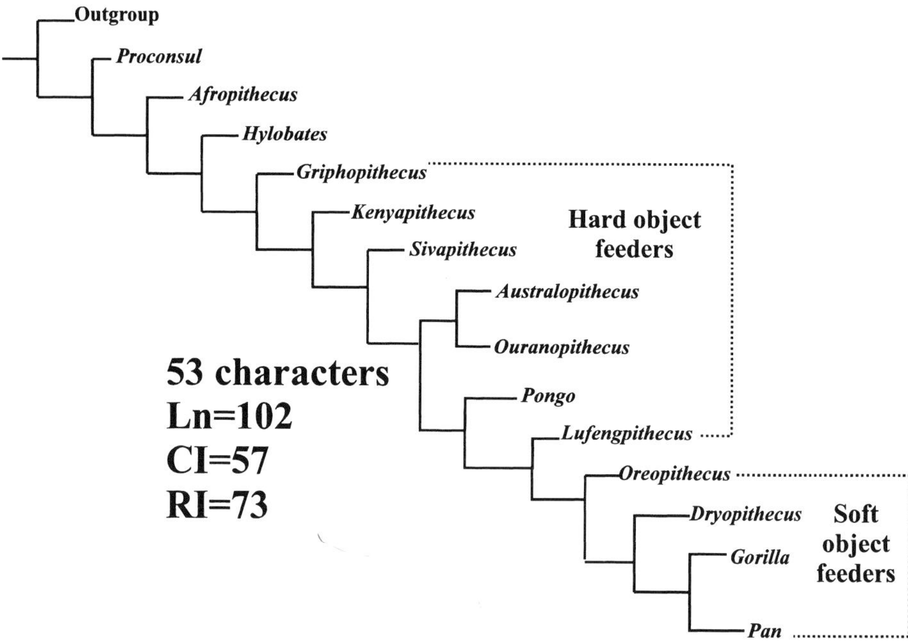

[Figure 10.2]
One of the five most parsimonious trees based on a data matrix in which all characters not known for *Kenyapithecus* are removed. These results are very strongly driven by character bias in gnathic morphology, and are not likely to represent the true phylogeny. Only 53 characters are included in the data matrix, as opposed to the 247 used to generate Figure 10.1. Among these, only 13 are postcranial and all of these are from the distal humerus. Differences amongst the most parsimonious cladograms involve positions amongst *Australopithecus, Ouranopithecus, Lufengpithecus* and *Pongo* and between *Dryopithecus* and each of the African apes. (See text for discussion.)

The placement of *Ouranopithecus* and *Australopithecus* as sister taxa is probably the result of character bias. No postcranial data are available for *Ouranopithecus*, so the link to *Australopithecus* is based exclusively on gnathic characters preserved in *Kenyapithecus*. There are many similarities between *Ouranopithecus* and *Australopithecus* in gnathic anatomy (Bonis & Koufos, 1993, 1997 and this volume; Koufos, 1993, 1995), but when all the data are considered these emerge as homoplasies (Begun, 1995; Begun & Kordos, 1997; Figure 10.1). All of these homoplasies are plausibly related to a single functional complex, mastication (see Kay & Ungar, 1997 for a similar conclusion based on other, related data). Convergence in the morphology of *Ouranopithecus* and *Australopithecus*, which can be demonstrated on the basis of parsimony, can be explained on the basis of functional demands on hominid masticatory structures related to the development of powerful chewing. The same features (large molars with thick enamel and flat occlusal surfaces, robust mandibles, structurally reinforced anterior faces, reduced canines) appear in varying degrees independently in many primates (e.g. *Gigantopithecus, Sivapithecus, Ankarapithecus, Kenyapithecus, Griphopithecus, Afropithecus, Theropithecus, Cebus, Hadropithecus, Megaladapis*). *Sivapithecus, Australopithecus, Ouranopithecus, Lufengpithecus,* and *Pongo* are all 'united' in this cladogram by primitive features of the dentition all functionally related to hard object feeding. *Kenyapithecus* is missing from this group. *Kenyapithecus* retains a number of primitive features of the dentition, such as molar proportions, not found in other hominids. *Oreopithecus, Dryopithecus* and living African apes are 'united' by primitive features of that 'clade' all related to soft object feeding – i.e. folivory and/or frugivory.

McCrossin & Benefit (1997) have argued on the basis of their analysis of what they refer to *Kenyapithecus,* which is referred here to *Griphopithecus* and in Ward *et al.* (1999) to *Equatorius,* that this genus is primitive relative to *Hylobates* and great apes. This interpretation is consistent with other recent analyses of *Griphopithecus* (Rose *et al.*, 1996; Ward & Brown, 1986; Nakatsukasa *et al.*, 1998; Ward *et al.*, 1999; Güleç & Begun unpublished data; P. Andrews, personal communication). This is also consistent with the interpretation depicted in Figure 10.1, although McCrossin & Benefit (1997) go on to conclude that no Miocene hominoid is more closely related to any living great ape than is *Hylobates.* Their bold statement that *Sivapithecus* and *Dryopithecus* postcrania share no more derived characters with living hominoids than does the postcrania of *Proconsul* is difficult to explain or evaluate, being in stark contrast to virtually every single published analysis of these fossils (Begun, 1992; summarised recently in Rose, 1997 and S. Ward, 1997). Even more curious is the more recent assertion by the same authors that *Griphopithecus* does have modern hominoid characters (McCrossin, 1997) and is the sister taxon to the African apes and humans (McCrossin *et al.*, 1998). This most recent interpretation is based on a newly discovered humerus that appears to be quite distinct from the previously described humerus of *Griphopithecus* from Maboko (Kenya), as well as a fragmentary radius. The distal radius, which is only briefly described in an abstract (McCrossin *et al.*, 1998), supposedly adds to evidence from other postcranial specimens and from the craniodental sample, that now suggest to these authors that *Kenyapithecus* (*Griphopithecus*) is a member of the African ape and human clade, although all but the radius was previously interpreted by the same authors as evidence that *Kenyapithecus* (*Griphopithecus*) is primitive relative to all living hominoids (Benefit & McCrossin, 1995; McCrossin & Benefit, 1997). Even when the few characters of the radius that can be gleaned from the brief description in McCrossin, *et al.* (1998) are added to the data matrix used to generate Figure 10.1, the hypothesis that *Griphopithecus* is a member of the African ape and human clade is extremely unparsimonious, requiring 26 more steps in a diversity of anatomical complexes throughout the cranial, dental and postcranial regions. In summary, it seems clear overall that until the anatomy of *Griphopithecus* is adequately described and subjected to a rigorous phylogenetic analysis the most reasonable interpretation is the consensus view that *Griphopithecus* is primitive relative to *Hylobates* and great apes.

The jury is still out on *Kenyapithecus.* Although a more definitive interpretation of the evolutionary relations of *Kenyapithecus* will have to await the recovery of more specimens, even if *Kenyapithecus* is a hominid, the recent suggestion by Ward *et al.* (1999) that a *Kenyapithecus*-like taxon is present at Paşalar, Turkey is consistent with the Eurasian origins hypothesis. Paşalar is about 15 Ma (Bernor & Tobien, 1990), or between 1 and 2.5 Ma older than Fort Ternan. If both localities contain *Kenyapithecus* or an eventual *Kenyapithecus* clade (e.g. two sister genera), then this would imply that *Kenyapithecus* at Fort Ternan is a migrant or a descendant of a migrant from Western Asia. Intriguingly, this is very similar to the scenario proposed by Tong & Jaeger (1993) to explain the rodents of Eurasian character present in the fauna from Fort Ternan.

Samburupithecus

It has been recently suggested that the earliest identifiable member of the African ape and human clade is *Samburupithecus* from Samburu Hills, Kenya (Ishida & Pickford, 1997). The evidence to support this is slight. None of the features proposed by Ishida & Pickford (1997) link *Samburupithecus* to australopithecines to the exclusion of all Miocene hominoids, and as noted by these authors, the specimen retains many primitive features found in early and middle Miocene hominoids. A low root of the zygomatic processes, a strongly inclined nasal aperture edge, the retention of molar cingula and thick enamel with high dentine relief are all characters found in *Proconsul, Afropithecus* and *Morotopithecus* but not in fossil and living great apes. *Samburupithecus* is in fact most likely to be a terminal member of one lineage of early and/or middle Miocene hominoid (Table 10.1). It has unusual dental characters of the type that are often found in terminal lineages with long, distinct evolutionary histories. For example, unusual occlusal morphology and dental proportions are found in *Oreopithecus, Gigantopithecus* and *Paranthropus* among hominoids, but also in Malagasy subfossil prosimians, *Daubentonia*, and the late surviving Miocene omomyid *Ekgmowechashala*. The distinctive features of *Samburupithecus* that give it a superficial resemblance to *Gorilla* may be related to the effects of long branch attraction in a late Miocene taxon rooted in the early or middle Miocene primitive hominoid radiation.

Because it is only known from a single maxillary fragment, it is impossible to be confident in determining the phylogenetic relations of *Samburupithecus*. But, even if additional evidence were found to support inclusion of *Samburupithecus* in the African ape and human clade, there is no necessary impact on the Eurasian origins hypothesis. Nothing in the anatomy of *Samburupithecus* links it phyletically to *Griphopithecus* or *Kenyapithecus* to the exclusion of other middle and late Miocene hominoids. A phylogenetic relationship between *Samburupithecus* and African apes would in fact imply a close relationship between *Samburupithecus* and Eurasian fossil great apes (consistent in part with Andrews, 1992), which all share apomorphies of the great apes not found in *Kenyapithecus* or *Griphopithecus*. At approximately 9.5 Ma, *Samburupithecus* could have been the result of the dispersal event coinciding with the extinction of various western European great apes (see below). However, at this point the bulk of the evidence suggests that *Samburupithecus* is not a hominid.

Miocene hominoid paleobiogeography revisited

As noted above, a challenge to the Eurasian origin hypothesis is the phyletic position of African Miocene taxa, particularly *Kenyapithecus* and *Griphopithecus*, but also *Afropithecus*, *Proconsul* and *Samburupithecus*. A review of the evidence available to date indicates that there are no strong reasons to reject the most parsimonious hypothesis presented in Figure 10.1, keeping in mind that the position of *Kenyapithecus* in Figure 10.1 is very unclear, given the small and limited nature of the sample. It does seem reasonable to suggest that the euhominoids form a clade that excludes all exclusively early and middle Miocene hominoids, with the possible exception of *Kenyapithecus*. No Miocene hominoid from Africa prior to the latest Miocene or Pliocene is convincingly attributable to this clade. While absence of evidence is not evidence of absence, hypotheses have to be based on currently available evidence. This evidence suggests that euhominoids were absent from Africa between 12.5 and 14 Ma and the appearance of the African ape and human clade. The evidence further suggests that even if *Kenyapithecus* is a hominid, it is excluded from the clade that includes *Dryopithecus*, *Ouranopithecus*, living great apes and fossil and living humans. Many researchers feel compelled to present *ad hoc* explanations such as sampling bias to account for the absence of hominoids where they are expected to be, such as East Africa in the late Miocene, while the palaeobiogeographic implications of known temporal and geographic distributions of other mammals are more readily accepted at face value. The observation that numerous types of latest Miocene to modern African mammals are absent from early or middle Miocene African faunas, ranging from proboscideans to rodents, is routinely accepted as evidence for a non-African origin of these lineages (e.g. Barry *et al.*, 1985; Thomas, 1985; Tong *et al.*, 1993; Bernor *et al.*, 1996; Gentry & Heizmann, 1996; Leakey *et al.*, 1996). Anticipated data cannot be used to falsify existing hypotheses. In addition, although it is frequently claimed that very few localities are known in this time period in Africa, this is simply not the case. Many localities are known between 5 and 12 Ma from Africa and Arabia (Bou Hanifia, Sahabi, Wadi Natrun, Menacer, Oued Zra, Oued el Atteuch, Khendek-el-Ouaich, Bled Douarah, Douaria, El Hamma du Djerid, M'dilla, Amama 1 and 2, the Aït Kandoula localities, Nchorora, Ngorora, Nakali, Samburu, Lothagam, Lukeino, Middle Awash, Baringo Basin, Manonga Valley, Sinda Basin, Nkondo, Abu Dhabi, and Langebaanweg), and only three have yielded scrappy hominoid remains (Thomas *et al.*, 1982; Benammi *et al.*, 1996; Leakey *et al.*, 1996). Lukeino and Lothagam are younger than 6 Ma (Leakey *et al.*, 1996) and the Samburu specimen, discussed above, is not diagnostically hominid. Many of these localities have strong Eurasian faunal affinities, and many sample forested or humid biotopes from which hominoids could be expected, if they were present (Thomas *et al.*, 1982; Leakey *et al.*, 1996; P. Andrews, personal communication). These localities and their contained fauna have been used to support various hypotheses of faunal exchanges between Eurasia and Africa in the late Miocene (e.g. Thomas *et al.*, 1982; Barry *et al.*, 1985; Thomas, 1985; Tong *et al.*, 1993; Bernor *et al.*, 1996; Benammi *et al.*, 1996; Gentry & Heizmann, 1996; Leakey *et al.*, 1996). Despite intensive survey and analysis, hominids are simply not among them, most probably because they had not arrived in Africa until about 8–9 Ma and were not sufficiently abundant to be represented in the fossil record until after 6 Ma.

The phylogenetic hypothesis presented in Figure 10.1 is consistent with the following palaeobiogeographic scenario. Hominoids first appear in Europe approximately 16.5 Ma, at Engelsweis in Germany (Heizmann, *et al.*, 1996; Andrews *et al.*, 1996; Heizmann & Begun, unpublished data). The thickly enamelled, low dentine penetrance hominoid molar from Engelsweis has affinities with *Griphopithecus* but is too fragmentary to attribute to a taxon with confidence (Heizmann and Begun, unpublished data).

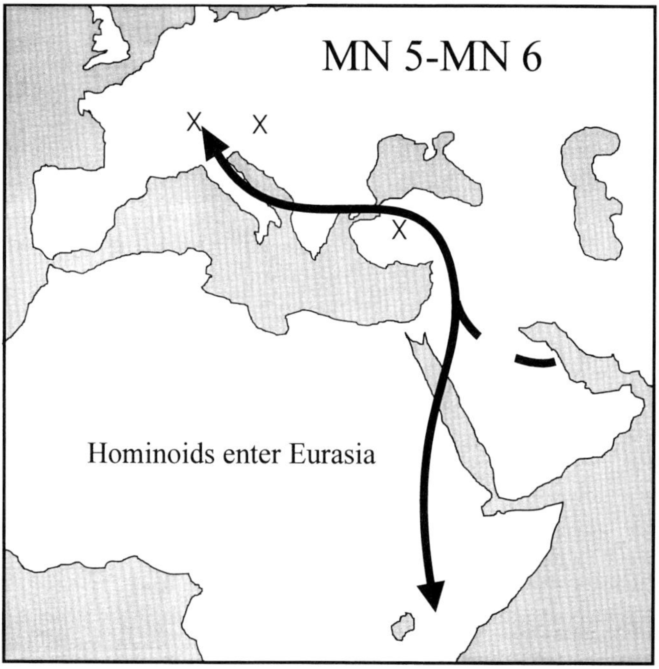

[Figure 10.3]
Early dispersal of hominoids into Eurasia. The ancestors of the earliest European thickly enamelled hominoids may have come directly from Africa (solid line) or they may have come from Saudi Arabia (dashed line), from which a similar but poorly preserved hominoid (*Heliopithecus*) is known. *Griphopithecus* is roughly contemporaneous in Eurasia and Africa and it is not clear in which direction the dispersal took place. (See text.)

Engelsweis pre-dates all African and Asian middle Miocene hominoid localities, and it is possible that the taxa from these sites, which are all thickly enamelled, originated in Europe (Figure 10.3).

There is a significant temporal gap between the earliest Eurasian and African middle Miocene hominoids (*c.* 16.5–15 Ma; Steininger *et al.*, 1996) and the first record of the euhominoids. This occurs by about 12.5 to 12 Ma both in Europe (*Dryopithecus* from St. Gaudens, St. Stefan, La Grive and Can Vila) and Asia (Chinji formation *Sivapithecus*) (Kappelman *et al.*, 1991; Steininger *et al.*, 1996). One sample that appears briefly to occupy this gap is Fort Ternan, dated to about 14–12.5 Ma (Pickford, 1986). *Kenyapithecus* from Fort Ternan and possibly Paşalar is derived relative to *Griphopithecus africanus* from Maboko (Ward *et al.*, 1999; Begun, 2000). As noted above, *Kenyapithecus wickeri* is poorly known but does retain primitive characters relative to euhominoids. It may be the sister taxon to the euhominoids, or a

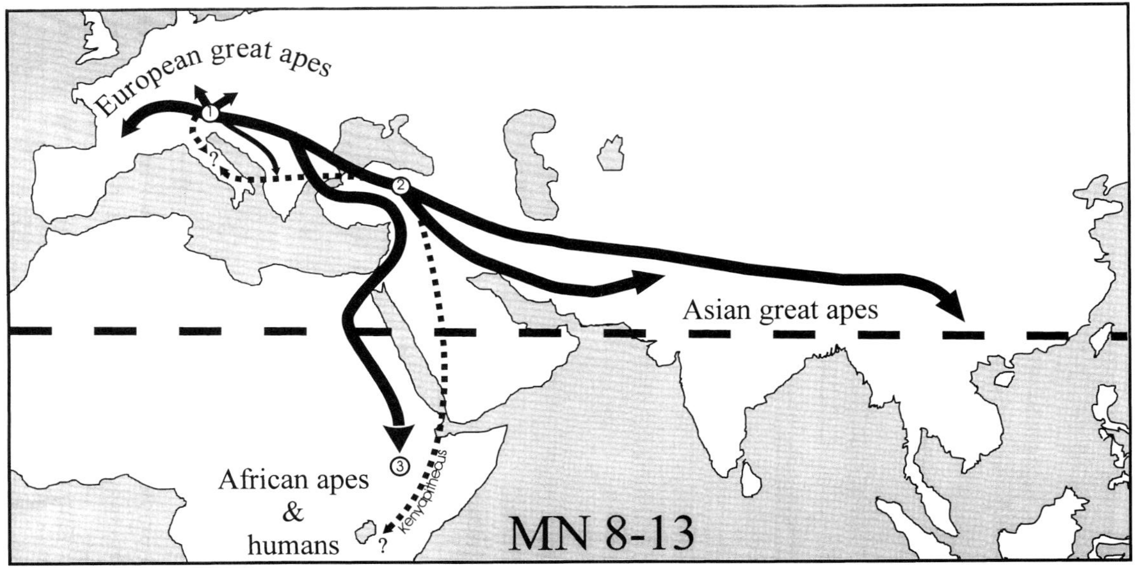

[Figure 10.4]
The evolution of Eurasian hominids. The first major division (1), between western and eastern Eurasian hominids may have occurred in Western Asia, as represented here, or in Central Europe. Both regions contain *Griphopithecus*, a potential ancestor of later occurring taxa. If the Asian great apes do not form a clade, as suggested in Figure 10.1, then at least two separate dispersal events into South and East Asia are required. Another major event (2) represents the diversification of European hominids into several species of *Dryopithecus* and *Ouranopithecus*. *Oreopithecus*, represented by the dashed lines leading to Italy, may be a part of this clade (Harrison & Rook, 1997) or it may represent a separate dispersal event (Begun, *et al*., 1997). If *Kenyapithecus* is also a member of the Eurasian hominid clade, as suggested in Figure 10.1, then it may have dispersed into Africa at this time as well. Ward *et al*. (1991) suggest that *Kenyapithecus sensu stricto* may be present in Turkey before it appears in Kenya. The third major event (3) represented here is the diversification of the African ape and human clade, the origins of which are very probably Eurasian, whether from a dryopithecini (Begun *et al*., 1997) or from some other Eurasian late Miocene hominine. By MN 13 all hominids living north of the Tropic of Cancer (thick dashed line) are extinct. The earliest extinctions tend to be in the northwest (Europe) and the latest extinctions tend to be more to the south and east (South Asia and East Asia). Hominids crossed back into the tropics at the end of the Miocene and evolved into *Pongo* in the east and African apes and humans in the west. (See text for discussion.)

Table 10.2. *New characters and character states used in this analysis*

			Taxa[a]									
Characters	**Character states**		**O**	**P**	**A**	**G**	**K**	**S**	**D**	**L**	**Or**	**Aust. & E.H.**
Olecranon process	projecting	reduced	0	0	0	0	–	–	–	–	1	1
Radial notch orient.	anterior	lateral	0	0	0	1	–	–	1	–	1	1
Radial notch shape	shallow	deep	0	0	0	0	–	–	1	–	1	1
Coranoid process	concave	beaked	0	0	0	0	–	–	1	–	1	1
Coranoid inclination	distal	anterior	0	0	0	0	–	–	1	–	1	1
P. p. p. arctic[b]	dorsal	proximal	0	0	0	0	–	0	1	1	0	1
Dist. hum. shaft[c]	deep	broad	0	0	0	0	1	1	1	–	1	1

[a] Taxonomic abbreviations, from left to right: **O**utgroup, ***P****roconsul*, ***A****fropithecus*, ***G****riphopithecus*, ***K****enyapithecus*, ***D****ryopithecus*, ***L****ufengpithecus*, ***O****reopithecus*, ***Aust****ralopithecus* and **e**xtant **h**ominoids. None of these characters have been published for *Ouranopithecus*.

[b] Proximal phalangeal proximal articular surface orientation.

[c] Distal humeral shaft cross section.

true hominid, and also it may have originated in Eurasia. Hylobatids are presumed to have evolved some time after the entrance of hominoids into Eurasia, the earliest current date for which is about 16.5 Ma, and before the divergence of Eurasian great apes, currently dated to about 13 Ma (Kappelman, *et al.*, 1991). Unfortunately, direct evidence of the evolutionary history of gibbons and siamangs continues to evade discovery.

The first known unambiguous hominids are Eurasian – *Dryopithecus* to the west and *Sivapithecus* to the east (or possibly a relation of *Kenyapithecus* in Turkey, see above). While a vicariance model may explain this distribution of these descendants from a more centrally restricted Central European–Western Asian ancestor (e.g. Andrews & Bernor, 1999), it is complicated by the fact that other clades without a direct relationship to *Dryopithecus* or *Sivapithecus* segregate geographically in a similar pattern. *Oreopithecus*, a late surviving and insular early hominid taxon from Italy, is the western counterpart to *Lufengpithecus*, a late surviving and isolated early hominid clade from southern China. In sum, the pattern of evolution of euhominoids in Eurasia is very complicated, which is reflected in the inability of researchers to reach a consensus regarding their phylogenetic relationships. Regardless of the precise pattern of relationship, this diversity indicates a true adaptive radiation, and is analogous to the radiation of platyrrhines upon their sudden introduction into South America in the early Oligocene. Relatively conservative early middle Miocene hominoids with one or more key innovations (?thickly enamelled, low dentine penetrance molars), radiate into an impressive array of forms with a level of diversity in body size, dietary strategy and positional behaviour equivalent to that of *all anthropoids* today.

Consistent with Figure 10.1 is the implication that the African ape and human clade evolved in Europe from a *Dryopithecus*-like ancestor. However, even if the relations among great apes represented in Figure 10.1 are inaccurate, the fact remains that *all* diagnostic Miocene hominids are Eurasian. Nothing in the anatomy of *Kenyapithecus* links that taxon to the African apes and humans to the exclusion of *Dryopithecus* and *Ouranopithecus*. The African ape and human clade may have evolved from an *Ouranopithecus*-like ancestor (Andrews, 1992), or even a *Sivapithecus*-like ancestor (Ward & Kimbel, 1983), but that ancestor appears to have been from Eurasia (Figure 10.4).

Acknowledgements

I am grateful to the organizers of the Hominoid Evolution and Climatic change in Europe workshop and to the editors of this volume for inviting me to attend the workshop and submit a chapter, and to the European Science Foundation and NSERC for providing financial support. Thank you also to Dr. Salvador Moyà-Solà for useful comments in his review of the first draft of this manuscript.

References

Abel, O. (1902). Zwie neue Menschenaffen aus den Leithakalkbildingen des wiener Bekkens. *S. Ber. Akad. Wiss. Wien, math-nat.* **1**: 1171–202.

Agustí, J., Köhler, M. & Moyà-Solà, S. (1996). Can Llobateres: the pattern and timing of the Vallesian hominoid radiation reconsidered. *J. Hum. Evol.* **31**: 143–55.

Andrews, P. (1992). Evolution and environment in the Hominoidea. *Nature*: **360**: 641–46.

Andrews, P. & Bernor, R. L. (1999) Vicariance biogeography and paleoecology of Eurasian Miocene hominoid primates. In Agusti, J., Rook, L. & Andrews, P. (eds.), *The Evolution of Neogene Terretrial Ecosystems in Europe*, pp. 454–87. Cambridge: Cambridge University Press.

Andrews, P. & Cronin, J. (1982). The relationships of *Sivapithecus* and *Ramapithecus* and the evolution of the orang-utan. *Nature*, **297**: 541–6.

Andrews, P., Harrison, T., Delson, E., Bernor, R. L. & Martin, L. (1996). Distribution and biochronology of European and Southwest Asian Miocene Catarrhines. In Bernor, R. L., Fahlbusch, V. & Mittmann, H.-W. (eds.), *The Evolution of Western Eurasian Neogene Mammal Faunas*, pp. 168–295. New York: Columbia University Press.

Barry, J. C., Johnson, N. M., Raza, S. M. & Jacobs, L. L. (1985). Neogene mammalian faunal change in southern Asia: correlations with climatic, tectonic, and eustatic events. *Geology* **13**: 637–40.

Begun, D. R. (1992). Phyletic diversity and locomotion in primitive European hominids. *Am. J. Phys. Anthrop.*: **87**: 311–40.

Begun, D. R. (1994). Relations among the great apes and humans: New interpretations based on the fossil great ape *Dryopithecus*. *Yrbk. Phys. Anthrop.* **37**: 11–63.

Begun, D. R. (1995). Late Miocene European orang-utans, gorillas, humans, or none of the above. *J. Hum. Evol.* **29**: 169–80.

Begun, D. R. (1996). Events in European hominoid evolution. *Europal.* **10**: 16–20.

Begun, D. R (1997). A Eurasian origin of the Hominidae. *Am. J. Phys. Anthrop.* **24** (Suppl.): 73–4.

Begun, D. R. (2000). Middle Miocene hominoid origins. *Science* **287**: 2375a.

Begun, D. R. & Kordos, L. (1997). Phyletic affinities and functional convergence in *Dryopithecus* and other Miocene and living hominids. In Begun, D. R., Ward, C. V. & Rose, M. D. (eds.), *Function, Phylogeny and Fossils: Miocene Hominoid Evolution and Adaptations*, pp. 291–316. New York: Plenum Press.

Begun, D. R., Ward, C. V. & Rose, M. D. (1997). Events in hominoid evolution. In Begun, D. R., Ward, C. V. & Rose, M. D. (eds.), *Function, Phylogeny and Fossils: Miocene Hominoid Evolution and Adaptations*, pp. 389–415. New York: Plenum Press.

Benammi, M., Calvo, M., Prévot, M. & Jaeger, J.-J. (1996). Magnetostratigraphy and paleontology of Aït Kandoula Basin (High Atlas, Morocco) and the African–European late Miocene terrestrial faunal exchanges. *Earth Planet. Sci. Lett.* **145**: 15–29.

Benefit, B. R. & McCrossin, M. L. (1995). Miocene hominoids and hominid origins. *Ann. Rev. Anthrop.* **24**: 237–56.

Bernor, R. L., Fahlbusch, V. Andrews, P. Bruijn, H. de Fortelius, M. & Rögl, F. (1996). The evolution of Western Eurasian Neogene mammal faunas: a chronologic, systematic, biogeographic, and paleoenvironmental synthesis. In Bernor, R. L., Fahlbusch, V. & Mittmann, H.-W. (eds.), *The Evolution of Western Eurasian Neogene Mammal Faunas*, pp. 449–69. New York: Columbia University Press.

Bernor R. L. & Tobien, H. (1990). The mammalian geochronology and biogeography of Paşalar (Middle Miocene, Turkey). *J. Hum. Evol.* **19**: 551–68.

Bonis, L. de. & Koufos, G. (1993). The face and mandible of *Ouranopithecus macedoniensis:* description of new specimens and comparisons. *J. Hum. Evol.* **24**: 469–91.

Bonis, L de. & Koufos, G. (1997). The phylogenetic and functional implications of *Ouranopithecus macedoniensis.* In Begun, D. R., Ward, C. V. & Rose, M. D. (eds.), *Function, Phylogeny and Fossils: Miocene Hominoid Origins and Adaptations.*, pp. 317–26. New York: Plenum Press.

Brown, B. (1989). The Mandibles of *Sivapithecus.* Ph.D. thesis, Kent State University.

Brown, B. & S. Ward (1988). Basicranial and facial topography in *Pongo* and *Sivapithecus.* In Schwartz, J. H. (ed.), *Orang-utan Biology*, pp. 247–60. New York: Oxford University Press.

Gebo, D. L., MacLatchy, L., Kityo, R., Deino, A., Kingston, J. & Pilbeam, D. (1997). A hominoid genus from the Early Miocene of Uganda. *Science* **276**: 401–4.

Gentry, A. & Heizmann, E. P. J. (1996). Miocene ruminants of the Central and Eastern Paratethys. In Bernor, R. L., Fahlbusch, V. & Mittmann, H.-W. (eds.), *The Evolution of Western Eurasian Neogene Mammal Faunas*, pp. 378–91. New York: Columbia University Press.

Harrison, T. (1988). A taxonomic revision of the small catarrhine primates from the early Miocene of East Africa. *Folia Primatol.* **50**: 59–108.

Harrison, T. (1992). A reassessment of the taxonomic and phylogenetic affinities of the fossil catarrhines from Fort Ternan, Kenya. *Primates* **33**: 501–22.

Harrison, T. & Rook, L. (1997). Enigmatic anthropoid or misunderstood ape: the phylogenetic status of *Oreopithecus bambolii* reconsidered. In Begun, D. R., Ward, C. V. & Rose, M. D. (eds.), *Function, Phylogeny and Fossils: Miocene Hominoid Origins and Adaptations*, pp. 327–62. New York: Plenum Press.

Heizmann, E., Duranthon, F. & Tassy, P. (1996). Miozäne Grossäugetiere. *Stuttg. Beitr. Naturk., Serie C* **39**: 1–60.

Ishida, H. & Pickford, M. (1997). A new late Miocene hominoid from Kenya: *Samburupithecus koptalami* gen. et sp. nov. *C. R. Acad. Sci. Ser. D* **325**: 823–9.

Kappelman, J., Kelley, J., Pilbeam, D., Sheikh, K. A., Ward, S., Anwar, M., Barry, J. C., Brown, B., Hake, P., Johnson, N. M., Raza, S. M. & Shah, S. M. I. (1991). The earliest occurrence of *Sivapithecus* from the middle Miocene Chinji Formation of Pakistan. *J. Hum. Evol.* **21**: 61–73.

Kay, R. F. & Ungar, P.S. (1997). Dental evidence for diet in some Miocene catarrhines with comments on the effects of phylogeny on the interpretation of adaptation. In Begun, D. R., Ward, C. V. & Rose, M. D. (eds.), *Function,*

Phylogeny and Fossils: Miocene Hominoid Origins and Adaptations, pp. 131–51. New York: Plenum Press.

Koufos, G. D. (1993). Mandible of *Ouranopithecus macedoniensis* (Hominidae, Primates) from a late Miocene locality of Macedonia. *Am. J. Phys. Anthrop.* **91**: 225–34.

Koufos, G. D. (1995). The first female maxilla of the hominoid *Ouranopithecus macedoniensis* from the late Miocene of Macedonia, Greece. *J. Hum. Evol.* **29**: 385–9.

Leakey, M. & Walker, A. (1997). *Afropithecus*: function and phylogeny. In Begun, D. R., Ward, C. V. & Rose, M. D. (eds.), *Function, Phylogeny and Fossils: Miocene Hominoid Evolution and Adaptations*, pp. 225–39. New York: Plenum Press.

Leakey, M. G., Feibel, C. S., Bernor, R. L., Harris, J. M., Cerling, T. E., Stewart, K. M., Storrs, G. W., Walker, A., Werdelin, L. & Winkler, A. J. (1996). Lothagam: a record of faunal change in the Late Miocene of East Africa. *J. Vert. Paleontol.* **16**: 556–70.

Le Gros Clark, W. E. & Leakey, L. S. B. (1951). The Miocene Hominoidea of East Africa. Brit. Nat. Hist. *Foss. Mamm. Afr.* **1**: 1–117.

McCrossin, M. L. (1997). New postcranial remains of *Kenyapithecus* and their implications for understanding the origins of hominoid terrestriality. *Am. J. Phys. Anthrop.*: **24** (Suppl.): 164.

McCrossin, M. L. & Benefit, B. R. (1993). Recently recovered *Kenyapithecus* mandible and its implications for great ape and human origins. *Proc. Natl. Acad. Sci. USA* **90**: 1962–6.

McCrossin, M. L. & Benefit, B. R. (1997). On the relationships and adaptations of *Kenyapithecus*, a large-bodied hominoid from the middle Miocene of eastern Africa. In Begun, D. R., Ward, C. V. & Rose, M. D. (eds.), *Function, Phylogeny and Fossils: Miocene Hominoid Origins and Adaptations*, pp. 241–67. New York: Plenum Press.

McCrossin, M. L., Benefit, B. R. & Gitau, S. N. (1998). Functional and phylogenetic analysis of the distal radius of *Kenyapithecus* with comments on the origin of the African ape and human clade. *Am. J. Phys. Anthrop.* **26** (Suppl.): 158–9.

Moyà-Solà, M. & Köhler, M. (1995). New partial cranium of *Dryopithecus* Lartet, 1863 (Hominoidea, Primates) from the upper Miocene of Can Llobateres, Barcelona, Spain. *J. Hum Evol.* **29**: 101–39.

Moyà-Solà, S., Köhler, M. & Alba, D. M. (1999). Primate evolution – in and out of Africa. *Curr. Biol.* **9**: R547–50.

Nakatsuksa, M., Ishida, H. Kunimatsu, Y. & Nakano, Y. (2000). Whole skeleton of the large hominoid from Nachola, northern Kenya. *Anthrop. Sci.* **108**: 92.

Pickford, M. (1982). New higher primate fossils from the Middle Miocene deposits at Majiwa and Kaloma, Western Kenya. *Am. J. Phys. Anthrop.* **58**: 1–19.

Pickford, M. (1986). Geochronology of the Hominoidea: a summary. In Else, J. G. & Lee, P. C. (eds.), *Primate Evolution*, pp. 123–8. Cambridge: Cambridge University Press.

Pickford, M. (1991). What caused the first steps towards the evolution of walkie-talkie primates. In *Origine(s) de la Bipédie chez les Hominidés*, ed. Cahiers de Paléontologie, pp. 275–93. Paris: CNRS.

Pilbeam, D. (1996). Genetic and morphological records of the Hominoidea and hominid origins: a synthesis. *Molec. Phylogenet. Evol.* **5**: 155–68.

Pilbeam, D. R. (1982). New hominoid skull material from the Miocene of Pakistan. *Nature* **295**: 232–4.

Rae, T. (1997). The early evolution of the hominoid face. In Begun, D. R., Ward, C. V. & Rose, M. D. (eds.), *Function, Phylogeny and Fossils: Miocene Hominoid Evolution and Adaptations*, pp. 59–77. New York: Plenum.

Rose, M. D. (1997). Functional and phylogenetic features of the forelimb in Miocene hominoids. In Begun, D. R., Ward, C. V. & Rose, M. D. (eds.), *Function, Phylogeny and Fossils: Miocene Hominoid Evolution and Adaptations*, pp. 79–100. New York: Plenum Press.

Rose, M. D., Nakano, Y. & Ishida, H. (1996). *Kenyapithecus* postcranial specimens from Nachola, Kenya. *African Study Monographs* **24**: 3–56.

Rose, M. D., Leakey, M. G., Leakey, R. E. F. & Walker, A. C. (1992). Postcranial specimens of *Simiolus enjiessi* and other primitive catarrhines from the early Miocene of Lake Turkana, Kenya. *J. Hum. Evol.* **22**: 171–237.

Senut, B. (1991). Origine(s) de la bipédie humaine: Approche Paléontologique. In Cahiers de Paléontologie, *Origin(s) de la bipédie chez les hominidés*. pp. 245–57. Paris: CNRS.

Steininger, F. F., Berggren, W. A., Kent, D. V., Bernor, R. L., Sen, S. & Agusti, J. (1996). Circum-Mediterranean Neogene (Miocene and Pliocene) marine-continental chronologic correlations of European mammal units. In Bernor, R. L., Fahlbusch, V. & Mittmann, H.-W. (eds.), *The Evolution of Western Eurasian Neogene Mammal Faunas*, pp. 7–46. New York: Columbia University Press.

Stewart, C.-B. & Disotell, T. R. (1998). Primate evolution – in and out of Africa. *Curr. Biol.* **8**: 582–8.

Thomas, H. (1985). The early and middle Miocene land connection of the Afro-Arabian plate and Asia: a major event for hominoid dispersal? In Delson, E. (ed.), *Ancestors: The Hard Evidence*, pp. 42–50. New York: Alan R. Liss.

Thomas H., Bernor R. & Jaeger J.-J. (1982). Origines du peuplement mammalien en Afrique du Nord durant le Miocène terminal. *Geobios* **15**: 283–97.

Tong, H. & Jaeger, J.-J. (1993). Muroid rodents from the middle Miocene Fort Ternan locality (Kenya) and their contribution to the phylogeny of muroids. *Palaeontographica* **229**: 51–73.

Ward, C. S. & Brown, B. (1986). The facial skeleton of *Sivapithecus indicus*. In Swindler, D. R. & Erwin, J. (eds.), *Comparative Primate Biology*, pp. 413–52. New York: Alan R. Liss.

Ward, C. V. (1997). Functional anatomy and phyletic implications of the hominoid trunk and hindlimb. In Begun, D. R., Ward, C. V. & Rose, M. D. (eds.), *Function, Phylogeny and Fossils: Miocene Hominoid Evolution and Adaptations*, pp. 101–30. New York: Plenum Press.

Ward, S. (1997). The taxonomy and phylogenetic relationships of *Sivapithecus* revisited. In Begun, D. R., Ward, C. V. &. Rose, M. D. (eds.), *Function, Phylogeny and Fossils: Miocene Hominid Origins and Adaptations*, pp. 269–90. New York: Plenum Press.

Ward, S., Brown, B., Hill, A., Kelley, J. & Downs W. (1999). *Equatorius*: a new hominoid genus from the middle Miocene of Kenya. *Science* **285**: 1382–6.

Ward, S. C. & Kimbel, W. H. (1983). Subnasal alveolar morphology and the systemic position of *Sivapithecus*. *Am. J. Phys. Anthrop.* **61**: 157–71.

Ward, S. C. & Pilbeam, D. R. (1983). Maxillofacial morphology of Miocene Hominoids from Africa and Indo-Pakistan. In Corruccini, R. L. & Ciochon, R. S. (eds.), *New Interpretations of Ape and Human Ancestry*, pp. 211–38. New York: Plenum Press.

11 Phylogenetic relationships of *Ouranopithecus macedoniensis* (Mammalia, Primates, Hominoidea, Hominidae) of the late Miocene deposits of Central Macedonia (Greece)

Louis de Bonis and George D. Koufos

The genus *Ouranopithecus* is based on the species *Ouranopithecus macedoniensis*, which has been described from specimens recovered in late Miocene layers of Macedonia, northern Greece. The first specimen, a mandible of a young individual, the type-specimen of the species, was first published under the name *Dryopithecus macedoniensis* (Bonis *et al.*, 1974). At that time, following a recent revision of Miocene hominoids (Simons & Pilbeam, 1965), all the Miocene genera, except '*Ramapithecus*', were included in the genus *Dryopithecus*. Some years later, the discovery of several jaws and isolated teeth led to naming the new genus *Ouranopithecus* (Bonis & Melentis, 1977). All the material was recovered from a locality named Ravin de la Pluie, 25 km west of the city of Thessaloniki, Macedonia, Greece. Later, a nearly complete face of *Ouranopithecus* was recovered from the locality of Xirochori close to the Ravin de la Pluie (Bonis *et al.*, 1990) and a mandible and a maxilla were found at Nikiti 1, 100 km east of Thessaloniki (Koufos, 1993; 1995). All the specimens are of similar geological age (Bonis & Koufos, 1999) and the three localities had similar palaeoenvironment. Nikiti 1, however, although closely resembling the open environments of Ravin de la Pluie and Xirochori 1 (Bonis *et al.*, 1992) by the presence of large numbers of giraffids, differs from them by the presence of boselaphines and the suid *Microstonyx* which indicate a more forested environment (Kostopoulos *et al.*, 1996; Kostopoulos & Koufos, 1996).

Dating and environment

The fossil vertebrate bearing localities of the Central Macedonian basin were discovered in 1915 (Arambourg & Piveteau, 1929; Bonis & Koufos, 1994). They are situated in a tectonic depression called the Axios (or Vardar) groove, between the Païkon mountains to the west and the Serbo-Macedonian mountains to the east (Mercier, 1973). The oldest Miocene layers are deposited on the Mesozoic or the marine Eocene basement and they belong to the Nea Messimbria Formation. This Formation consists of an anticlinal on a horst which is the western prolongation of the Khortiatis mountain. The thickness of the Nea Messimuria Formation is 1000 m on the north side and 600 m on the south side (Rapport IFP, 1967). There are conglomerates at the base and then red clays and red silts with gravels. This Formation is overlain by the greyish sandy marls of the Vathylakkos Formation (Bonis *et al.* 1987).The fauna of the Nea Messimuria Formation has been dated to late Vallesian age of mammals or MN 10 (Bonis *et al.*, 1987). The faunas of the other Formations of the basin have been dated to the Turolian age of mammals or MN 11, 12 and 13 (Bonis *et al.*, 1987). The palaeomagnetic data indicate the Chron 4A (about 9 Ma) for the Vallesian hominoid bearing localities (Kondopoulou *et al.*, 1992; Sen *et al.*, 2000).

When they are compared to recent African and Asian mammalian faunas, the Macedonian Vallesian faunas fit very well with the open environment faunas. They contain several genera and species of antelopes and giraffes as well as open area rodents, which all indicate a savannah-like environment even if the quality of grass was different from that of recent African savannah (Quade *et al.*, 1994).

Main characters of *Ouranopithecus*

Ouranopithecus is known through several upper and lower jaws, isolated teeth and a quite complete face. Its size can reach that of a female gorilla for the larger specimens but there is great sexual dimorphism (Bonis & Melentis 1978).

Upper dentition

The difference in size between the upper incisors I1 and I2 is large. The incisors index I1/I2 is smaller in humans, a little smaller in some australopithecines and African apes, but higher in *Pongo* and *Sivapithecus* (Bonis & Koufos, 1993). It seems difficult to use this index from a phylogenetic point of view. The relative homeomorphy of upper incisors might be considered to be a good synapomorphy linking chimpanzee and humans, but on the one hand, *Laccopithecus* (pliopithecid or hylobatid), from the locality of Lufeng (China), seems to have homeomorphic incisors (Wu & Pan, 1984; 1985), and on the other hand *Australopithecus afarensis* has heteromorphic upper incisors (Johanson *et al.*, 1982). The crown of I1 is quite low and shovel-shaped. The height difference between a fresh and a worn I1 is low because the wear is principally on the lingual face of the crown. When worn this face has a basal cusp which is pear-shaped. This pattern is very similar to that of *Gigantopithecus blacki* from the early Pleistocene of China (Wu, 1962) and it might indicate some link between both genera.

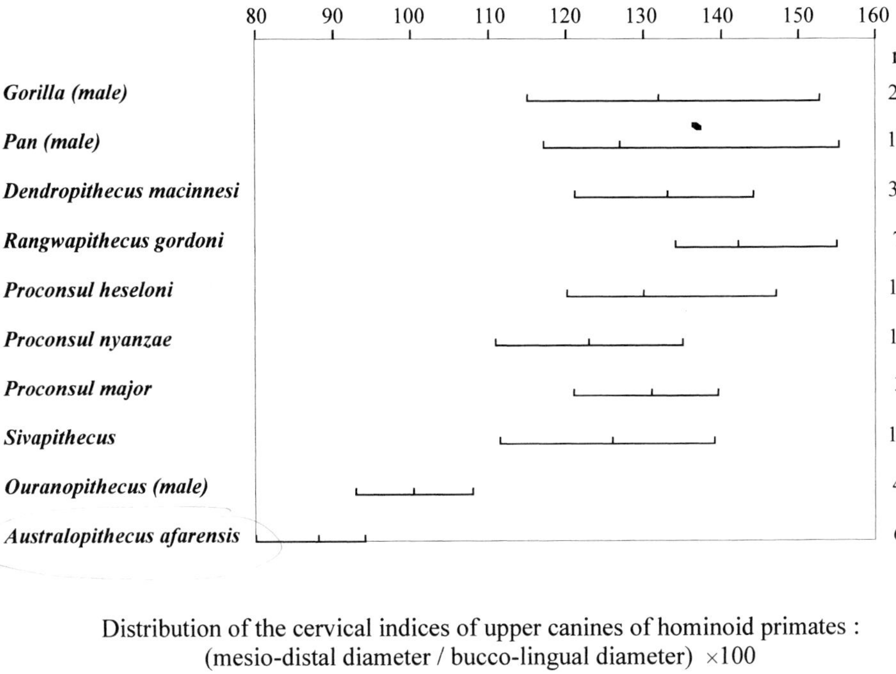

[Figure 11.1]
Repartition of cervical indices of the upper canines in some hominoid primates.

The size and the shape of the canine is important for estimating phylogenetic trends (Figure 11.1). In the plesiomorphic state, as observed in most cercopithecoids and in most early Miocene or recent hominoids, the crown is a little flattened buccolingually and there is a sharp distal crest. The cervical section is elliptic, the major axis being mesiodistal and the length/breadth index is high. *Ouranopithecus* index is lower and closer to the index in hominines which in section is rounded.

This is a derived character the function of which is unclear. The crown is low relative to the size of the cheek teeth and it can be considered as reduced when it is compared to primitive early or middle Miocene primates as are the species of *Proconsul* (Bonis & Melentis, 1978; Bonis & Koufos, 1993). On the other hand, it is less reduced than those of Pliocene hominines, *Australopithecus afarensis* (Johanson *et al.*, 1982) or *A. anamensis* (Leakey *et al.*, 1995; Leakey *et al.*, 1998). There is a small wear facet on the mesial face due to the lower canine. Young individuals also have two facets on the lingual face. The first is situated on the distal crest and the second, larger one, on the anterior part of the crown, close to the neck. Both correspond to the lower P_3 and fuse rapidly so that dentine is exposed on the whole linguo-distal face.

We shall see below the consequence of this type of wear.

The cingulum is lacking or vestigial on all the cheek teeth. This character is derived when compared with propliopithecids or with *Proconsul*. The premolars have two rounded cusps of about the same size and the crowns are symmetrical. P3 is very different from fossil or recent apes where the P3 is asymmetrical. The anterior half of the buccal face is higher and more developed than the posterior one and the cusps, especially the buccal, are higher and sharper. The premolars of *Ouranopithecus* are more similar to those of the Pliocene hominines. This change in shape, and probably in function, of P3 is another derived character found in the Macedonian genus as well as in *Australopithecus*. P4 is similar to P3, *Ouranopithecus* is homeomorphic like the australopithecines, while fossil and recent apes are heteromorphic for upper premolars. The molars are large relative to the premolars. M1 and M3 are smaller than M2. The cusps are low and not sharp. There are several distal accessory cusps on the unworn M3 exactly as on the M3 of *Australopithecus afarensis* (Johanson *et al.*, 1982). The multiplication of accessory cusps is also a derived character linked to the function of the teeth. All the cheek teeth are thick enamelled (Andrews & Martin, 1991). The thickness could be compared to that of some australopithecines. The enamel thickness is not an univocal character. Recent studies have concluded that most of the primates share almost the same enamel thickness relative to the body weight (Beynon *et al.*, 1998). Some species among the hominoids have thicker enamel than average: for example, *Proconsul heseloni, P. nyanzae, P. africanus, Sivapithecus* sp., and *Homo sapiens*, and especially *Australopithecus africanus* or the species of *Paranthropus*. The enamel thickness of *Ouranopithecus* is similar to that of *Paranthropus, Australopithecus africanus* and Paşalar primate, but is thicker than that of *Proconsul africanus, P. major, Dryopithecus* and *Sivapithecus* (Andrews & Martin, 1991, Fig. 4). The relatively thick enamel of the Rusinga and Mfangano species of *Proconsul* might be linked to the palaeoenvironment which is recently described as a semi-arid seasonal climate (Bestland & Krull, 1999). *Ardipithecus ramidus* has thin-enamelled cheek teeth (White *et al.*, 1994) but this species is not fully described and its relationships with the hominines are still unclear. There is also, on *Ouranopithecus* cheek teeth, a well-marked wear gradient similar to that found in australopithecines and which is more marked than in sivapithecines (Bonis *et al.*, 1998). This gradient depends not only on the thickness of the enamel but also on the time necessary for the posterior molars to fully erupt. The time passing between eruptions of two successive molars was longer than in recent apes and more similar to hominines. The timing of dental eruptions has been calculated for living primates (Dean, 1998; Shellis *et al.*, 1998; Ramirez Rozzi,

Table 11.1. *Crown formation and eruption times (years) in Pan* and *Homo*

	Time of crown growing		Age of dental eruption (average)	
	Pan	*Homo*	*Pan*	*Homo*
P3	3.8–6	2.9–3.9	6.1	10.4
P4	3.8–6	2.9–3.9	5.6	11
M1	2.7–4	2.3–2.4	3	6.4
M2	2.7–4	2.3–2.4	6.5	12.5
M3	2.7–4	2.3–2.4	10.5	20
Difference M1–M2			3.5	6.1
Difference M2–M3			4	7.5

1999) and correlated with increment enamel (striae of Retzius and cross striations) and crown completion. The conclusion reached is that fossil apes and fossil humans had the same growing pattern as that of recent apes and humans for molars, but have a reduced growing time for premolars. These data do not falsify the data given for the wear gradient. Studies on recent primates (Schultz in Vallois, 1955, fig. 1996) show that crown formation time is not directly linked to crown eruption (Table 11.1).

Lower dentition

The incisors are smaller than those of recent apes with a small difference in size between I1 and I2, the second one being a little larger than the first one. In cervical dimensions, the canines are smaller, relative to the cheek teeth, than those of fossil or recent apes and can be considered as reduced. The female canines do not rise above the level of the incisors or the premolars. The male canines are larger than the female ones. A comparison using the height of the lower canine versus the body size seems to reach different results (Kelley, Chapter 12). In this case the height would be better than cervical dimensions in estimating the reduction because the canines would be used as weapons by males and the height of the crown would be the most significant measurement. The canines of two mandibles from Ravin de la Pluie locality are compared with the face found in the Xirochori locality (Bonis & Koufos, 1993). The body size is estimated from the surface of the right orbit to be 40 to 60 kg. The ratio canine/body weight did not correspond to the reduced canines when it was compared with those of recent or fossil primates and *Ouranopithecus* would be macrodont concerning the cheek teeth. We can discuss several points of this hypothesis. First, it would be simpler to use directly the ratio between orbital surface and canine

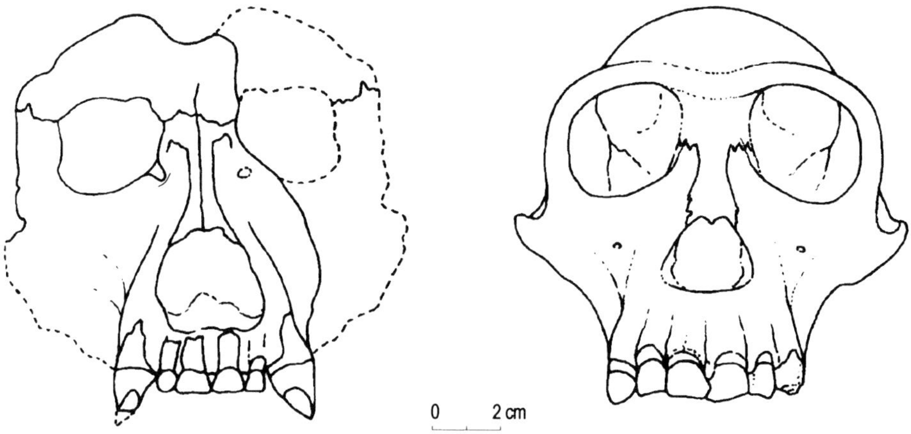

[Figure 11.2]
Face of *Ouranopithecus macedoniensis* (left) and *Pan paniscus* (right). The external bi-orbital breadth are 12 cm and 11 cm respectively. We can conclude that the skull (and probably the body weight) of *Ouranopithecus* is bigger than that of *Pan* despite the orbital surfaces are 9.16 and 11.30 cm^2 respectively.

height than to use an estimated body size with an increase of the total variance, but it does not matter. If we look at the results obtained for the ratio of the occlusal surface of M1/M_1 versus body size (Gingerich *et al.*, 1982), the weight of *Ouranopithecus* is 72.6 kg (51 to 91 kg) for M1 and 84.3 kg (70 to 101 kg) for M_1 far above the weight estimated from the orbit surface. It is difficult to infer exactly the body size from any skull measurement (Smith, 1981). It would be possible, for instance, to say that *Ouranopithecus* is not macrodont for the cheek teeth but that it is 'micro-orbital'. In other respects it appears that the orbit surface of *Ouranopithecus* is smaller than that of a male chimpanzee, which has a smaller skull, and that of a male orang-outan when the whole face is compared (Figure 11.2).

From another point of view, the canines of male primates are not really weapons, although we can be reasonably sure that the upper canines of *Smilodon* were used as weapons to kill prey, but hominoid primates are not carnivores that kill by biting. *Gorilla* is vegetarian despite its large canines and although sometimes chimpanzees eat meat, they are not specialised hunters and their prey is much smaller than they are themselves. Male catarrhines frequently use their canines to threat display (Rowe, 1996). Large canines can be very impressive and they are used for instance as an intraspecific intrasexual message between males to avoid actual combat and the risk of injury. Gorillas also use hooling, chest beating and slapping and throwing branches or leaves. When a male shows its canines to another male, their robustness is as significant as their heights and both are estimated relative to the size of the cheek teeth. The fists and the robustness of Mike Tyson are more impressive that his height.

The lower anterior premolar, P_3, is very significant for estimating phyletic relationships among hominoid primates and to separate hominines from apes. In the latter this tooth is elongated, a little flattened laterally, asymmetric and with an oblique protocristid. The anterior part of the buccal face has, particularly on male premolars, a clear vertical wear facet, called honing facet, due to the upper canine. This facet is a primitive character of all the catarrhine primates. The shape of the hominines P_3 is clearly different. The tooth is not elongated, the occlusal outline is more rounded, the protocristid is much less oblique with generally a lingual cuspid, the buccal face is more symmetric without any honing facet. The *Ouranopithecus* P_3 differs from those of fossil or recent apes by the same characters except the lingual cuspid. The honing facet is absent and this tooth is far more hominine-like than ape like. Linked probably to the absence of honing facet and to a different function of the canine, the wear of the crown is similar to that found in *Australopithecus afarensis* (Figure 11.3). The enamel in *Ouranopithecus* teeth wears in such a way that the dentine can be seen on the anterior part of the protocristid, on the top and on the posterior part of the same cristid. This striking similarity in the wear pattern to *Australopithecus* means certainly a similarity in the way of chewing and probably a similarity in the diet. It is probably the result of a similarity in the way of life, of a similar dwelling in an open environment. All these characters are certainly derived *vis-à-vis* the primitive hominoid way of life.

The cuspids of the lower P_4 are not equal, the metaconid being higher than the protoconid. Both foveae, anterior and posterior, are deep and well developed. The size of the lower molars increases from M_1 to M_3. The cingulum is absent or very weak, weaker than that of the contemporaneous species *Dryopithecus fontani*. As in the upper jaw, the last one has accessory cuspids similar to that of *Australopithecus afarensis*. We can make similar statements concerning the wear gradient for the upper molars. The thick enamel is also present but to a lesser extent in *Kenyapithecus, Equatorius, Griphopithecus, Afropithecus, Sivapithecus* and to an even lesser extent in *Pongo* (Andrews & Martin, 1991). But opinions differ regarding the thickness of *Pongo* dental enamel. Shellis and co-workers (1998) claim that the primate dental enamel thickness is allometric (despite the thin enamel of *Gorilla*) and that *Pongo* 'has . . . the greater average enamel thickness in the large teeth . . .' which in fact is not very different from Andrews & Martin (1991).

A multivariate metric approach of dental measurements focused on

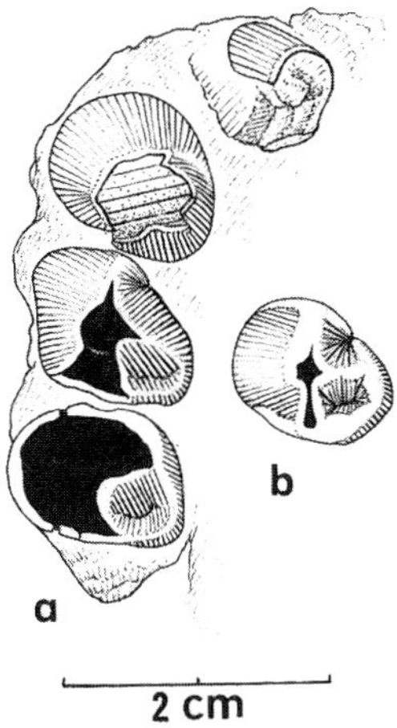

[Figure 11.3]
Mandible of *Ouranopithecus macedoniensis* (RPI 196) and P_3 of *Australopithecus afarensis* (AL 277-1). There is a striking similarity of both P_3 wear patterns.

Gigantopithecus, and including also *Ouranopithecus* (Gelvin, 1980), gives significant data on the morphometric similarities of the latter. A canonical variate analysis based on logarithmically transformed measurements of the lower dentition shows that *Ouranopithecus* plots close to *Gigantopithecus blacki* and *G. bilaspurensis* (or *Indopithecus giganteus*) and also to *Australopithecus*, *Paranthopus* and *Homo erectus* on both plans defined by the first and second canonical variates and the first and third canonical variates, which is the opposite to *Proconsul* or *Sivapithecus*. A cluster analysis based on the same logarithmically transformed measurements gives similar results (Gelvin, 1980). The metric analysis, therefore recognises some links between *Ouranopithecus* and *Gigantopithecus* on one hand, as well as *Ouranopithecus* and Plio-Pleistocene hominines on the other hand. '*Ouranopithecus* follows a hominid dental pattern more closely than does *Gigantopithecus*' (Gelvin, 1980). We reach the same conclusion from the morphological study.

Dental microwear

Dental microwear is greatly dependent on the biomechanics of the jaw and on the diet of the animal, with both being linked. Several studies on mammals, and particularly on primates, show that it is possible to deduce the kind of food eaten by an animal some hours before its death from the pattern of microwear on the teeth. The ratio and direction of pits and striations on the surface of enamel make it possible to know the last few meals of an animal and, with several specimens, the feeding of a species. Frugivores, folivores and hard object feeders have different microwear patterns. A study of *Ouranopithecus* tooth microwear by Ungar (1996) came to the conclusion that the Macedonian primate, with a high ratio of pits versus striations on the cheek teeth and special wear of the incisors, was a hard object feeder feeding on seeds, nuts, roots or tubers directly on the ground. This diet corresponds to an open environment dweller. This is also the conclusion we reached when studying the palaeoenvironment of *Ouranopithecus* through the faunal associations found in the hominoid primate bearing localities (Bonis *et al.*, 1992). These results fit well with the presence of very thick enamel on the cheek teeth of *Ouranopithecus*. The microwear pattern as well as the thick enamel are derived characters for hominoid primates. The whole super-family originated certainly from forest dwelling primates with thinner enamel such as *Proconsul africanus* or *P. major* (Andrews & Martin, 1991). The microwear pattern as well as the thickness of enamel are shared by australopithecines.

Face

The face of *Ouranopithecus* is known from a couple of specimens. An upper maxilla (RPl 128) unearthed from the locality Ravin de la Pluie which shows well the nasal area (Bonis & Melentis, 1978), and an almost complete face (XIR 1) from the locality Xirochori 1 (Bonis & Koufos, 1993). Both can be compared with other late Miocene Eurasiatic taxa – *Sivapithecus, Dryopithecus, Ankarapithecus* and *Lufengpithecus*. The face is large with a large interorbital distance. The right orbit, the shape of which does not seem distorted, is relatively small, low and horizontally quadrangular. These characters might be derived. The inter-orbital distance of *Sivapithecus* is very narrow and the orbits are tall and vertically oval-shaped (Pilbeam, 1982; Pilbeam & Smith, 1982; Preuss, 1982; Ward & Kimbel, 1983; Ward & Pilbeam, 1983) , all are pongid derived features. The orbits of *Dryopithecus* are D-shaped and more or less taller depending on the reconstruction (Kordos, 1987; Begun, 1994; Moyà-Solà & Köhler, 1993). The shape of the orbital region of *Ankarapithecus* does not seem very different (Alpagut *et al.*, 1996). The skull of *Lufengpithecus* is crushed but the orbits seem rounded and horizontally rectangular. The bulging supra-orbital torus (brow ridge) is well defined and is considered to be a derived feature relative to other Miocene hominoids. The glabella is swollen but less than the brow ridge. *Lufenpithecus* is very different with a wide and sunken glabella region (Kordos, 1988; Wu, 1987;Wu *et al.*, 1983). The lateral outline of the upper face is quite vertical and not oblique as it is in middle Miocene hominoid primates. This is a derived feature as is also another different derived feature – the concave lateral outline of the pongids *Pongo* and *Sivapithecus*. The structure of the naso-alveolar area is primitive and is similar to *Dryopithecus* or *Gorilla*, but it is not as primitive as *Proconsul* or *Morotopithecus* (Ward & Kimbel, 1983; Ward & Pilbeam, 1983). On the other hand, the naso-alveolar morphology is less derived than in *Australopithecus afarensis* or the recent chimpanzee. *Sivapithecus* and *Pongo* are clearly more derived than the other hominoids. A maxilla attributed to *Ankarapithecus* was described as pongid-like (Andrews & Tekkaya, 1980) but later it was considered as more primitive (Begun & Gülec, 1995). This opinion was reinforced by a new specimen attributed to the same genus (Alpagut *et al.*, 1996; Andrews & Alpagut, Chapter 9).

Mandible

The mandible is known from several specimens. It also shows sexual dimorphism – the male mandibles are deeper than the female ones. The symphysis is primitive with a long planum alveolare and two well-marked tori – superior and inferior. In the gonial area, a well indicated crest (tuberosites massetericus) corresponds to a well developed muscle masseter superficialis (lamina prima). This character is linked to a powerful chewing capacity. The mandibular condyle gives some data on the functional jaw mechanism. Apes have a large slightly convex condyle which is articulated on a quite flat glenoid fossa without a well marked anterior process. The temporomandibular joint (TMJ) of *Homo* is clearly different. The anterior glenoid process is well developed and the glenoid fossa is deeper and narrower than that of apes. The mandibular condyle is also narrower and more convex. Australopithecines are intermediate for these characters between apes and humans (James, 1960; Picq, 1990) and *Australopithecus afarensis* is more ape-like than are recent hominines for the TMJ morphology (Picq, 1990). The narrowness of the *Ouranopithecus* mandibular condyle is more hominine-like than ape-like (Bonis & Koufos, 1997). The *Ouranopithecus* TMJ was probably not very different from that of australopithecines. The morphology of the TMJ is linked to the method of chewing and we can assume that this does not differ between both groups.

Conclusion: homologies or homoplasies

Several characters are present in *Ouranopithecus* that are also present in Pliocene australopithecines. Some are plesiomorphic as are the large interorbital distance, naso-alveolar structure or shape of the mandibular symphisis. But some other shared features are apomorphic. These are:

- Bulging brow ridge with small glabella depression.
- Vertical lateral outline of upper face.
- Canine reduction.
- Shape of the canines.
- Symmetric upper P3.
- Shape of the paracone of P3.
- Low size difference between paracone and protocone of P3.
- Accessory cusps-cuspids on M3/M_3.
- Not elongated P_3.
- Protocristid (or metalophid or metacristid) of P_3 less oblique.
- Buccal wall of P_3 less asymmetric.
- Absence of honing facet.
- Very thick cheek teeth enamel.
- Shape of the wear pattern of P_3.
- Well marked molar wear gradient.

Some of these characters are known independently in other lineages (homoplasy) such as the enamel thickness which is present in *Sivapithecus* and to a lesser extent in *Pongo*, or the canine reduction (Kelley, Chapter 12), which could be also homoplasic between *Ouranopithecus* and the Pliocene hominines. But it would be strange and not parsimonious if all the features were homoplasic even if some of them are linked together. So we think that they testify to a sister group relationships between *Ouranopithecus* and hominines. A hominoid primates radiation took place in Europe and Asia during the whole late Miocene. At the same time, there is little data known from Africa (Ishida & Pickford, 1997; Ishida *et al.*, 1984). In Eurasia, two genera, such as *Ouranopithecus* and *Dryopithecus*, had developed the same ecological adaptations as those of later hominines and African apes. It would be possible to consider both genera together as the sister group of the recent African set – apes and australopithecines (Begun, Chapter 10) but it seems to us that it would more parsimonious if *Ouranopithecus* alone was the sister group of hominines, and *Dryopithecus* alone was a sister taxum of African apes. Until now one of the strongest arguments against such a hypothesis was based on the results of the 'molecular clock'. Most molecular biologists, faithfully followed by most palaeoanthropologists, claimed that the splitting between African apes and humans, and of course australopithecines, could not be older than 6 Ma. So the place of *Ouranopithecus* as a sister group of *Australopithecus* was impossible and all its characters shared with the latter are homoplasic. But recently (Arnason *et al.*, 1996) an analysis of primate relationships based on the complete set of mitochondrial DNA leads to a different dating. The splitting between humans and apes is older than 10 Ma. This dating corresponds to a significant climatic change in Africa with the development of more open environment dwelling mammals, for instance antelopes, and an appearance of new kinds of primate more adapted to this new landscape and to different food. *Kenyapithecus* in Africa or *Griphopithecus* in Eurasia are good examples of this kind of primate. The *Ouranopithecus* ancestry could be located in this ancestral stem. Other findings that fit between middle Miocene primates and *Ouranopithecus* or between the latter and the oldest australopithecines will be necessary to either confirm or deny this hypothesis.

Acknowledgements

We thank the European Science Foundation to allow the organisation of the workshop 'Phylogeny of Eurasian Neogene Hominoid Primates' in Nikiti. The field campaigns in Greece have been supported by grants of the Foundation Singer Polignac, the Leakey Foundation and the French CNRS (programme Paléoenvironnements et Evolution des Hominidae) to the senior author. The second author thanks the Leakey Foundation, the Community of Nikiti, and the Aristotle University of Thessaloniki for supporting the excavations at Nikiti. We thank D. Begun for helpful comments and criticisms on the manuscripts and all the people present in the Nikiti's workshop for exiting and helpful discussions.

References

Alpagut, B., Andrews, P., Fortelius, M., Kappelman, J., Temizoy, J., Celebi, H. & Lindsay, W. (1996). A new specimen of *Ankarapithecus meteai* from the Sinap formation of central Anatolia. *Nature*, **382**: 349–51.

Andrews, P. & Martin, L. (1991). Hominoid dietary evolution. *Phil. Trans. Roy. Soc. London*, B, **334**: 199–209.

Andrews, P. & Tekkaya, I. (1980). A revision of the Turkish Miocene hominoid *Sivapithecus meteai*. *Paleontology*, **23**: 86–95.

Arambourg, C. & Piveteau, J. (1929). Les vertébrés du Pontien de Salonique. *Ann. Paléontol.* **18**: 59–138.

Arnason, U., Gullgerg, A., Janke, A. & Xu, X. (1996). Pattern and timing of evolutionary divergences among hominoids based on analysis of complete mtDNAS. *J. Mol. Evol.* **43**: 650–61.

Begun, D. R. (1994). Relations among the great apes and human: New interpretation based on the fossil great ape *Dryopithecus*. *Ybk Phys. Anthrop.* **37**: 11–63.

Begun, D. R. & Güleç E. (1995). Restoration and reinterpretation of the facial specimen attributed to *Sivapithecus meteai* from Kayancak (Yassören), central Anatolia, Turkey. *Am. J. Phys. Anthrop.* **20** (Suppl.): 63–4.

Bestland, E. A. & Krull, E. S. (1999). Palaeoenvironments of Early Miocene Kisingiri volcano *Proconsul* sites: evidence from carbon isotopes, palaeosols and hydromagmatic deposits. *J. Geol. Soc. London* **156**: 965–76.

Beynon, A. D., Dean, M. C., Leakey, M. G., Reid D. J. & Walker A. (1998). Comparative dental development and microstructure of *Proconsul* teeth from Rusinga Island, Kenya. *J. Hum. Evol.* **35**: 163–209.

Bonis, L. de, Bouvrain, G., Geraads, D. & Koufos, G. D. (1990). New hominid skull material from the late Miocene of Macedonia in northern Greece. *Nature* **345**: 712–14.

Bonis, L. de, Bouvrain, G., Geraads, D. & Koufos, G. D. (1992). Diversity and paleoecology of Greek late Miocene mammalian faunas. *Palaeogeogr. Palaeoclimatol. Palaeoecol.* **91**: 99–121.

Bonis, L. de, Bouvrain, G., Geraads, D. & Melentis J. (1974). Première découverte d'un primate hominoïde dans le Miocène supérieur de Macédoine (Grèce). *Comp. Rend. Acad. Sci.* Paris **278**: 3063–6.

Bonis, L. de, Bouvrain, G. & Koufos, G. (1987). Late Miocene mammal localities of the lower Axios valley (Macedonia, Greece) and their stratigraphic significance. *Mod. Geol.* **13**: 141–7.

Bonis, L. de & Koufos, G. D. (1993). The face and the mandible of *Ouranopithecus macedoniensis*: description of new specimens and comparisons. *J. Hum. Evol.* **24**: 469–91.

Bonis, L. de & Koufos, G. D. (1994). Our ancestors'ancestor: *Ouranopithecus* is a Greek link in human ancestry. *Evolut. Anthrop.* **3**: 75–83.

Bonis, L. de & Koufos, G. (1997). The phylogenetic and functional implications of *Ouranopithecus macedoniensis*. In Begun, D. R., Ward, C. V. & Martin, L. (eds.), *Function, Phylogeny and Fossils, Miocene Hominoid Evolution and Adaptations*, pp. 317–26. New York: Plenum Press.

Bonis, L. de & Koufos, G. D. (1999). The Miocene large mammal succession in Greece. In Agustí, J., Rook, L. & Andrews, P. (eds.), *Hominoid Evolution and Climatic Change in Europe*, vol. 1, *The Evolution of Neogene Terrestrial Ecosystems in Europe*, pp. 205–37. Cambridge: Cambridge University Press.

Bonis, L. de, Koufos, G., Guy, F., Peigné S. & Sylvestrou I. (1998). Nouveaux restes du primate hominoïde *Ouranopithecus* dans les dépôts du Miocène supérieur de Macédoine (Geèce). *Comp. Rend. Acad. Sci.* Paris **327**: 141–6.

Bonis, L. de & Melentis; J. (1977). Un nouveau genre de primate hominoide dans le vallesien (Miocène supérieur) de Macedoine. *C.R. Acad. Sci. Paris* **284**: 1393–6.

Dean, M. C. (1998). A comparative study of cross striation spacing in cuspal enamel and on four methods of estimating the time taken to grow molar cuspal enamel in *Pan, Pongo* and *Homo. J. Hum. Evol.* **35**: 449–62.

Gelvin, B. R. (1980). Morphometric affinities of *Gigantopithecus. Am. J. Phys. Anthrop.* **53**: 541–68.

Gingerich, Ph. D., Smith, B. H. & Rosenberg, K. (1982). Allometric scaling in the dentition of primates and prediction of body weight from tooth size in fossils. *Am. J. Phys. Anthrop.* **58**: 81–100.

Ishida, H. & Pickford, M. (1997). A new late Miocene hominoid from Kenya, *C.R. Acad. Sc. Paris* **325**: 823–9.

Ishida, H., Pickford, M., Nakaya, H. & Nakano, Y. (1984). Fossil anthropoids from Nachola and Samburu Hills, Samburu District, Kenya. *Afr. Study Monogr. Suppl. Iss.* **2**: 73–85.

James, W. W. (1960). *The Jaws and Teeth of Primates*. London: Pitman Medical Publishing.

Johanson, D. C., White, T. D. & Coppens, Y. (1982). Dental remains from the Hadar Formation, Ethiopia : 1974–1977 collections. *Am. J. Phys. Anthrop.* **57**: 545–603.

Kondopoulou, D., Sen, S., Koufos, G. and Bonis, L. de (1992). Magneto and biostratigraphy of the late Miocene mammalian locality of Prochoma (Macedonia, Greece). *Paleontol. Evol. Barcelona* **24–25**: 135–9.

Kordos, L. (1987). Description and reconstruction of the skull of *Rudapithecus hungaricus* Kretzoï (Mammalia). *Ann. Hist. Nat. Mus. Ntl. Hung.* **79**: 151–3.

Kordos, L. (1988). Comparison of early primate skulls from Rudabanya (Hungary) and Lufeng (China). *Anthrop. Hung.* **20**: 9–22.

Kostopoulos, D., Koliadimou, K. & Koufos, G. D. (1996). Giraffids of Nikiti. Palaeontographica **239**: 61–88.

Kostopoulos, D. & Koufos, G. D. (1996). Late Miocene bovids (Mammalia, Artiodactyla) from the locality of Nikiti 1, Macedonia, Greece. *Ann. Paléontol.* **87**: 251–300.

Koufos, G. D. (1993). A mandible of *Ouranopithecus macedoniensis* from the late Miocene of Macedonia (Greece). *Am. J. Phys. Anthrop.* **91**: 225–34.

Koufos, G. D. (1995). The first female maxilla of the hominoid *Ouranopithecus macedoniensis* from the late Miocene of Macedonia (Greece). *J. Hum. Evol.* **29**: 385–99.

Leakey, M. G., Feibel, C. S., McDougall, I. & Walker, A. (1995). New four million year old hominid species from Kanapoi and Allia Bay, Kenya. *Nature* 365–71.

Leakey, M. G., Feibel, C. S., McDougall, I., Ward, C. & Walker, A. (1998). New specimens and confirmation of an early age for *Autralopithecus anamensis*. *Nature* **393**: 62–6.

Mercier, J. (1973). Etude géologique des zones internes des Hellénides en Macédoine centrale. *Ann. Géol. Pays hellén.* **1**: **20** (B): 1–596.

Moyà-Solà, S. & Köhler, M. (1993). Recent discoveries of *Dryopithecus* shed new light on evolution of great apes. *Nature* **365**: 543–5.

Picq, P. (1990). L'articulation temporo-mandibulaire des hominidés. *Cah. Paléoanthrop.* Paris: CNRS.

Pilbeam, D. (1982). New hominoid skull from the Miocene of Pakistan. *Nature* **348**: 237–9.

Pilbeam, D. & Smith, R. (1982). New skull remains of *Sivapithecus* from Pakistan. *Mem. Geol. Surv. Pakistan* **II**: 1–13.

Preuss, T. M. (1982). The face of *Sivapithecus indicus*: description of a new, relatively complete specimen from the Siwaliks of Pakistan. *Folia Primatol.* **38**: 141–57.

Quade, J., Solounias, N. & Cerling, T. E. (1994). Stable isotop evidence from palaeosol carbonates and different teeth in Greece for forests or woodlands over the past 11 Ma. *Palaeogeogr. Palaeoclimatol. Palaeoecol.* **108**: 41–53.

Ramirez Rozzi, F. (1999). L'horloge dentaire et la croissance des hominidés. In Les origines de l'humanité, Pour la Science (Dossier hors série), pp. 60–3.

Rapport de l'Institut Français du Pétrole (1967). *Bassin tertiaire de Salonique. Etude Géologique et Pétrolière*. (Unpublished, 128 pp.)

Rowe, N. (1996). *The Pictorial Guide to the Living Primates*. East Hampton, NY: Pogonias Press.

Sen, S., Koufos, G. D., Kondopoulou, D. & Bonis, L. de (2000). Magnetostratigraphy of the late Miocene continental deposits of the lower Axios valley, Macedonia,

Greece. *Bull. Geol. Soc. Greece* No. 9, pp. 197–206.

Simons, E. & Pilbeam, D. (1965). Preliminary revision of the Dryopithecinae (Pongidae, Anthropoidea). *Folia Primatol.* **3**: 81–152.

Shellis, R. P., Beynon, A. D., Reid, D. J. & Hiiemar K. M. (1998). Variation in the molar enamel thickness among primates. *J. Hum. Evol.* **35**: 307–22.

Smith, R. J. (1981). On the definition of variables in studies of primate dental allometry. *Am. J. Phys. Anthrop.* **55**: 323–9.

Ungar, P. S. (1996). Dental microwear of European Miocene catarrhines: evidence for diet and tooth use. *J. Hum. Evol.* **31**: 335–66.

Vallois, H. (1955). Ordre des Primates. In Grasse, P.-P. (ed.), *Traité de Zoologie*, vol. 17, pp. 1854–2206. Mammifères. Paris: Masson.

Ward, S. & Kimbel, W. (1983). Subnasal alveolar morphology and the systematic position of *Sivapithecus*. *Am. J. Phys. Anthrop.* **61**: 157–71.

Ward, S. & Pilbeam, D. (1983). Maxillofacial morphology from hominoid primates from Africa and Indo-Pakistan. In Ciochon, R., & Corruccini, R. (eds.), *New Interpretations on Apes and Human Ancestry*, pp. 210–19. New York: Plenum Press.

White, T., Suwa, G. & Asfaw B. (1994). *Australopithecus ramidus*, a new species of early hominid from Aramis, Ethiopia. *Nature* **371**: 306–12.

Wu, R. (1962). The mandibles and dentition of *Gigantopithecus*. *Palaeontol. Sin.* Beijin **146** (D): 1–94.

Wu, R. (1987). A revision of the classification of the Lufeng great apes. *Acta Anthrop. Sin.* **6**: 263–71.

Wu, R. & Pan, Y. (1984). A late Miocene gibbon-like primate from Lufeng, Yunnan province. *Acta Anthrop. Sin.* **3**: 193–200.

Wu, R. & Pan, Y. (1985). Preliminary observation on the cranium of *Laccopithecus robustus* from Lufeng, Yunnan, with reference to its phylogenetic relationships. *Acta Anthrop. Sin.* **4**: 7–12.

Wu, R., Xu, Q. & Lu, Q. (1983). Morphological features of *Ramapithecus* and *Sivapithecus* and their phylogenetic relationships – morphology and comparison of the crania. *Acta Anthrop. Sin.* **2**: 1–10.

12 Phylogeny and sexually dimorphic characters: Canine reduction in *Ouranopithecus*

Jay Kelley

Degrees and patterns of morphological sexual dimorphism are problematic characters for the purpose of phylogenetic analysis. Since sexual dimorphism is a product of at least partly independent selection in males and females – although the underlying genetics of certain characters can produce a correlated response (Plavcan, 1998) – the relative contributions of male and female values to any measure of sexual dimorphism will vary. For example, species of *Hylobates* and *Callicebus* both have very minimal canine size dimorphism (Plavcan & van Schaik, 1992), but in *Hylobates* this is achieved through female canine hypertrophy, whereas in *Callicebus* it results from male canine reduction (Plavcan *et al.*, 1995). Therefore, similar or identical measures of sexual dimorphism will not necessarily be homologous and can instead reflect homoplasy. Moreover, even sexual dimorphisms that reflect similar contributions from male and female values in the different species can still result from homoplasy; species of *Hylobates* and the Pitheciinae all express low levels of canine size dimorphism as a result of female canine hypertrophy, but this has clearly evolved independently in the two lineages. While the possibility of homoplasy exists for any character, it has quite clearly been especially common with respect to sexual size dimorphism, including canine size dimorphism, both within and between higher primate lineages.

While sexual dimorphisms as a class of characters offer poor prospects for phylogenetic reconstruction, it is nevertheless legitimate to explore the phylogenetic significance of morphological characters that happen to be sexually dimorphic. However, this can only be done in males and females separately since, by definition, males and females express different states for sexually dimorphic characters.

Among the most obvious sexually dimorphic characters that preserve in the fossil record is canine size. Of particular interest here is canine size reduction, especially male canine reduction, since it is one of the most frequently cited apomorphies of the human lineage and can be reliably documented in the earliest well-known Pliocene members of the lineage (Johanson *et al.*, 1982; White & Johanson, 1982; Plavcan & van Schaik, 1997; Asfaw, *et al.*, 1999). It has been claimed that the late Miocene hominoid *Ouranopithecus macedoniensis*, known from several localities in northern Greece, also had reduced canines, especially male canines, and that this is a synapomorphy warranting placement of *Ouranopithecus* in the human lineage (Bonis & Melentis, 1987; Bonis *et al.*, 1990, 1998; Bonis & Koufos, 1993, 1994).

Here I will evaluate both the claim of canine reduction in this species, as well as the overall phylogenetic utility of this character. I will demonstrate that the male canines of *Ouranopithecus* are probably as large relative to body mass as those of most extant great apes. I will also show that there has been homoplasy in male canine reduction within the African ape and human clade.

Evidence for canine reduction in *Ouranopithecus*

The claim for male canine reduction in *Ouranopithecus* is based on ratios of canine size to molar size (Bonis, *et al.*, 1981; Bonis & Koufos, 1993), with canine size calculated as either the cross-sectional area or mesiodistal length at the base of the crown. Using these measures of relative canine size, Bonis & Koufos (1993) demonstrated that two presumed *Ouranopithecus* males, RPl 128 from Ravin de la Pluie (Bonis & Melentis, 1978) and XIR 1 from Xirochori (Bonis *et al.*, 1990), lie within the female ranges but outside the male ranges of extant African apes. They concluded from this that *Ouranopithecus* males have reduced canines. A similar analysis using a more diverse comparative sample and also including mandibular canines confirms these results (Table 12.1). Andrews (1990) has argued that, rather than demonstrating reduced male canines, a more reasonable interpretation of these results is that the two specimens are instead females and that *Ouranopithecus* does not therefore have reduced canines. However, the absolute size of these two individuals coupled with sex-specific features of canine shape strongly suggest that they are in fact males (Kelley, 1995a). Thus, using the measures of relative canine size employed by Bonis and Koufos, reduced male canines would appear to be characteristic of this species.

Ratios of canine to molar size as an indication of genuine canine reduction can, however, be misleading. An obvious problem with such ratios is that *apparent* canine reduction can result from having absolutely very large molars (megadontia) rather than diminished canines. Molar megadontia has been suggested to characterise a number of Miocene hominoids, including *Ouranopithecus* (Kelley, 1986, 1995a). That *Ouranopithecus* is in fact highly megadont in the molar dentition is evident in a comparison of molar size with the extant apes, especially *Gorilla*. The anterior molars of *Ouranopithecus* males are only marginally smaller on average than those of

Table 12.1. *Male canine/molar ratios*[a] *in extant and fossil hominoids*

	Upper canine	Lower canine
Gorilla	1.19–1.60 (10)	1.02–1.46 (10)
Pan	1.17–1.66 (10)	1.02–1.42 (10)
Pongo	1.39–1.58 (10)	1.13–1.35 (10)
Proconsul	1.42–1.61 (4)	1.26–1.52 (9)
Sivapithecus	1.14–1.32 (5)	0.96–1.29 (5)
Lufengpithecus	1.21–1.34 (5)	0.90–1.15 (7)
Ouranopithecus	1.02–1.07 (2)	0.89–1.00 (3)

[a] Canine maximum length/first molar mesiodistal length; extant ape data from Kay (1982). Sample sizes in parentheses.

Table 12.2. *Molar occlusal area*[a] *in male* Ouranopithecus *and* Gorilla

	M^1	M^2	M^3	M_1	M_2	M_3
Ouranopithecus	194 (3)	240 (3)	240 (3)	208 (4)	242 (6)	297 (8)
Gorilla	231(20)	266(20)	236(20)	212(20)	269(20)	270(20)
Ouranopithecus/Gorilla (%)	84	90	102	98	90	110

[a] Occlusal area (mm^2) calculated as maximum mesiodistal length × maximum buccolingual breadth; *Gorilla* data from Pilbeam (1969), *Ouranopithecus* data from Bonis *et al.* (1998) and L. de Bonis (personal communication). Sample sizes in parentheses. Right and left sides averaged for *Ouranopithecus* where antimeres present.

Gorilla while the third molar is actually slightly larger (Table 12.2; see also Bonis & Melentis, 1984, figure 2; Koufos, 1995, figure 7). However, even the most liberal estimate of body mass for *Ouranopithecus* males is less than half the approximately 170 kg average male mass of *Gorilla* (see discussion below), revealing that *Ouranopithecus* is exceptionally megadont in the molar dentition. Therefore, a large portion, if not all, of the apparent canine reduction in *Ouranopithecus* deduced from canine/molar ratios is certainly due to molar megadontia rather than the possession of genuinely reduced canines.

The real issue in canine reduction is not canine size in relation to molar size but, rather, canine size in relation to body size. The point of biological significance in canine reduction is that, not only is there a departure from the general isometry between canine size and body size evident in primates as a whole, but also a departure from the negatively allometric canine/body size relationship that characterises the extant great apes (Plavcan *et al.*, 1995).

Additionally, with respect to the selective pressures governing canine size, I would also argue that any measure of relative canine size based on canine cervical crown dimensions is largely irrelevant. It has been convincingly demonstrated that the primary functional correlate of relative canine size in extant primates is aggressive behaviour (Plavcan & van Schaik, 1992, 1994; Plavcan *et al.*, 1995). Canines are first and foremost weapons and, whether used in display or for actual fighting, the canine features of consequence are most likely to be crown height and sharpness. This is perhaps less obvious with respect to canine displays than to fighting but, intuitively, it seems more likely that the visual impact of such displays lies with the projection of the canine rather than with some impression of the mesiodistal length of the canine at the gum line. In the following comparative analysis, therefore, relative canine size in *Ouranopithecus* and the extant great apes will be expressed as canine height in relation to body mass.

Relative canine size in *Ouranopithecus*

Estimates of body mass

The most reliable estimates of body mass derive from postcrania. Unfortunately, with the exception of an unpublished phalanx, there are no postcranial remains of *Ouranopithecus* by which to estimate body mass. Body mass estimates based on tooth size give inconsistent results and would be particularly suspect for *Ouranopithecus* given the indications of molar megadontia in this taxon.

Another means for estimating body mass in fossil primates relies on measures of orbital size. Aiello & Wood (1994) and Kappelman *et al.* (2000) have each explored the relationship between orbital size and body mass in primates using a variety of orbital measures. The most reliable of these regressions, in terms of mean percentage prediction error as well as percentage of individuals with an estimated mass within ± 20% of actual mass, was derived by Kappelman *et al.* utilizing a measure of orbital area (Table 12.3).

The partial skull from Xirochori (XIR 1) preserves a nearly intact right orbit. Based on overall dental size, particularly canine cervical crown area, this individual is almost certainly male as noted above (see Bonis *et al.*, 1990; Koufos, 1995). Using the digitized area of the right orbit, Kappelman *et al.* (2000) estimated the mass of this individual at 48 kg, with lower and upper 95% confidence limits of 34 and 69 kg respectively (Table 12.3). The 48 kg estimate is substantially less than an earlier estimate of 72 kg by Bonis & Koufos (1994) based on molar size, but it is the largest of the various mass estimates based on orbital size (Table 12.3).

Table 12.3. *Body mass estimates from orbital size in extant hominoids and XIR 1*

Measure	Mean % prediction error [a]	% Within ± 20% actual body mass [a]	Estimate for XIR 1 (kg)
Aiello & Wood (1994) regressions:			
Orbital breadth	24	42	40.0
Orbital height	19	58	26.2
Orbital area	16	50	31.0
Kappelman et al. *(2000) regressions:*			
Orbital height	22	56	–
Orbital area	14	78	48.2

[a] Figures for extant hominoid comparative samples.

Although the reliability of species or sex mean mass estimates from single individuals is low, a comparison of molar size of XIR 1 with two other probable *Ouranopithecus* males (RPl 193 and RPl 128) reveals that XIR 1 is somewhat smaller than the former and slightly larger than the latter (Table 12.4; see also Bonis *et al.*, 1998). It is not unreasonable, therefore, to assume that XIR 1 would have at least approximated the average male mass for the species. Nevertheless, some allowance will be made in the following analysis of relative canine size to accommodate the possibility that 48 kg underestimates the actual mean male mass in *Ouranopithecus*. On the other hand, as 48 kg is the highest estimate for XIR 1 from the various orbitally-based regression equations, an analysis based on this body mass estimate is perhaps somewhat conservative since a higher body mass produces a lower value of relative canine size.

Mandibular relative canine height

All of the male maxillary canines of *Ouranopithecus* are very heavily worn, making attempts to estimate original, unworn height difficult and of questionable reliability. In contrast, two male mandibles preserve canines that are either virtually unworn (RPl 55) or very minimally worn (RPl 75) (Bonis & Melentis, 1977). Since the analysis of relative canine size is based on canine height, the initial focus will therefore be on the mandibular canines.

RPl 55 and RPl 75 were determined to be male based on the size and shape of their canines (Kelley, 1995a). Maximum buccal crown height of the canines was measured on scanning electron microscope (SEM) quality epoxy-resin replicas made from silicone-rubber moulds of the original

Table 12.4. *Molar Size*[a] *in* Ouranopithecus *maxillary specimens*

	Sex	M^1	M^2	M^3
XIR 1	M	191	232	237
RPl 128	M	185	235	228
RPl 193	M	206	254	256
RPl 775	F	221[b]	253[b]	214
NKT 89	F	149	169	189

[a] Molar area calculated as mesiodistal length × bucco-lingual breadth; left and right sides averaged where antimeres present.
[b] Molar area increased by post-depositional water infiltration and subsequent expansion.

specimens. For both specimens, the antimere showing the least wear was measured. The measured canine height of RPl 55 was 20.9 mm, while the estimated unworn height of the slightly more worn RPl 75 was 21.5 mm, for an average height of 21.2 mm.

Relative canine height is expressed as the calculated residual from a least squares regression of extant great ape canine height (Kelley, 1995b) against body mass (Smith & Jungers, 1997). Expected canine height values were determined for *Ouranopithecus,* and for the early hominid *Australopithecus afarensis,* by entering body mass estimates for these taxa into the great ape regression equation. Residual canine height values were then calculated by subtracting actual canine heights from the expected values. The extant great ape comparative sample includes separate canine height and body mass values for all included subspecies (Table 12.5). Body mass estimates for *A. afarensis* were taken from McHenry (1992) and canine heights were measured on reference casts of unworn or minimally worn canines of this species at the Cleveland Museum of Natural History (Ohio, USA).

Residual canine height values for *Ouranopithecus, A. afarensis* and the extant great apes are shown in Figure 12.1. These results reveal that average lower canine height in *Ouranopithecus* is almost exactly what would be expected for a male ape with a body mass of 48 kg. Residual canine height in *Ouranopithecus* is substantially greater than in either *Pan paniscus,* the only extant ape with reduced male canines, or *A. afarensis.*

Since the errors associated with body mass estimates based on orbit size are large, it might be argued that the 48 kg estimate for XIR 1 is too low, or, even if accurate, that XIR 1 was a small male. However, even when the average body mass estimate for *Ouranopithecus* is increased by 25% to 60 kg, residual male canine height is still within the extant great ape range

Table 12.5. *Male canine height (mm) and body mass (kg) in extant great apes*

	Upper canine	Lower canine	Body mass
Pan t. troglodytes	23.5	21.0	59.7
Pan t. schweinfurthii	23.5	21.9	42.7
Pan t. verus	22.9	20.8	46.3
Pan paniscus	17.4	14.6	45.0
Gorilla g. gorilla	32.9	27.4	170.4
Gorilla g. graueri	33.2	27.0	175.2
Gorilla g. beringei	29.9	25.2	162.5
Pongo p. pygmaeus	26.5	25.0	78.5
Pongo p. abelli	27.0	25.3	77.9

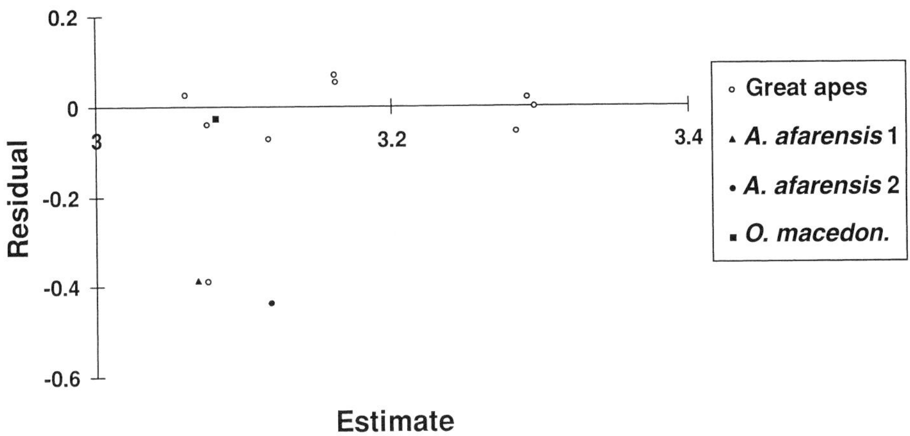

[Figure 12.1]
Male residual mandibular canine height in great apes, *Ouranopithecus*, and *Australopithecus afarensis*, with average male body mass in *Ouranopithecus* estimated at 48 kg. The lowest great ape value is for *Pan paniscus*. The two values for *A. afarensis* correspond to the two different estimates for average male mass for this species (44.6 and 60.1 kg) given in McHenry (1992). All data were natural log transformed.

and is virtually identical to the value for *Pan troglodytes troglodytes* (Figure 12.2).

Judging by the two mandibular specimens, therefore, *Ouranopithecus* does not have reduced male canines. Canine height in this taxon is very close to what would be expected for an extant male ape of this size.

Maxillary relative canine height

While the above analysis of mandibular canines does not support the notion of canine reduction in *Ouranopithecus*, characterisations of this taxon as

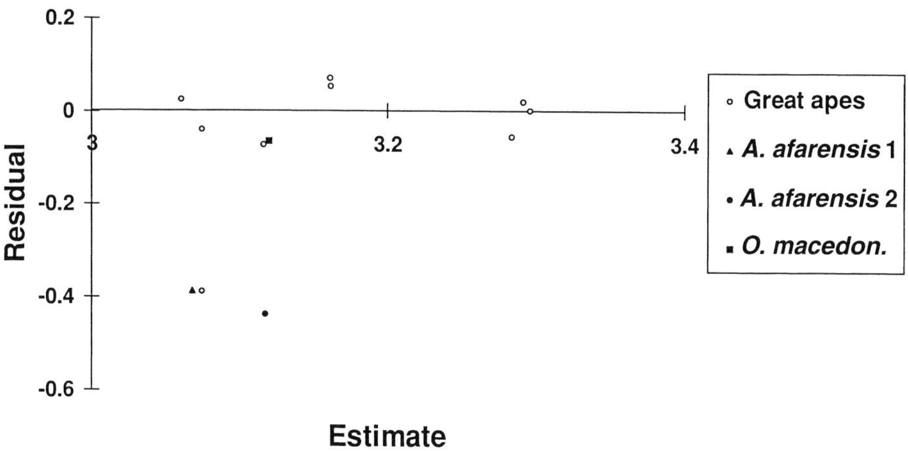

[Figure 12.2]
Male residual mandibular canine height as in Figure 12.1 but with average male body mass in *Ouranopithecus* estimated at 60 kg. All other data as in Figure 12.1.

having reduced canines have been based for the most part on maxillary specimens. Further, while the analyses that suggest canine reduction in *Ouranopithecus* were based on canine/molar ratios as noted, maxillary specimens of *Ouranopithecus* have also been figured by Bonis & Koufos (1993, 1994) to demonstrate similar canine height in the latter and *A. afarensis*. As figured by Bonis and Koufos, however, the *Ouranopithecus* canines have not been corrected for the effects of heavy wear (Figure 12.3), making them appear similar in terms of relative height to the genuinely reduced canines of *A. afarensis*. Based on such comparisons and given the foregoing analysis of the mandibular canines, one might be misled into concluding that perhaps only the maxillary canines of *Ouranopithecus* were reduced while the mandibular canines remained high crowned. However, the necessity of proper occlusion would seemingly rule out such a morphological chimera and, in fact, such a pattern is not observed among extant primates, including those with reduced canines.

A new palate from Ravin de la Pluie preserving a partially damaged but minimally worn canine, RPl 775, is also relevant to assessment of maxillary relative canine size in *Ouranopithecus* (Bonis *et al.*, 1998). The specimen was described as a male by Bonis *et al.* based on the relatively large size of its postcanine teeth, which are mostly comparable to those of other presumed male specimens from this site and Xirochori (Table 12.4). The specimen is noteworthy because the canine is comparatively small and obviously relatively low-crowned, although canine height was not reported.

As a male, this specimen might be considered to provide additional

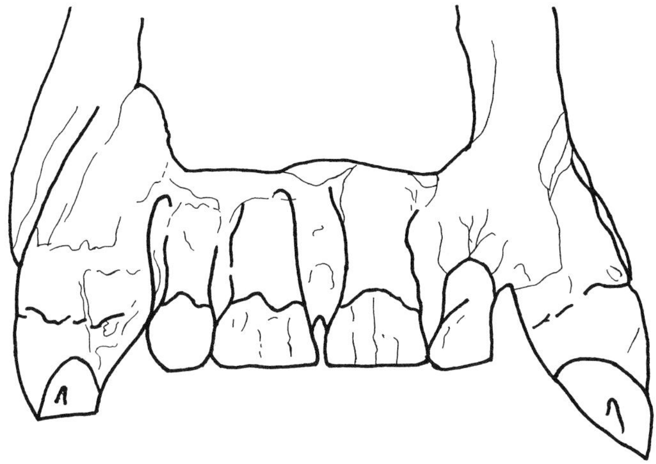

[Figure 12.3]
Anterior view of the alveolar portion of the maxilla of XIR 1 from Xirochori (redrawn from Bonis & Koufos, 1993). Note that both canines are worn to the base of the mesial groove.

evidence for reduced male canines in *Ouranopithecus*. Unfortunately, damage to the mesial portion of the canine crown cervically precludes calculating the canine metric indices by which this individual could be reliably sexed (Kelley, 1995a,b). However, other features of the canine of this specimen strongly suggest that it is female, in spite of the apparently large size of the cheek teeth (see below). Most notably, the canine has a massive lingual cingulum, a feature that has frequently been noted to distinguish female canines from those of males among many fossil and extant anthropoid primates (e.g., Remane, 1927; Hooijer, 1948; Harrison, 1988). A large cingulum is absent on the other presumed male canines of *Ouranopithecus*, but a wide dentine hollow, the remnant of a large lingual cingulum which has been worn away, is present on RPl 208, an obvious and undisputed female canine of this species (Bonis & Melentis, 1978; Kelley, 1995a; Koufos, 1995). In addition, the conformation of the crown dentine and the root at the mesial cervix in RPl 775 is such that, if the enamel were intact and the surface over the mesial groove unworn, the enamel would have projected mesial to the line of the root (Kelley, personal observation). This is perhaps the most consistent identifying feature of female canines among extant great apes (Kelley, 1995b). This morphology would be inconsistent with that of the other male canines of *Ouranopithecus* (Koufos, 1995), but is similar to that seen in the female canine, RPl 208 (Bonis & Melentis, 1978; Bonis *et al.*, 1990; Koufos, 1995). Given the preserved features of the undisputed male canines of *Ouranopithecus*, if RPl 775 is male there would need to be two distinct male canine morphologies in *O. macedoniensis*.

It is also necessary to comment briefly on the large postcanine teeth in RPl 775, since it was the size of the postcanine teeth that led Bonis *et al.* (1998) to regard RPl 775 as a male. As described by Bonis *et al.* (1998), the postcanine teeth of RPl 775 are substantially larger than those of a presumed female, NKT 89, but are similar in size to those of the likely males (Table 12.4). However, most of the postcanine teeth of RPl 775 have been affected by post-depositional infiltration of water and subsequent expansion (Kelley, personal observation; see also Bonis *et al.*, 1998, figure 1). The only completely intact molar teeth in RPl 775, the M^3s, are intermediate in size between those of NKT 89 and the likely males (Table 12.4). Moreover, the M^3 of RPl 775 is substantially smaller than M^2 and is even slightly smaller than M^1, whereas in the other four associated maxillary dentitions of *Ouranopithecus* M^3 and M^2 are subequal in size and both are much larger than M^1 (Table 12.4). This comparison reveals the extent to which water infiltration has increased the size of M^1 and M^2 of RPl 775. RPl 775 is certainly a large female but not unusually so when post-depositional alteration is taken into account, and when the entire *Ouranopithecus* sample is viewed in the context of molar size ranges within extant primate species. Even the most highly dimorphic primates, such as *Pongo* and *Gorilla*, exhibit substantial overlap in male and female postcanine tooth size (Pilbeam & Zwell, 1972; Swindler, 1976; Uchida, 1998). The only primate in which this may not be so is *Lufengpithecus lufengensis* (Kelley & Plavcan, 1998). RPl 775 simply represents the upper end of the female tooth size range, a part of the range that would be expected to be filled in as sample size increased.

In spite of the heavy wear on all upper canines, there is a means by which average unworn upper canine height in *Ouranopithecus* can be estimated. In all extant great apes, and very probably in virtually all primates, upper canine height is greater than lower canine height. This holds not only for species averages but, with very few exceptions, for upper and lower canine heights in individuals as well. Using just the unworn canines from the extant ape canine data base of Kelley (1995b), only in *Pongo* were there individuals (3 out of 24) in which the lower canine was higher-crowned than the upper canine, and in each case only minimally so. In all the remaining individuals of *Pongo*, and in all individuals of the other great apes including *Pan paniscus*, the upper canine was higher-crowned (Table 12.6). These observations not only reveal that it is highly improbable that only the maxillary canines would be reduced in *Ouranopithecus*, but, when combined with the *Ouranopithecus* lower canine data, they can be further used to model

Table 12.6. *Comparative upper and lower canine height in great apes*

Taxon	*N*	No. of Individuals Lower > Upper	Average % Upper > Lower
Pan troglodytes	36	0	11.8
Pan paniscus	7	0	19.6
Gorilla gorilla	23	0	20.2
Pongo pygmaeus	24	3	5.5

Based on unworn and minimally worn canines (Kelley, 1995b).

average upper canine height in this taxon.

The average percentage by which individual upper canine height exceeds lower canine height varies among extant great ape species, with the lowest value (5.5%) in *Pongo pygmaeus* and the highest (20.2%) in *Gorilla gorilla*; *Pan troglodytes* (11.8%) is intermediate (Table 12.6). Using *P. pygmaeus* and *P. troglodytes* as models (to be conservative), the average lower canine height of 21.2 mm determined from RPl 55 and RPl 75 leads to a predicted average maxillary canine height in *Ouranopithecus* of 22.4 mm and 23.7 mm respectively.

Judging by the degree of wear and the height of the remaining portions of the canine crowns in XIR 1, the above predictions would appear to be reasonable, perhaps even somewhat conservative. Both canines of XIR 1 are worn down to the base of the mesial groove, but the left canine preserves somewhat more of the crown, measured at 15.5 mm between the cervix and the worn edge apically (Figure 12.3). Using unworn maxillary canines of the Miocene hominoids *Sivapithecus* and *Lufengpithecus* as guides, the amount of wear evident in XIR 1 would represent an approximately 30–40% reduction of original, unworn canine height. Accordingly, unworn crown height in this individual would be estimated to have been in the range of about 22 to 26 mm, which is consistent with the above estimates of average maxillary canine height in *Ouranopithecus* based on the lower canines.

An analysis of male relative maxillary canine height was carried out using the lower of the two predicted average unworn canine height values, again to be conservative. The procedures and extant ape comparative samples were the same as for the mandibular canine analysis. Not surprisingly, the results are similar to those for the mandibular canines. Residual maxillary canine height is shown to be only slightly less than expected, even for the 60 kg body mass estimate (Figure 12.4). Residual maxillary canine height in *Ouranopithecus* is substantially greater than in either *P. paniscus* or *A. afarensis*, regardless of which canine height or body mass estimates are used.

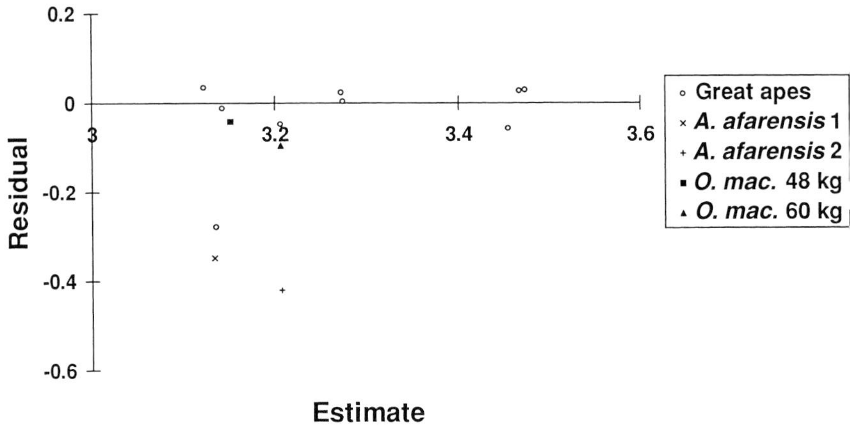

[Figure 12.4]
Male residual maxillary canine height in great apes, *Ouranopithecus* (at both 48 and 60 kg), and *Australopithecus afarensis*, with average male canine height in *Ouranopithecus* estimated at 22.4 mm (see text). Other data sources as for Figure 12.1.

While the analysis of maxillary relative canine height is not as reliable as that for the mandibular canines – being based on only one individual and on predicted rather than measured canine height for *Ouranopithecus* – it is nevertheless consistent with the latter in that there is no evidence for canine reduction in *Ouranopithecus*.

Canine reduction in *Ouranopithecus*?

As the estimates of male relative canine height in *Ouranopithecus* are based on only three individuals, and given statistical error associated with the body mass estimates based on orbital area, the accuracy of these estimates can certainly be questioned. In spite of these limitations, however, it cannot be claimed that the available canine sample supports the notion of canine reduction, or even incipient canine reduction, in *Ouranopithecus*. What data there are indicate that canine height in *Ouranopithecus* was within the range expected for an ape of its body size. Even allowing for the possibility of an underestimate of average male body mass for *Ouranopithecus* by as much as 20% does not alter the results in favour of reduced canines. By any reasonable estimate, relative canine height in this taxon was substantially greater than in either *P. paniscus*, the only extant great ape with reduced male canines, or *A. afarensis*. Apparent canine reduction in *Ouranopithecus* based on cervical crown dimensions rather than canine height is almost certainly an artifact of molar megadontia.

Utility of relative canine size as a character in phylogenetic analyses

As described above, *P. paniscus* males display nearly the same degree of canine reduction as *A. afarensis* males. This represents: (1) a shared derived character, in which case *A. afarensis* and other members of the human lineage would share a last common ancestor with *P. paniscus* to the exclusion of *P. troglodytes*; (2) a shared primitive character, in which case *P. troglodytes* would have undergone a character reversal, from having reduced canines to having high-crowned canines; or (3) a homoplasy. Only in the first alternative, which can be dismissed based on considerable genetic data (e.g., Caccone & Powell, 1989; Ruvolo *et al.*, 1993; Morin *et al.*, 1994), does canine reduction have any phylogenetic significance. As a shared primitive character, canine reduction is of no phylogenetic value for establishing a phylogenetic link specifically with the human lineage. More importantly, the presence of homoplasy among three closely-related species should give pause to attaching much significance to the presence of the character in other species of presumed close relationship. Therefore, evidence of genuine canine reduction in any Miocene hominoid species should be viewed with caution as an indicator of a phylogenetic link specifically with humans to the exclusion of chimpanzees.

Acknowledgments

I would like to thank Louis de Bonis and George Koufos for inviting me to participate in the workshop on 'Phylogeny of Eurasian Neogene Hominoid Primates'. I also thank the European Science Foundation and the Coordination Committee of the Network on Hominoid Evolution and Environmental Change in the Neogene of Europe. Thanks to Peter Andrews and David Pilbeam for discussions on this topic, and to Bernard Wood for commenting on the manuscript. The research reported here and my participation in the workshop were supported by NSF grant SBR 9408664.

References

Aiello, L. C. & Wood, B. A. (1994). Cranial variables as predictors of hominine body mass. *Am. J. Phys. Anthrop.* **95**: 409–26.

Andrews, P. (1990). Lining up the ancestors. *Nature* **345**: 664–5.

Asfaw, B, White, T., Lovejoy, O., Latimer, B., Simpson, S. & Suwa, G. (1999). *Australopithecus garhi*: a new species of early hominid from Ethiopia. *Science* **284**: 629–35.

Bonis, L. de & Koufos, G. D. (1993). The face and the mandible of *Ouranopithecus macedoniensis*: description of new specimens and comparisons. *J. Hum. Evol.* **24**: 469–91.

Bonis, L. de & Koufos, G. D. (1994). Our ancestor's ancestor: *Ouranopithecus* is a Greek link in human ancestry. *Evol. Anthrop.* **3**: 75–83.

Bonis, L. de & Melentis, J. (1977). Les Primates hominoïdes du Vallésien de Macédoine (Grèce). Étude de la mâchoire inférieure. *Géobios* **10**: 849–85.

Bonis, L. de & Melentis, J. (1978). Les Primates hominoïdes du Miocène supérieur de Macédoine. Étude de la mâchoire supérieure. *Ann. Paléontol.* **64**: 185–202.

Bonis, L. de & Melentis, J. (1984). La position phylétique d'*Ouranopithecus*. *Cour. Forsch. Inst. Senckenberg* **69**: 13–23.

Bonis, L. de and Melentis, J. (1987). Intérêt de l'anatomie naso-maxillaire pour la phylogénie des Hominidae. *C. R. Acad. Sci. Paris, Série II* **304**: 767–9.

Bonis, L. de, Bouvrain, G., Geraads, D. & Koufos, G. (1990). New hominid skull material from the late Miocene of Macedonia in northern Greece. *Nature* **345**: 712–14.

Bonis, L. de, Johanson, D., Melentis, J. & White, T. (1981). Variations métriques de la denture chez les Hominidés primitifs: comparaison entre *Australopithecus afarensis* et *Ouranopithecus macedoniensis*. *C. R. Acad. Sci. Paris, Série II* **292**: 373–6.

Bonis, L. de, Koufos, G. D., Guy, F., Peigné, S. & Sylvestrou, I. (1998). Nouveaux restes du primate hominoïde *Ouranopithecus* dans les dépôts du Miocène supérieur de Macédoine (Grèce), *C. R. Acad. Sci., Sci. terre planètes* **327**: 141–6.

Caccone, A. and Powell, J. R. (1989). DNA divergence among hominoids. *Evolution* **43**: 925–42.

Harrison, T. (1988). A taxonomic revision of the small catarrhine primates from the early Miocene of East Africa. *Folia Primatol.* **50**: 59–108.

Hooijer, D. A. (1948). Prehistoric teeth of man and of the orang-utan from central Sumatra, with notes on the fossil orang-utan from Java and southern China. *Zool. Meded.* **29**: 175–301.

Johanson, D. C., White, T. D. & Coppens, Y. (1982). Dental remains from the Hadar formation, Ethiopia: 1974–1977 Collections. *Am. J. Phys. Anthrop.* **57**: 545–604.

Kappelman, J., Richmond, B., Ryan, T., Seiffert, E., Begun, D. & Güleç, E. (2000). Primate fossils from the Sinap formation. In Fortelius, M., Kappelman, J., Sen, S., and Bernor, R. L. (eds.), *Geology and Paleontology of the Miocene Sinap Formation in Central Turkey*. New York: Columbia University Press.

Kay, R. F. (1982). *Sivapithecus simonsi*, a new species of Miocene hominoid, with comments on the phylogenetic status of the Ramapithecinae. *Int. J. Primatol.* **3**: 113–73.

Kelley, J. (1986). *Paleobiology of Miocene Hominoids*. PhD Thesis, Yale University.

Kelley, J. (1995a). Sex determination in Miocene catarrhine primates. *Am. J. Phys. Anthrop.* **96**: 391–417.

Kelley, J. (1995b). Sexual dimorphism in canine shape among extant great apes. *Am. J. Phys. Anthrop.* **96**: 365–89.

Kelley, J. & Plavcan, J. M. (1998). A simulation test of hominoid species number at Lufeng, China: implications for the use of the coefficient of variation in paleotaxonomy. *J. Hum. Evol.* **35**: 577–96.

Koufos, G. D. (1995). The first female maxilla of the hominoid *Ouranopithecus macedoniensis* from the late Miocene of Macedonia, Greece. *J. Hum. Evol.* **29**: 385–99.

McHenry, H. M. (1992). Body size and proportions in early hominids. *Am. J. Phys. Anthrop.* **87**: 407–31.

Morin, P. A., Moore, J. J., Chakraborty, R., Jin, L. Goodall, J. & Woodruff, D. S. (1994). Kin selection, social structure, gene flow, and the evolution of chimpanzees. *Science* **265**: 1193–201.

Pilbeam, D. R. (1969). Tertiary Pongidae of East Africa: Evolutionary relationships and taxonomy, *Peabody Mus. Bull.* **31**: 1–185.

Pilbeam, D. & Zwell, M. (1972). The single species hypothesis, sexual dimorphism, and variability in early hominids. *Yb. Phys. Anthrop.* **16**: 69–79.

Plavcan, J. M. (1998). Correlated response, competition, and female canine size in primates, *Am. J. Phys. Anthrop.* **107**: 401–16.

Plavcan, J. M. & van Schaik, C. P. (1992). Intrasexual competition and canine dimorphism in anthropoid primates, *Am. J. Phys. Anthropol.* **87**: 461–77.

Plavcan, J. M. & van Schaik, C. P. (1994). Canine dimorphism, *Evol. Anthrop.* **2**: 208–14.

Plavcan, J. M. & van Schaik, C. P. (1997). Interpreting hominid behavior on the basis of sexual dimorphism. *J. Hum. Evol.* **32**: 345–74.

Plavcan, J. M., van Schaik, C. P. & Kappeler, P. M. (1995). Competition, coalitions and canine size in primates. *J. Hum. Evol.* **28**: 245–76.

Remane, A. (1927). Studien über die Phylogenie des menschlichen Eckzahns. *Z. Anat. Entw. Gesch.* **82**: 391–481.

Ruvolo, M., Zehr, S., von Dornum,, M., Pan, D., Chang, B. & Lin, J. (1993). Mitochondrial COII sequences and modern human origins. *Mol. Biol. Evol.* **10**: 1115–35.

Smith, R. J. & Jungers, W. L. (1997). Body mass in comparative primatology. *J. Hum. Evol.* **32**: 523–59.

Swindler, D. R. (1976). *Dentition of Living Primates.* London: Academic Press.

Uchida, A. (1998). Variation in tooth morphology of *Gorilla gorilla. J. Hum. Evol.* **34**: 55–70.

White, T. D. & Johanson, D. C. (1982). Pliocene hominid mandibles from the Hadar Formation, Ethiopia: 1974–1977 Collections. *Am. J. Phys. Anthrop.* **57**: 501–44.

13 Heterochrony and the cranial anatomy of *Oreopithecus*: some cladistic fallacies and the significance of developmental constraints in phylogenetic analysis

David M. Alba, Salvador Moyà-Solà, Meike Köhler and Lorenzo Rook

One of the several aims of evolutionary biology is to recognize the evolutionary branching order of organisms by means of phylogenetic reconstruction methods. In the last 20 years, other fields of evolutionary biology have undergone a dramatic progress. This is especially true for the study of the developmental basis of evolution (Raff & Kaufman, 1983; Raff, 1996), which has stressed the significance of development for evolution and vice versa. However, this new synthesis between developmental and evolutionary biology, prompted by the publication of Gould's (1977) *Ontogeny and Phylogeny*, has not yet been applied to phylogenetic studies of hominoid evolution. In our opinion, this situation cannot be maintained any longer, because morphological evolution cannot be properly understood without considering the constraining effect of development upon evolution. This raises the question of whether or not current widely used methods of phylogenetic reconstruction (i.e. cladistics), based on the use of extensive datasets instead of a deep understanding of individual characters, are reliable. After reviewing the significance of developmental constraints in evolution, we will argue that the reasons underlying the appearance of characters are absolutely relevant for phylogenetic analysis, and will try to exemplify their significance through the example of *Oreopithecus bambolii*, an extinct orthograde great ape of the late Miocene (*c*. 8 Ma) of the Tusco-Sardinian paleobioprovince. It is our contention that cladistic methodology is seriously flawed in several aspects and can give quite erroneous insights of phylogenetic relationships when evolutionary constraints as well as morpho-functional considerations are not taken into account.

Heterochrony and evolutionary constraints

'Why do we observe certain forms in nature and no others?'

Alberch (1989, p. 22).

'There are many factors that act to constrain evolution and to channel it down a certain number of broad, and sometimes not so broad, pathways ... The ontogenetic development of an organism is a built-in constraining factor in the first place.'

McNamara (1997, pp. 133–4).

There are two different approaches to morphological evolution and adaptation (Alberch, 1989): first, the 'classical' externalist/functionalist approach that emphasises the role of natural selection while disregarding internal factors such as constraints, so that morphological evolution is considered to be driven mainly by external (ecological) factors; and second, the internalist approach, which, instead of focusing on adaptation, emphasises the limiting and channeling effect of constraints upon adaptation as well as their significance in determining the course of evolution. In fact, however, the distinction between internal and external factors is more academical than real. Since both are intimately linked (Alberch, 1989), it is necessary to take both approaches into account simultaneously to be able to investigate how evolution really proceeds (see also Wake, 1991).

Albeit traditionally disregarded by most evolutionary biologists, constraints play an important role in evolution, since they limit the potential spectrum of variability available to be sorted by whatever mechanism, thus restricting the potential adaptive changes under the action of natural selection (Gould & Lewontin, 1979). Ecological factors, of course, determine fitness and, therefore, control the sorting of variability during evolution. Nevertheless, besides these external factors, adaptation can be also restricted prior to the action of sorting mechanisms. In other words, adaptation is constrained because some variability never appears (or appears less frequently) and, consequently, it cannot be selected, either positively or negatively – in Alberch's (1985: p. 430) words: 'if phenotypic variation is not present, selection cannot operate on it'. This is the reason why some investigators have even asserted that organismal evolution could be so constrained that 'the constraints themselves become more interesting and more important in delimiting pathways of change than the selective force that may mediate change when it occurs' (Gould & Lewontin, 1979, p. 252).

Two basic types of evolutionary constraints can be distinguished (Gould, 1989): (1) structural (= architectural, formal) constraints 'are universal properties of matter' (Gould, 1989, p. 517), i.e. constraints due to the fact that chemical and physical laws (including the geometrical properties of form), to which everything is subject, do not allow the appearance of certain morphologies; and (2) phylogenetic (= historical) constraints 'are particular consequences of contingent histories' (Gould, 1989: p. 517). In other words, every living organism is loaded by its phyletic heritage (resulting from its own evolutionary history), so that future evolution is constrained by the limited set of possibilities upon which natural selection can act. Unlike structural constraints, phylogenetic ones can sometimes be overcome (though indirectly) by means of alternative possibilities, as illustrated by the principle of the panda's thumb (Gould, 1980).

Developmental constraints are possibly the most frequent phylogenetic constraints. 'A developmental constraint is any bias on the expression of phenotypic variability caused by the structure of the developmental interactions that generate global patterns' (Alberch, 1985, p. 430; see also Alberch, 1982). In other words, evolution works from what is available, and since development offers only a limited range of possible morphologies to be sorted, the directional action of natural selection is limited – and oriented – by developmental constraints (Alberch, 1982). Although evolution is mainly a matter of adaptation to environmental conditions, it actually results from the interaction between constraints and functional adaptation (Gould, 1997). If morphological variability was generated at random, almost endless possibilities would be available for natural selection. However, since this is not true due to the canalisation of developmental processes, adaptation is constrained by development.

It is beginning to be recognised (Klingenberg, 1998) that heterochrony – 'the phenomenon of morphological evolution due to selection upon variability resulting from shifts in developmental timing, and causing parallelisms between ontogeny and phylogeny' (Alba, in press) – can explain the generation of some developmental constraints, understood as biases in the direction of morphological evolution (i.e. 'organismal morphologies that evolve more or less easily than others'; Klingenberg, 1998: p. 113) due to biases on the generation of variability (Alberch, 1982, 1985). It should be noted that developmental constraints are not only limits imposed upon the action of natural selection, but also factors channeling (directing) the course of evolution (Gould, 1989), irrespective – or in spite – of immediate adaptation. This second, positive view of developmental constraints is especially true for heterochronic ones (i.e. for constraints due to the selection of variability produced by the alteration of developmental timing), which could be the most common for probabilistic reasons. As suggested by Raff & Kaufman (1983: p. 179), 'Given the stability of developmental pathways, heterochrony provides the course of least resistance for evolutionary change'. In other words, if changes in developmental timing are one of the easiest ways to produce variability, heterochrony would be very pervasive in evolution (as long suspected by some authors, e.g. Gould, 1977; McKinney & McNamara, 1991; McNamara, 1997), and consequently the role of developmental constraints in evolution (and in phylogenetic reconstruction) should not be ignored. Gould (1989: p. 11) already noted that 'Heterochrony is perhaps our best empirical mode for the study of developmental constraint'.

In Alba's (in press) words, the pervasive nature of heterochrony 'means that constraints must have played an important role in evolution because of limiting the range of potentially adaptive variation, so that instead of strolling freely through a universe of almost endless possibilities, evolution more often travels along roads that, to a certain extent, were already predetermined (channelled and constrained) by the ancestral developmental pathways'.

The role of evolutionary constraints in phylogenetic analysis

Although usually considered the most objective and reliable methodology of phylogenetic reconstruction, cladistics suffers from different problems. In addition to the subjective description and definition of character states, there are two potential sources of mistake: (1) the pervasive nature of homoplasy[1] ('false homology', i.e. similarity among taxa due to independent evolution); and (2) the lack of clear-cut criteria to undertake a priori[2] character analysis and weighting. In fact, both are different aspects of the same problem, since the latter point causes an improper understanding of characters that, among others, makes possible the confusion of homoplasy and homology. Since the advent of cladistics, it has been recognised that only synapomorphies (shared derived homologies) are useful to decipher the order of evolutionary branching, so that they have to be distinguished from symplesiomorphies (shared primitive homologies) (Hennig, 1966). This is not an easy task, however, because truly homologous characters have to be firstly distinguished from homoplastic ones.

To solve the problem of homoplasy, cladistic methodology relies on the principle of parsimony (or 'Ockham's razor') to decide among competing hypotheses. 'The principle of parsimony states that the simplest explanation consistent with a data set should be chosen over more complex explanations' (Stewart, 1993, p. 603), so that the hypothesis with fewer steps (the so-called 'most parsimonious one') is chosen as the correct one. However, in spite of minimising the absolute number of evolutionary transitions (and, thus, of homoplasies), cladistic analyses always detect a posteriori a significant minimum degree of homoplasy. In fact, it is widely recognised by most

[1] Although some authors prefer the term 'convergence' instead of 'homoplasy' (e.g. Moore & Willmer, 1997), we prefer 'homoplasy' and use 'convergence' in a more restricted sense (see later).

[2] We use the term 'a priori' only to refer to temporal priority, irrespective of logical priority, since character weighting would be in fact 'a posteriori' if justified by previous experience (see discussion in Gosliner & Ghiselin, 1984).

biologists (e.g. Gosliner & Ghiselin, 1984; Alberch, 1989; Willmer, 1990; Wake, 1991; Moore & Willmer, 1997; Shubin, 1998), including palaeoprimatologists (Fleagle, 1999) and other palaeontologists (Carroll, 1988; Conway Morris, 1998), that homoplasy seems to be very common. Hominoid primates are not an exception, with homoplasies being very pervasive and affecting both craniodental and postcranial characters (Begun & Kordos 1997; see also Larson, 1998).

Undoubtedly, parsimony remains as an allowable assumption on philosophical grounds for proposing hypotheses that are open to testing, since the validity of any particular hypothesis depends on independent testing, and not on the methodology used to derive it. However, parsimony-based criteria are currently used not only to propose phylogenetic hypotheses, but also to decide among competing ones without further testing. This is not justified. It should not be forgotten that cladograms are hypotheses that can be tested not only by future, more refined cladistic studies, but also by all the other available independent sources of information. Thus, independent testing could be carried out by means of any kind of independent information not included in the parsimony-based analysis (morphological characters previously not considered, DNA sequences, (palaeo)biogeography and biochronology), as well as by the adequacy of future discoveries to be made in the fossil record with the predictions of pre-existing cladograms. It seems now that phylogenies based on molecular evidence, although not without problems, offer the best morphology-independent source of phylogenetic information, with the advantage that powerful statistical methods not relying on parsimony can be more readily employed with this kind of data (Sidow, 1994; Moore & Willmer, 1997). The great reliance of most morphology-based phylogenetic analyses on the principle of parsimony is understandable given the scarcity of independent sources of evidence (especially in the fossil record). Nevertheless, once the great preponderance of homoplasy in evolution has been demonstrated (even by cladistic studies themselves), one should wonder whether parsimony is a reliable criterion to decide among competing phylogenetic hypotheses or not. Although the cladistic use of the principle of parsimony does not imply that homoplasy is rare in evolution, it should be also clear that the hypothesis with fewer steps is not necessarily the 'most parsimonious' one in a philosophical sense, provided that there are good reasons to think that homoplasy is very common. In fact, despite claims to the contrary, it has been already noted by several authors that if homoplasy is really pervasive in evolution, as suspected by some investigators (e.g. Gosliner & Ghiselin, 1984), then 'existing cladistic methods fall' (Wake, 1991: p. 544). Even compromised adherents of the principle of parsimony have accepted that 'Homoplasy ... is the ultimate trickster of parsimony [and] can cause historically incorrect trees to be most parsimonious' (Stewart, 1993: p. 606), so that 'The most parsimonious tree will be the correct phylogeny only if the number of shared-derived characters is high enough and the number of homoplasies low enough' (p. 604).

Obviously, all these problems are especially important when working with fossils, since the lack of many data sources and the scarcity of available evidence (in comparison with living species) makes phylogenetic inferences even more insecure, and definitive statements almost impossible by definition. In fact, Wood and Collard (Chapter 6) have shown that a cladistic analysis of living and fossil hominoids fails to generate the nowadays well established molecular phylogeny of living hominoids, thus casting doubt on the phylogenetic conclusions derived for fossil taxa using the cladistic methodology. There have been some attempts to circumvent the problem of homoplasy in the case of hominoid and hominid evolution. For example, to account for the fact that Nature is not parsimonious, Begun & Kordos (1997) have proposed that all cladograms with a consistency index (CI) up to 5% lower than that of the most parsimonious cladogram should be considered equally likely (see also Begun *et al.*, 1997). We agree with them that there is no objective reason to assume that the most parsimonious cladogram reflects real phylogeny more accurately that other slightly less parsimonious cladograms. However, their solution does not seem a good one, since the great number of cladograms fitting the 5% criterion (as found by Begun & Kordos, 1997) precludes any secure conclusion using only these criteria. Moreover, since parsimony maximises character congruence, *the degree of homoplastic evolution detected by cladistic studies will be only a minimum estimate, being no objective reason to think that it could not be even greater* (*cf.* Moore & Willmer, 1997). Consequently, it is useless to accept or reject cladograms on the basis of a given degree of homoplasy (e.g. a 5% difference in relation to the most parsimonious one), since the real (not the minimum) degree of homoplasy has been never consistently evaluated in Nature. Although Moore & Willmer (1997: p. 1) assert that 'The question "How common is convergence [= homoplasy]?" remains unanswered and may be unanswerable', they conclude that in general it has been greatly underestimated, having a higher prevalence than homologous similarity, at least in some groups like the Opistobranchia (Gosliner & Ghiselin, 1984), or even being so high as to preclude the proposal of a robust phylogeny for plethodontid salamanders (Wake, 1991), among many other examples.

An alternative solution to the problem of homoplasy has been proposed by Lieberman *et al.* (1996), who try to circumvent it by means of a posteriori re-evaluation of characters. Again, we agree with them in that a closer investigation of characters is necessary because of the fact that parsimony is an unsatisfactory criterion for those situations in which homoplasy is frequent. In our opinion, however, characters should be investigated *prior* to their use in any kind of phylogenetic analysis, because of the nonoperational nature of the definition of homology (Bock, 1974) as similarity of organisms due to the descent from a common ancestor. In our opinion, homologies and homoplasies must be distinguished without considering any phylogenetic hypothesis (irrespective whether it is the most parsimonious or not), at least on preliminary grounds. Although it seems that circularity is always involved to certain degree when discerning between homology and homoplasy (Moore & Willmer, 1997), the usual cladistic procedure of assuming characters to be homologous a priori, and identifying homoplasies only a posteriori (once the most parsimonious phylogeny has been derived), relies on a methodology that has to assume minimal homoplastic evolution precisely to be able to evaluate the prevalence of homoplasy, thus resulting in an absolutely circular reasoning (Bock, 1974; Gosliner & Ghiselin, 1984; Moore & Willmer, 1997). This explains why homoplasies are only perceived as *ad hoc* hypothesis under a cladistic framework ('parsimony criterion consists of nothing other than avoiding unnecessary *ad hoc* hypotheses of homoplasy, those not required by observation': Farris, 1982: p. 330; see also Siebert, 1992), where they are invoked a posteriori to explain incongruent characters and thus preclude refutation of the preferred (most parsimonious) phylogenetic hypothesis (Neff, 1986). Unfortunately, a priori character weighting has been rejected by many strict cladists, who, however, have not proposed good alternatives to overcome the problem of homoplastic evolution. Perhaps this responds to the fact that, for cladists, character incongruence is only 'noise' (and therefore problematic), whereas for those biologists interested not only in phylogeny but also in adaptation and the evolution of morphology, homoplasies are something very interesting and deserve further explanation.

Of course, a priori character weighting (which can be used simultaneously with parsimony, Farris, 1982) requires some additional hypotheses about adaptation and constraints. In fact, characters themselves should be treated 'as hypotheses subject to test and possible refutation prior to their use at higher levels, such as in cladistic analysis' (Neff, 1986: p. 115), which can only be accomplished by means of some a priori weighting method. In our opinion, it is better to rely on a few deeply understood characters than on an extensive dataset full of potentially homoplastic characters, and this is the reason why, as pointed out by Köhler & Moyà-Solà (1997a), the current use of long lists of discrete non-weighted characters cannot solve problems such as homoplastic evolution (due to heterochrony, allometry, etc.), high adaptive and/or ecophenotypical plasticity of characters and unknown polarity of change. In this way the major discrepancies found among the cladistic studies of, say, Rose (1997) and Ward (1997) (based on the forelimb and the trunk and hindlimb, respectively) can be explained. (And the same could be said about the above-mentioned discrepancies found by Wood & Collard, Chapter 6, between phylogenies obtained by means of cladistic methodology for living and extinct hominoids, and phylogenies for living organisms derived using molecular data.) Instead, we should investigate the origin itself of the characters to understand 'the biological basis of the features that we use to reconstruct evolutionary history' (Shubin, 1998: p. 13), so that 'before conducting any cladistic analysis, the usefulness of a character should be independently assessed in the light of its underlying functional and structural constraints and developmental processes' (Köhler & Moyà-Solà, 1997a: p. 328).

Since Darwin (1859), homoplasy has been mainly attributed to convergence, i.e. independent evolution of similar characters in different lineages. However, it can also be due to parallelism, i.e. independent evolution of similar characters in related lineages as a result of their common ancestry (or, in other words, as a result of historical constraints; see Simpson, 1961; Futuyma, 1986). Concisely stated, '"Parallelism" means that taxa began with the same initial conditions, and independently underwent the same changes', whereas '"Convergence" means that the taxa began with different initial conditions and, by different pathways, arrived at a similar condition' (Gosliner & Ghiselin, 1984). Convergence is usually the result of adaptation, like for example the colonisation of the same environment (as in the classic example between dolphins and ichthyosaurs, or between birds, pterosaurs and bats). Parallelism, however, can be the result of different causes: although the same kind of adaptive reasons can be involved (e.g. the independent evolution of the same ecomorphs of *Anolis* lizards on different islands of the Greater Antilles: Losos *et al.*, 1998; see also Harvey & Partridge, 1998; Vogel, 1998), parallelism is also due to design limitations because of developmental constraints, as proposed for example by Hetch (1989) in the case of neoteny of the urodele cranium, or by Wake (1991) in the evolution of plethodontid salamanders. If these kinds of evolutionary constraints (especially heterochronic ones) have effectively played an important role in evolution, this would explain why cladistic analysis (which does not take them into account) always detects a posteriori such a high degree of homoplasy *in spite of* using the principle of parsimony.

The current status of the phylogenetic position of *Oreopithecus*

Oreopithecus is in many aspects a very particular catarrhine primate, with a typical hominoid postcranial, a very specialized dentition and a quite unusual cranial morphology. Craniodental anatomy has always played an important role in the discussion of the phylogenetic relationships of this fossil primate. Monkey, ape or hominid, the systematic position of *Oreopithecus* has been discussed for more than a century since Gervais described the type specimen in 1872. Among others, *Oreopithecus* has been considered to be an aberrant cercopithecoid (Delson, 1979; Szalay & Delson, 1979; Rosenberger & Delson, 1985), a relative of the primitive anthropoid genus *Apidium* (Simons, 1960), a generalised catarrhine and basal hominoid genus related to the African *Nyanzapithecus* and classified in its own family Oreopithecidae (Harrison, 1985, 1986a,b,c), a great ape of unclear affinities (Sarmiento, 1987; Harrison, 1991), a primitive hominid *s.s.* (Hürzeler, 1954, 1956, 1958, 1960; Straus, 1963), a stem great ape probably related to the genus *Dryopithecus* (Harrison & Rook, 1997) and a member of the Eurasian great ape clade (Ponginae) descended from *Dryopithecus* (Moyà-Solà & Köhler, 1997). (See Delson, 1986, for a comprehensive treatment of the old literature.)

Hürzeler (1958) and Straus (1957, 1958a, 1963) have shown that *Oreopithecus* was undoubtedly a hominoid (ape) instead of a cercopithecoid (monkey), although it was considered by Hürzeler (1954, 1956, 1958, 1960; see also Straus, 1963), on the basis of cranial as well as postcranial remains, an ancient collateral extinct branch of the Hominidae. The hominoid hypothesis was corroborated by the postcranial studies of Sarmiento (1983, 1987) and Harrison (1986b), although noticing the strikingly diverging dental morphology of *Oreopithecus* relative to the typical hominoid pattern (see also Harrison & Rook, 1997). Surprisingly, in spite of the derived dental status of *Oreopithecus*, it was considered to share many traits with the hypothetical ancestor of the great ape and human clade (Sarmiento, 1987), or even to fit quite well with the ancestral inferred catarrhine morphotype (Harrison 1986b, 1991). None the less, at the same time, other investigators maintained that *Oreopithecus* was a cercopithecid (Delson, 1979; Rosenberger & Delson, 1985) on the basis of some dental features and in spite of the obvious hominoid nature of the postcranium. However, since the dental morphology of *Oreopithecus* is extremely derived (Sarmiento, 1987; Harrison & Rook, 1997), as we will see later, and the dental homologies shared with hominoids are mainly symplesiomorphies (Sarmiento, 1987), this kind of evidence cannot be used to decipher whether *Oreopithecus* is a cercopithccoid, a hominoid or a primitive catarrhine. On the contrary, postcranial evidence clearly shows that *Oreopithecus* must be accepted as a hominoid beyond any reasonable doubt (Sarmiento, 1987; Harrison & Rook, 1997). Moreover, it has been recently recognized that *Oreopithecus* belongs to the great ape and human clade (the Hominidae *s.l.*), as shown among others by the lack of subarcuate fossa in the petrosal bone (Moyà-Solà & Köhler, 1997), which is a great ape synapomorphy (Moyà-Solà & Köhler, 1993, 1995; Moyà & Köhler, 1993).

The question remains about the concrete phylogenetic affinities of *Oreopithecus* among the hominoids. Amongst Miocene apes, only *Dryopithecus* has a comparable postcranial anatomy and intermembral proportions (Moyà-Solà & Köhler, 1997) to that of *Oreopithecus*. Moreover, they share the morphology of the zygomatic bone (flat, anteriorly oriented and with three foramina situated above the inferior rim of the orbit), which has been proposed as a synapomorphy of the Eurasian great apes or Ponginae (Schwartz, 1990; Moyà-Solà & Köhler, 1993, 1995; Moyà & Köhler, 1993). We therefore conclude, in agreement with Harrison & Rook (1997), that *Oreopithecus* was most probably derived from a *Dryopithecus*-like ancestor. This exclusively phylogenetic proposal, however, does not explain the *Oreopithecus* strikingly derived craniodental morphology that, during so much time, precluded any secure taxonomic assignment as a hominoid (not surprisingly, since fossil hominoid systematics has been traditionally based on cradiodental features; Pilbeam, 1996, 1997), in spite of being the most well-represented Eurasian fossil ape since the discovery of the strikingly complete (although crushed) skeleton IGF 11778 by Hürzeler in 1958 (Straus, 1958b; Hürzeler, 1960; Figure 13.1). The uniquely derived features of *Oreopithecus* can be only understood once the developmental constraints and insular conditions that took place during its evolution are taken into account.

In our opinion, as firstly proposed by Moyà-Solà & Köhler (1997), *Oreopithecus* must be considered a highly derived (both cranially and postcranially) endemic insular descendant of *Dryopithecus*, which evolved by means of a phenomenon of paedomorphic cranial heterochrony (probably neoteny). In fact, not only the European faunal affinities of the Maremma region (Harrison & Harrison, 1989; Andrews *et al.*, 1996) argue in favour of a European origin for *Oreopithecus* – and remember that *Dryopithecus* is distributed throughout this continent (Begun *et al.*, 1990; Andrews, 1992) – but also there is possibly only an insignificant temporal gap between the last appearance of *Dryopithecus* in Can Llobateres at 9.5 Ma (Agustí *et al.*, 1996) and the first occurrence of insular faunas in the Maremma region (which lasted from 9.5 to 6.5 Ma; Moyà-Solà *et al.*, 1999b). Therefore, if we are right, *Oreopithecus* belongs, along with living *Pongo* and all the other Eurasian

[Figure 13.1]
Photograph of the skeleton of *Oreopithecus bambolii* (IGF 11778) found in 1958 by J. Hürzeler.

great ape fossil genera (*Dryopithecus, Ouranopithecus, Ankarapithecus, Lufengpithecus, Sivapithecus* and *Gigantopithecus*), to the same adaptive radiation that took place during the middle and late Miocene of Eurasia (Pickford *et al.*, 1997) in the form of initial dispersion thought Europe and Asia and subsequent vicariant evolution (Agustí *et al.*, 1996). In this case, the early branching of *Oreopithecus* with respect to all the other great apes indicated by the cladistic studies of Begun *et al.* (1997) could be simply an artifact due to the unique highly autapomorphic and superficially primitive nature of many characters in this taxon. In fact, until recently *Oreopithecus* was still considered to possess some primitive catarrhine features, such as the relatively short face, which we believe to be secondarily derived, paedomorphic features only superficially resembling the inferred primitive condition.

This illustrates the significance of evolutionary constraints with regard to the recognition of homoplastic characters. Especially problematic are those due to parallelism, since in this case there can be no obvious (adaptive) reasons to explain why a given character should have evolved independently in two lineages or should have even repeatedly reversed to the primitive condition (as it is the case of the short face of *Oreopithecus* and other paedomorphic primates). Of course, this is due to phylogenetic constraints, and in particular to the possession of common developmental pathways that can be modified only in a limited number of ways. Amongst others, the easiness of changes in developmental timing makes evolution highly recurrent, thus increasing the likeliness of homoplasy. Perhaps characters should not be routinely assumed to be homologous when homoplasy is not unambiguously demonstrated, especially when no effort is made (at least in cladistic studies) to detect homoplasy. Therefore, in the following discussion, we would like to point out the need for a deep understanding of characters before accepting them as true homologies useful in constructing phylogenetic hypotheses. This understanding of characters, in fact a kind of a priori weighting, would be useful not only to eliminate homoplasy and to properly define character states, but also to understand how evolution works and what kind of inconsistencies can be found among different growth fields and/or functional complexes. Our aim here is to test and discuss these ideas on the basis of the evolution of *Oreopithecus bambolii* using cranial and dental anatomy as examples.

The phylogenetic position of *Oreopithecus* from a heterochronic perspective

– or neotenous

To test the hypothesis of paedomorphic evolution in *Oreopithecus*, we have mainly relied on the published description of Moyà-Solà & Köhler (1997), based on the cranium IGF 11778 and measures of other cranial specimens (BA 40, 43, 60, 61, 62, 63 and 78), some of which are currently being restored at the 'Institut de Paleontologia M. Crusafont' in Sabadell (Spain), as well as on the equally crushed cranium recently rediscovered in the 'Museo Geologico G. Capellini' of the Bologna University, Italy (Gentili *et al.*, 1998;

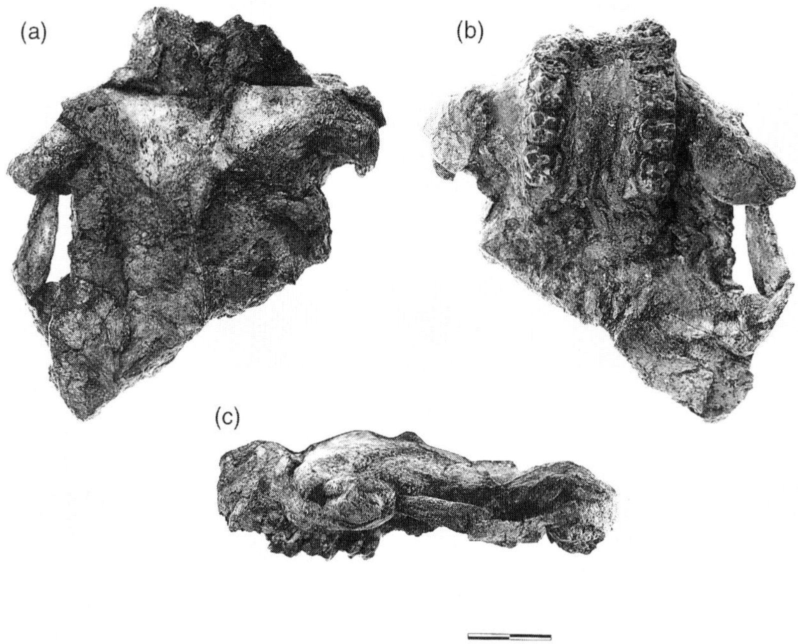

[Figure 13.2]
Photograph of a partial crushed cranium of *Oreopithecus bambolii* from Casteani, housed in the Museum 'G. Capellini' (University of Bologna). (a) Superior view; (b) Palatal view; (c) Left lateral view. Note the distal hallucial phalanx adhered to the braincase. (Scale bar = 1 cm.)

Figure 13.2). We have not taken into account the recently published reconstruction of the same cranium IGF 11778 (Clarke, 1998), because of some artifacts of the reconstruction due to an improper consideration of diagenetical plastic deformation (surprisingly, the reconstructed cranium is longer than the crushed one, which is absolutely impossible).

The craniodental anatomy of *Oreopithecus bambolii* can be summarised as follows (Moyà-Solà & Köhler, 1997; Figure 13.3):

- Extremely short face, with a wide upper face and a high vertically ascending mandibular ramus.
- Orbits and root of the zygomatic placed far forward on the face above the premolars, and root of the zygomatic very low and close to the alveolar margin of premolars.
- Flat and anteriorly oriented zygomatic bone, with three foramina situated above the inferior rim of the orbit, similar to that of *Dryopithecus* and *Pongo* although much slender.
- Prominent saggital crest on a small and globular braincase, with a degree

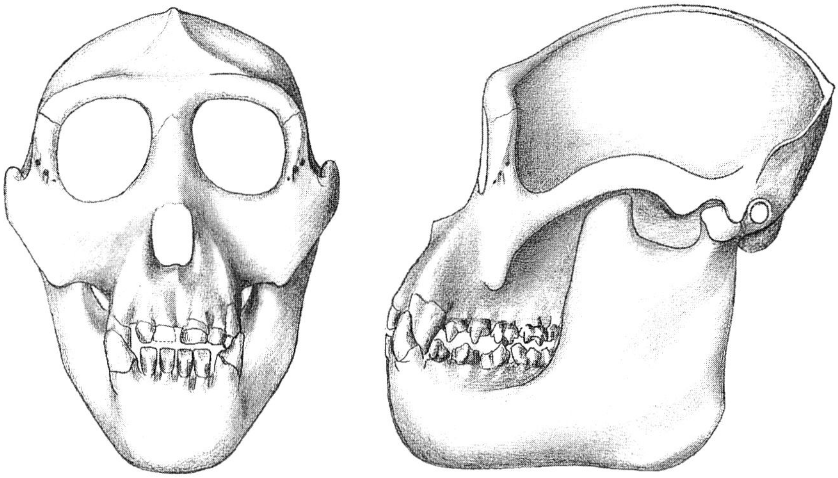

[Figure 13.3]
Reconstruction of the cranium of *Oreopithecus bambolii.*
(See text for further details.)

of encephalization probably below that of most anthropoids and other living hominoids (Jungers, 1987; Harrison, 1989; Martin, 1996), although not so low as Harrison (1989) suggested.

- Petrosal bone lacking subarcuate fossa.
- Small postcanine teeth in relation to body mass (Jungers, 1987, 1990), and also small canines, especially with regard to basal crown area (Alba *et al.*, unpublished data).
- Absence of diastema, vertically implanted incisors, M_3 frequently located behind the ascending ramus, and M^3 usually not fully grown and placed obliquely in adults.
- Teeth with very prominent cusps and supplementary cusps, crests and cingula.

Until recently, the cranial morphology of *Oreopithecus* was seen as a unique combination of some autapomorphies and basically primitive catarrhine features (Harrison, 1986b). In short, *Oreopithecus* was considered to fit quite well to the primitive catarrhine morphotype, on the basis of characters such as relatively short face, orbits situated far forward on the face above the premolars and as high as broad, anterior root of the zygomatic arch close to the alveolar margin of the cheek teeth, and globular but relatively low neurocranium, among others (Harrison, 1986b). Most of the cranial characters of *Oreopithecus*, however, and especially the shortening of the face, the vertically implanted orbits and the low zygomatic root above

Table 13.1. *Frequency distribution of the position of the anterior root of the zygomatic relative to the upper tooth row in* Oreopithecus *and some living hominoids*

Genus	N	Position of zygomatic arc above tooth row								
		M^2/M^3	Post M^2	Mid M^2	Ant M^2	M^1/M^2	Post M^1	Mid M^1	Ant M^1	P^4/M^1
Oreopithecus	2								50.0%	50.0%
Gorilla	26		3.9%	11.6%	34.7%	34.7%	15.4%			
Pan	25	12.0%	8.0%	8.0%	16.0%	44.0%	8.0%	4.0%		
Homo	20					5.0%	27.0%	40.0%	25.0%	3.0%
Hylobates	65	4.6%	12.3%	37.0%	29.2%	10.8%	3.1%	3.1%		
Symphalangus	15	27.0%	7.0%	40.0%	13.0%	13.0%				

Adapted from Harrison (1986b).

the premolars fit neither the reconstructed primitive catarrhine morphology nor the cranial pattern of any living or fossil adult catarrhine primate, but rather that of all infantile and juvenile great apes (Moyà-Solà & Köhler, 1997).

Making heterochronic inferences on fossil primates is not straightforward for several reasons, such as the scarcity and the fragmentary nature of the fossil remains, which preclude a quantitative assessment of many characters, not to say of ontogenetic change. For these reasons, and because the restoration of some cranial remains has not finished yet, we will concentrate only on one of the above-mentioned characters, the position of the zygomatic root, though a deeper analysis including more characters will be presented elsewhere in the future. Harrison (1986b) already noted that the zygomatic root in *Oreopithecus* was positioned far forward on the face (above P^4/M^1) in relation to other catarrhine primates (interestingly, a character also found in *Australopithecus* and *Homo*, although not in such an extreme condition; Table 13.1). Nevertheless, he attributed this trait to powerful mastication, instead of considering the role that heterochrony could have played in the evolution of this character. Interestingly, as pointed out by Moyà-Solà & Köhler (1997), the condition that adult specimens of *Oreopithecus* display (zygomatic root above P^4 or P^4/M^1) most closely resembles the typical infantile hominoid condition (above the dp^4 or P^4), although some variability exists. For example, it is situated above anterior M^1 I the male cranium IGF11778, but above the P^4-M^1 in the female specimen BAC61. The latter, which preserves the right orbit, zygomatic and maxilla, shows that the face is very flat. We can conclude, therefore, that we are dealing with a paedomorphic trait, which can be attributed to a phenomenon of neoteny because the rate of shape change (measured as position of the zygomatic root) seems to have been reduced in the descendant (*Oreopithecus*) in relation to the ancestral ontogenetic trajectory (as inferred from that of living *Pan* and *Pongo*, which are identical for this feature; Figure 13.4).

Interestingly, there is another endemic primate in the fossil record of Mediterranean islands, which shows paedomorphic features in the skull and dissociation between cranial and postcranial reduction. It is the 'dwarf' early Pleistocene *Macaca majori* from Sardinia (Azzaroli, 1946; Zanaga, 1998). This fossil macaque is considerably smaller than the supposed ancestor (the European *Macaca sylvana florentina*) and is comparable in size with the smallest extant East Asian macaque species. However, the facial dimensions of *Macaca majori* are relatively small and show an extreme reduction in length, with a consequent reduction of premolars (especially P3s), a lack of diastemata and also an extreme reduction in canine size (to such an

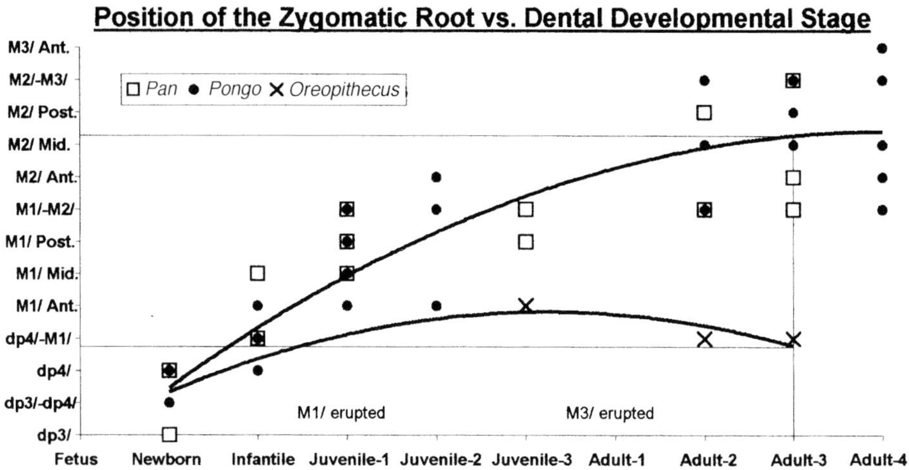

[Figure 13.4]
An example of paedomorphosis in *Oreopithecus bambolii.*
Shape (position of the zygomatic root) plotted *versus* maturation (dental developmental stages), assuming the ontogenetic trajectory shared by living *Pan troglodytes* and *Pongo* to be the ancestral one for great apes, and initial shape to be unchanged in *Oreopithecus*. The rate of shape changes seems to have been decreased, although the lack of chronological age precludes a firm conclusion.

extent that no sexual dimorphism is apparently detectable). In any case, the role of heterochronic phenomena in the evolution of *Macaca majori* is the object of ongoing studies.

Amongst living apes, cranial paedomorphosis has been documented in the pygmy chimpanzee (*Pan paniscus*), which differs both in cranial size and shape from the common chimpanzee (*Pan troglodytes*), but resemble more closely subadults of this species (Cramer, 1977; McHenry & Corruccini, 1981; Shea, 1983, 1984a, 1988). Shea (1983, 1984a,b, 1988) has shown that many traits are ontogenetically scaled in the three living species of African apes, and in particular that the paedomorphic cranium of the pygmy chimpanzee (taking the common chimp as a model of the ancestor) is attained through ontogenetic scaling of many (though not all) cranial dimensions followed by earlier truncation in size but not in age. According to Shea's terminology, this should be termed rate hypomorphosis, although it could be equally labelled neoteny (decreased rate of shape change) with ontogenetic scaling (Alba, in press). This should not be confused with previous claims of neoteny *sensu* Gould (1977) in the pygmy chimp (e.g. Shea, 1983, 1989), which were based on the dissociation found between the head and the rest of the body, and have been recently criticised on the basis of the claim (still controversial) that classical tools for heterochronic diagnosis cannot be reliably applied to dissociation between different growth fields (Godfrey & Sutherland, 1995, 1996; Alba, in press).

In *Oreopithecus*, for obvious reasons, there are still many uncertainties regarding the type of heterochrony involved, which will require further detailed analyses. For example, although paedomorphosis is clearly documented in the facial region, it is not clear whether it also applies to the rest of the head. The neurocranium of *Oreopithecus* is smaller than expected on the basis of body mass and reduced in comparison to *Dryopithecus*, which could explain the presence of a sagittal crest to compensate for the loss of surface available for muscular insertion (if the size of the neurocranium is reduced in relation to body mass, additional surface is needed in order to maintain the efficiency of the masticatory apparatus). However, it is not clear as to whether or not the small neurocranium should be considered a reflection of generalised cranial paedomorphosis. On the one hand, the negatively allometric trajectory of brain versus body mass during postnatal ontogeny indicates that a relatively small neurocranium should be considered a peramorphic instead of paedomorphic trait and vice versa; for example, this was the line of reasoning followed by Gould (1977) with regard to the hypothesis of human neoteny. On the other hand, however, there is an inherent relativism of heterochronic diagnoses, in the sense that it depends on the stage at which comparisons are made (Alba *et al.*, in press). When the polarity of ontogenetic change is reversed, as in the growth of the braincase versus the postcranium, the diagnosis of paedomorphosis or peramorphosis can be very problematical (see discussion in Alba, in press). For this reason, it would be very useful to know the type of paedomorphic heterochrony (e.g. neoteny, progenesis; see Figure 13.5) involved in the case of *Oreopithecus*. Unfortunately, this is beyond our current understanding of this case of paedomorphosis; in other words, we cannot be absolutely certain about the type of heterochrony involved in *Oreopithecus* (for example, we cannot reliably discern between neoteny and progenesis), since only developmental dental age (instead of chronological absolute age) is currently available for this fossil primate. This means that comparisons are made at homologous developmental stages, although we do not know whether these stages occurred at the same age in the ancestor and in the descendant, whereas heterochronic studies should ideally employ both types of information simultaneously (Alba, in press).

Be that as it may, the case of the pygmy chimpanzee clearly shows that a reduction in the size of the head, and especially of the neurocranium, when compared to the rest of the body, is fully compatible with paedomorphosis, as long as neoteny (or any other type of heterochrony involved) affects more

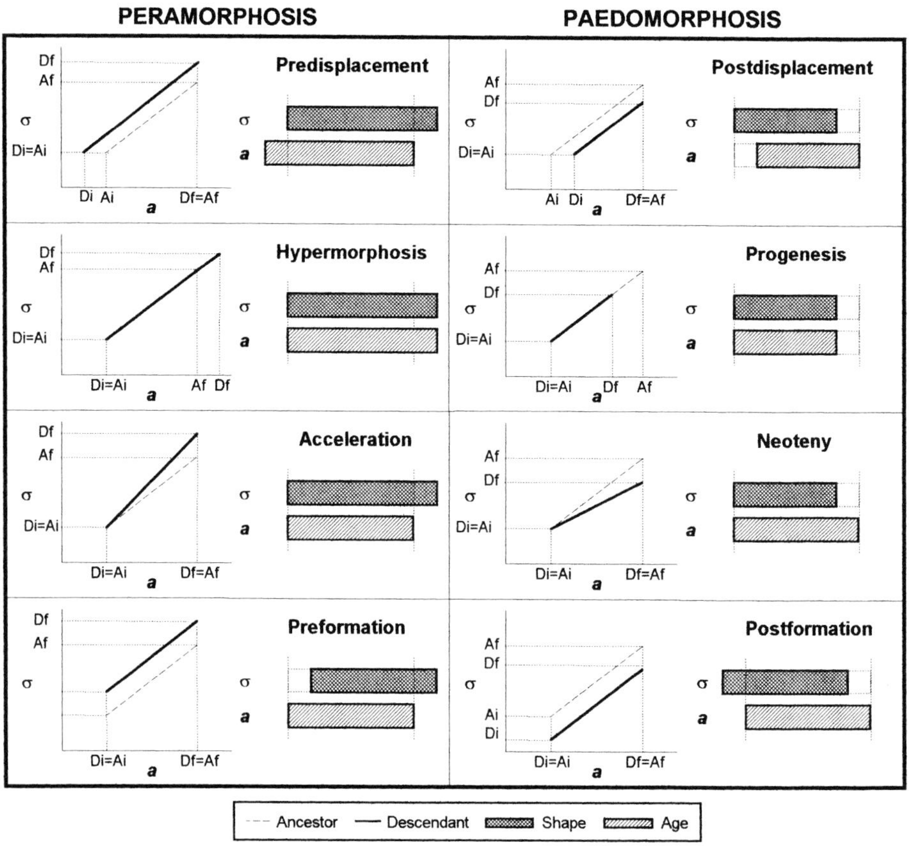

[Figure 13.5]
Different pure types of heterochrony, plotted in shape-age axes and in simplified clock models. (Redrawn from Alba, in press). Symbols: *σ*, shape; ***a***, age; A, ancestor; D, descendant; i, initial; f, final.

strongly the head (i.e. the head is at least partially dissociated from the rest of the body). Since heterochrony in *Oreopithecus* apparently has affected only the cranium, the relative size of the braincase with regard to the postcranium becomes irrelevant from a heterochronic point of view (although perhaps not from an adaptive one). Moreover, the fact that paedomorphosis has not affected equally the whole body suggests that, as proposed by Moyà-Solà & Köhler (1997), neoteny instead of progenesis is probably involved (Figure 13.6), although other different heterochronic changes (perhaps associated to life history evolution) cannot be entirely discarded for the moment. We thus conclude that the most likely hypothesis for the moment is that the reduction of prognathism in *Oreopithecus* is nothing more than the reflection of a generalised paedomorphosis affecting

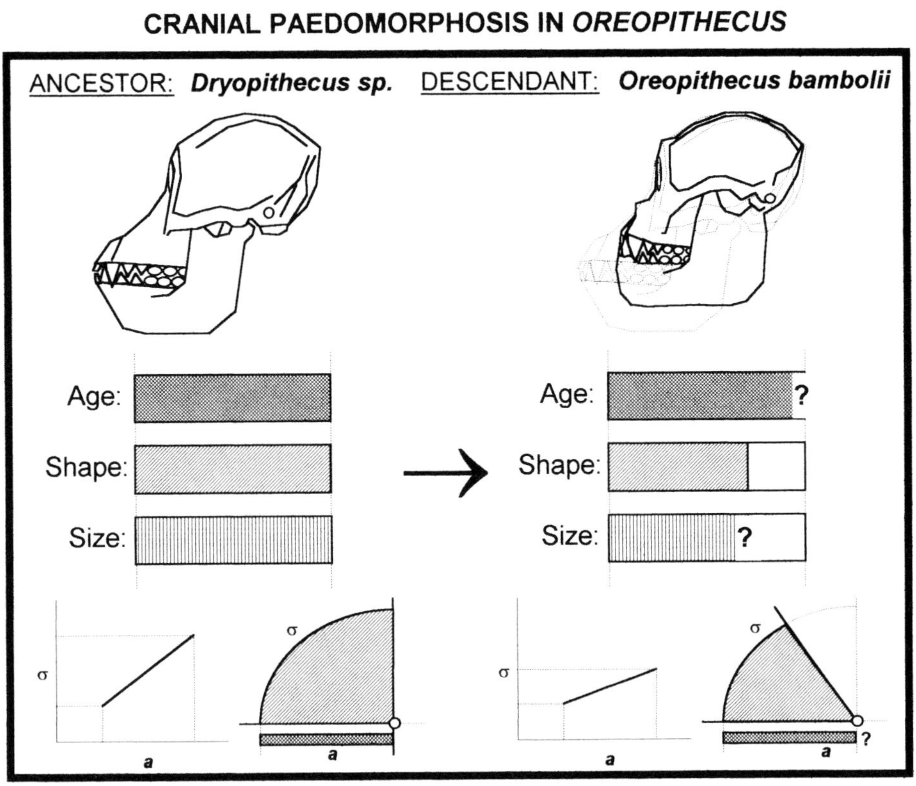

[Figure 13.6]
Schematic depiction of neoteny in *Oreopithecus bambolii*.
Question marks indicate some uncertainty. (Redrawn from Alba, in press).

the facial region (since prognathism increases during ontogeny in hominoids, the retention of a juvenile facial morphology implies a reduction of prognathism) and also, probably, the rest of the head.

Another different question is to decipher the target of selection involved in this case. As an hypothesis that needs further testing, we propose that the paedomorphic morphology of the cranium in general, and the shortened face in particular, could have been related to the biomechanical demands of bipedal standing and locomotion (Figure 13.7). This type of locomotion has been proposed for this fossil ape by Köhler & Moyà-Solà (1997b) on the basis of many postcranial adaptations to bipedalism, including the lumbar lordosis, the *genu valgum* (carrying angle in the knee joint), several hominid-like characters in the pelvis (short ischium and short pubic symphisis, with a wide subpubic angle, a great ischial spine and a well-developed anterior–inferior iliac spine), and the highly derived foot morphology with human-like proportions (increased power arm/load arm ratio). Moreover, this

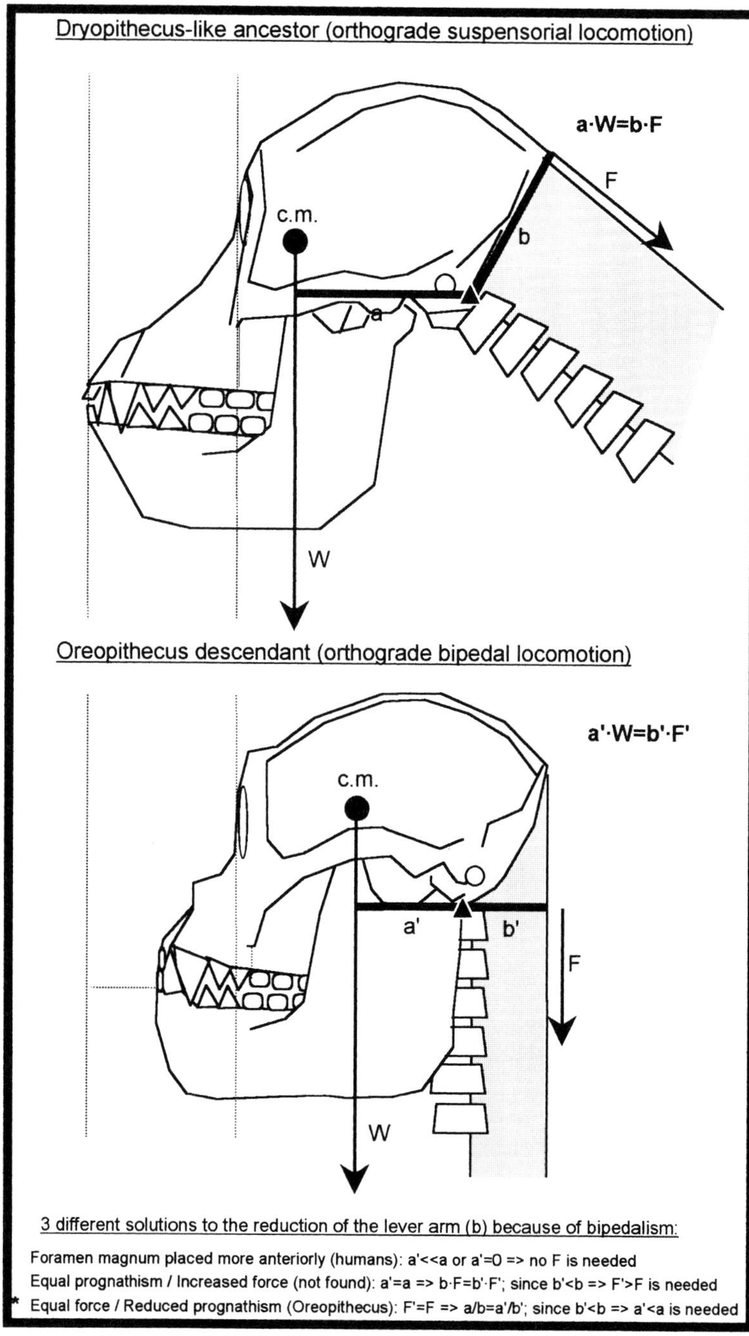

[Figure 13.7]
Schematic depiction of the biomechanical hypothesis to explain the reduction of prognathism by means of paedomorphic heterochrony in *Oreopithecus bambolii.* (See the text for further explanations.)

hypothesis has been recently confirmed by two independent lines of evidence: the discovery that *Oreopithecus* possessed enhanced manipulative capabilities (including hominid-like precission grip capability; Moyà-Solà *et al.*, 1999a), which is only possible when the hands have been freed from locomotor demands; and the study of the trabecular network of the cancellous bone of the pelvis (Rook *et al.*, 1999), which indicates that *Oreopithecus* not only was adapted to bipedalism but also habitually used this type of locomotion.

The acquisition of bipedal postures and locomotion, more economic and less risky for an insular primate (Köhler & Moyà-Solà, 1997b; Moyà-Solà *et al.*, 1999b) when compared to arboreal climbing and suspension, would have required some readjustments to sustain the weight of the head. Usually, the weight of the head is sustained by the force generated by the nuchal musculature, which is possible because the lever arm of this musculature is enough. In a bipedal primate, however, the vertebral column is placed vertically, so that the lever arm of nuchal musculature is decreased and, therefore, an increased force would be required to sustain the head. Alternative solutions are needed, however, since the amount of force that the nuchal muscles can generate is limited by the compressive stresses that the intervertebral disks can resist. In the case of the human lineage, this problem was solved by placing the *foramen magnum* (and, thus, the vertebral column) more anteriorly (nearer to the center of mass), thus decreasing (eliminating) the load arm of the head. Like humans, *Oreopithecus* also reduced the load arm of the head, but in this case by means of a decreased degree of prognathism. Therefore, in the case of *Oreopithecus*, cranial (especially facial) neoteny could have been simply the easiest way to readjust the positioning of the head relative to the vertebral column. This would mean that the small neurocranium relative to body mass of *Oreopithecus*, paralleling the condition found in the pigmy chimpanzee, could be merely a non-adaptive byproduct of dissociation between the cranial growth field (for the reasons outlined above) and the rest of the body, which could have been permitted thanks to the typical relaxation of predatory selection pressures on insular ecosystems.

An alternative hypothesis relates the particular craniodental morphology of *Oreopithecus*, and particularly the highly derived teeth and the shortened face, with an adaptation to extremely folivory. According to Harrison & Rook (1997), several characters of the face, mandible and neurocranium, coupled with some dental specialisations, would conform to a 'functional–behavioural' complex related to powerful chewing of leaves and other fibrous plant material. Certainly, it has been suggested by many authors, on the basis of dental shape (Szalay & Delson, 1979; Rosenberger & Delson,

1985; Harrison, 1986a; Delson, 1992; Ungar, 1996; Harrison & Rook, 1997; Kay & Ungar, 1997) as well as dental microwear (Ungar, 1996; King, Chapter 5), that *Oreopithecus* was a folivore primate. However, it is not so clear at all that *Oreopithecus* was a folivore (see Moyà-Solà & Köhler, 1997, for an alternative hypothesis), or at least such an extreme folivore as suggested by some authors. Although studies of dental microwear have shown that *Oreopithecus* was more folivore than the frugivorous *Dryopithecus*, extreme folivory in *Oreopithecus* is only sustained by the pronounced relieve of postcanine teeth (Ungar, 1996; Kay & Ungar, 1997). However, this pronounced relief, with the presence of accessory cusps, crests and cingula, does not resemble the lophodont pattern found in cercopithecids, but rather the dental shape of peccaries (Moyà-Solà & Köhler, 1997; Harrison & Rook, 1997), which in our opinion might indicate a more omnivorous diet. More significantly, many dental traits found in *Oreopithecus* (e.g. the vertically implanted incisors, the absence of diastemata, microdontia, not enough space for M3) are not expected adaptations to extreme folivory, but rather necessary consequences of the lack of space available for teeth, some of which would have been probably disadvantageous for a folivorous primate. Thus, microdontia could respond to the need of accommodating an adult dentition in juvenile-sized jaws, and the same structural constraint could explain the pronounced relief as an adaptive readjustment to prevent the reduction of masticatory surface (Moyà-Solà & Köhler, 1997), as shown by the molarised and bicuspid P_3, which Hürzeler (1958, 1968) considered to be a synapomorphy with humans, but that in fact would have evolved independently in both taxa (Harrison, 1986b). Alternatively (though not exclusively), some dental peculiarities of *Oreopithecus* could have been merely indirectly selected as heterochronic non-adaptive byproducts or spandrels (*sensu* Gould & Lewontin, 1979, 1982; Gould, 1997), although this issue cannot be further considered here because the influence of heterochrony upon dental evolution has never been extensively investigated (but see Jernvall, 1995, for a first approximation).

In any case, it should be noted that no evidence has been so far presented that contradicts the paedomorphic hypothesis first proposed by Moyà-Solà & Köhler (1997). Even if the folivore hypothesis is right, this would not invalidate the heterochronic hypothesis, since the shortened face of *Oreopithecus* could have evolved in the same way (and by means of the same modifications of the developmental programme) although for different adaptive reasons (folivory instead of bipedalism). Even other targets of selection, like behaviour and life-history, cannot be excluded. For example, neither *Macaca majori* nor pygmy chimpanzee support a biomechanical explanation related to bipedalism. Rather, they suggest a possible relationship between paedomorphosis and the reduction of sexual dimorphism in the facial region, which Shea (1983, 1984a, 1989) linked to juvenile behaviour and a decrease of male–male and male–female agonistic interactions in the case of *Pan paniscus*. On the other hand, the case of *Macaca majori* also suggests that some factor related to insularity could have been involved, given the similarities in the paedomorphic pattern between the two known insular forms. Life history evolution is one interesting possibility given the relationship proposed by Gould (1977) between neoteny and *K*-selection. The insular environments are characterised by high levels of inter- and intraspecific competition (due to the lack of predators and the severe limitation of resources), and therefore by density-dependent mortality, so that the evolution of *K*-selected life history strategies optimizing the carrying capacity of the environment is favoured. The possibility cannot be even discarded that each of these cases (*Oreopithecus*, *Pan paniscus* and *Macaca majori*), in spite of the similar developmental changes involved, is due to a somewhat or totally different adaptive explanation (Susman & Jungers, 1981). This suggests the intriguing possibility that, due to developmental constraints, very similar morphological patterns could evolve as a result of different adaptive reasons (Alba *et al.*, 1999). Although paradoxical at first sight, this should not be surprising, given the fact that heterochrony refers to the developmental changes involved in the genesis of the selected variability and not to the target of selection that ultimately was responsible for the selection of certain traits (perhaps only a few, with all the rest being merely dragged as non-adaptive consequences). This is precisely the reason why developmental and other phylogenetic constraints should not be discarded in phylogenetic analyses, since they can be responsible for a great deal of homoplasy that could not be easily explained or even recognised by traditional (adaptive) explanations disregarding internal factors.

Summary and conclusions

Although some traits in *Oreopithecus* might be consistent with powerful mastication associated with extreme folivory (the reduced prognathism, the high relief of postcanine teeth and the low degree of encephalisation), these traits can be also interpreted otherwise (biomechanical demands of bipedal locomotion, the need to maintain as high as possible the postcanine masticatory surface, and the low energy intake due to the scarcity of resources, respectively). Given the fact that other traits (especially microdontia and the subsequent loss of masticatory surface) are totally unexpected under the hypothesis of extreme folivory, we conclude that the biomechanical hypothesis proposed here is currently the most likely explanation of *Oreopithecus* heterochrony, although other alternative interpretations cannot be entirely discarded. In any case, even if the hypothesis of bipedalism proved to be wrong, another alternative explanation could be equally compatible with the heterochronic hypothesis that, after all, only explains how the selected variability was generated, but not why it was selected for. Moreover, at least we have learnt during the last decade that *Oreopithecus bambolii* is an extremely derived anthropoid primate, which has absolutely nothing to do with the primitive catarrhine or hominoid morphotype. No matter how this ancestral morphotype is reconstructed, the similarities found with *Oreopithecus* are basically homoplasies, the result of independent convergent evolution between ancestral catarrhines and a remarkably peculiar endemic insular fossil primate. In fact, not only the craniodental anatomy of *Oreopithecus* is derived, but also the postcranial features recently investigated in detail have proved to be autapomorphic (and, interestingly, convergent with the human lineage), including the bipedal (although not hominid-like) locomotion (Köhler & Moyà-Solà, 1997b; Rook *et al.*, 1999), and the hominid-like precision grip capability in the hand (Moyà-Solà *et al.*, 1999a). Therefore, the once widely held assumption that *Oreopithecus* reflected some kind of primitive morphotype (in spite of the eminently derived nature of the dentition), nowadays still partially substantiated with respect to the great ape and human clade by cladistic studies (Begun *et al.*, 1997), was only due to an incomplete analysis of morphology and the failure to understand the origin of the variability that was selected during the process of endemic insular evolution.

Only in the light of heterochrony and evolutionary constraits has it recently been possible to realize that these previous claims of primitivism in *Oreopithecus* were due to a merely superficial resemblance in cranial shape to the inferred primitive catarrhine morpotype, whereas in fact, this taxon is probably related to the continental middle and late Miocene great ape *Dryopithecus*. Although *Oreopithecus* can be fully understood only in the light of insularity (the particular set of conditions typical of insular ecosystems, including the absence of terrestrial predators and the scarcity of space and resources; see for example Sondaar, 1977; and Moyà-Solà *et al.*, 1999b), and even though it is the only known insular ape, the lessons taken from this example are of general applicability (the insular *Oreopithecus* is not an 'isolated' case). Certainly, the particular evolutionary history of *Oreopithecus* is a tale of endemic insular evolution, but the general narrative rules are the same as for its continental counterparts – pervasive homoplasy through extensive parallelism and convergence at the service of adaptation. Only a thoughful analysis of characters, as opposed to the blind and extremely reductionist, computer-assisted cladistics as nowadays practiced by some palaeoprimatologists, can potentially discern what characters should be discarded.

In other words, we conclude that, irrespective of our personally preferred systematic philosophy and of the methods used to infer phylogenetic relationships, morphology must be not only measured and compared, but also understood. This is due to the fact that, as this example shows, phylogenetic relationships cannot be properly assessed without trying to discern why and how characters evolved the way they did – as in the case of the cranial and dental features of *Oreopithecus* where neoteny had a powerful influence. An accurate investigation of characters, which requires the simultaneous consideration of both internal and external factors (evolutionary constraints and natural selection, respectively), is therefore absolutely essential prior to carrying out any kind of phylogenetic analysis. Otherwise, it would be impossible in many cases to distinguish, not only homoplasies from homologies, but also autapomorphies from symplesiomorphies, as shown by the example of *Oreopithecus*, in which highly derived cranial features were erroneously interpreted to reflect a generalised primitive morphology. In Pilbeam's (1996: p. 164) words, 'simple counting characters to estimate most parsimonious trees without examining carefully the quality or likely genalogical value of every character will no longer do. These interesting problems of character incongruence emphasize the need to engage much more carefully in an analysis of the developmental bases for complex morphologies in order to improve the quality of character selection and to raise the probability of identifying homoplasies'.

In our opinion, characters should only be used in phylogenetic reconstruction once they have been properly understood through careful investigation. It has been argued by many cladists that the a priori weighting of characters is erroneous as it is based on hypotheses constructed on the basis of theoretical concepts (e.g. functional morphology), whilst a phylogenetic analysis should be exclusively a methodological issue. This cannot be true, however, since phylogenetic analyses are made on the basis of hypotheses of polarity that, of course, are equally theoretical instead of methodological. Only the way of processing information is a methodological consideration, not the way in which the basic information (hypotheses of polarity) is obtained. The consequences of disregarding a priori analysis and weighting of characters is well exemplified not only by the particular case of *Oreopithecus*, but also by the high disparity of phylogenetic hypotheses on hominoid evolution currently under discussion, which have been constructed upon the same taxa, the same specimens, and in many cases (surprisingly) even the same characters. We hope to have shown how much a deep understanding of characters intended to be used in phylogenetic studies is needed, a task that can only be accomplished by means of exploring the interplay between adaptation (studied by functional morphology) and structural as well as historical constraints (studied by allometry and heterochrony, among others). Only with the help of these powerful tools it will be possible to ascertain the usefulness of the characters intended to be used in phylogenetic analysis. The tortuous scientific history of *Oreopithecus bambolii* exemplifies how, without applying these methodological principles, cladistic analysis must be considered a blind exercise, potentially misleading and nearly unrelated to the true underlying biological facts.

Acknowledgments

We want to express our gratitude to George D. Koufos and Louis de Bonis for inviting us to participate in the European Science Foundation workshop on 'Phylogeny of Eurasian Neogene Hominoid Primates', and also to contribute to this volume. We also gratefully thank Burkart Engesser for permission to study the *Oreopithecus* specimens from Basel and for the loan of material, Brian Shea for helpful discussion on heterochrony, and an anonymous referee for useful comments and suggestions. This work has been partially supported by a predoctoral fellowship (1999FI 00765) to D.M.A. from the Generalitat de Catalunya.

References

Agustí, J., Köhler, M., Moyà-Solà, S., Cabrera, L., Garcès, M. & Parès, J. M. (1996). Can Llobateres: the pattern and timing of the Vallesian hominoid radiation reconsidered. *J. Hum. Evol.* **31**: 143–55.

Alba, D. M. (in press). Shape and stage in heterochronic models. In Minugh-Purvis, N. & McNamara, K. J. (ed.), *Human Evolution through Developmental Change*. Johns Hopkins University Press.

Alba, D. M., Moyà-Solà, S. & Köhler, M. (1999). *Oreopithecus*, bonobos, heterochrony, and the paradox of form. In Fontdevila, A. (ed.) *Seventh Congress of the European Society for Evolutionary Biology. Barcelona, Spain 23–28 August 1999*, p. II-8. (Abstract).

Alberch, P. (1982). Developmental constraints in evolutionary processes. In Bonner, J. T. (ed.), *Evolution and Development*, pp. 313–32. Berlin: Springer-Verlag.

Alberch, P. (1985). Developmental constraints: why St. Bernards often have an extra digit and Poodles never do. *Am. Nat.*, **126**: 430–3.

Alberch, P. (1989). The logic of monsters: evidence for internal constraint in development and evolution. *Geobios* (mémoire spécial) **12**: 21–57.

Andrews, P. (1992). Evolution and environment in the Hominoidea. *Nature* **360**: 641–6.

Andrews, P., Harrison, T., Delson, E., Bernor, R. L. & Martin, L. (1996). Distribution and biochronology of European and Southwest Asian Miocene catarrhines. In Bernor, R. L., Fahlbusch, V. & Mittmann, H.-W. (ed.), *The Evolution of Western Eurasian Neogene Mammal Faunas*, pp. 168–207. New York: Columbia University Press.

Azzaroli, A. (1946). La scimmia fossile della Sardegna. *Riv. Sci. Preistor.* **1**: 68–76.

Begun, D. R. & Kordos, L. (1997). Phyletic affinities and functional convergence in *Dryopithecus* and other Miocene and living hominoids. In Begun, D. R., Ward, C. V. & Rose, M. D. (ed.), *Function, Phylogeny and Fossils: Miocene Hominoid Evolution and Adaptation*, pp. 291–316. New York: Plenum Press.

Begun, D. R., Ward, C. V. & Rose, M. D. (1997). Events in hominoid evolution. In Begun, D. R., Ward, C. V. & Rose, M. D. (ed.), *Function, Phylogeny and Fossils: Miocene Hominoid Evolution and Adaptation*, pp. 389–415. New York: Plenum Press.

Begun, D. R., Moyà-Solà, S. & Köhler, M. (1990). New Miocene hominoid specimens from Can Llobateres (Vallès Penedès, Spain) and their geological and paleoecological context. *J. Hum. Evol.* **19**: 255–68.

Bock, W. J. (1974). Philosophical foundations of classical evolutionary classification. *Syst. Zool.* **22**: 375–92.

Carroll, R. L. (1988). *Vertebrate Paleontology and Evolution*. New York: W. H. Freeman and Company.

Clarke, R. J. (1998). First complete restoration of the *Oreopithecus* skull. *Hum. Evol.* **12**: 221–32.

Conway Morris, S. (1998). *The Crucible of Creation. The Burgess Shale and the Rise of Animals*. New York: Oxford University Press.

Cramer, D. L. (1977). Craniofacial morphology of *Pan paniscus*. *Cont. Primatol.* **10**: 1–64.

Darwin, C. (1859, 1964). *On the Origin of Species*. A Facsimile of the First Edition with an Introduction by Ernst Mayr. Cambridge, MA: Harvard University Press.

Delson, E. (1979). *Oreopithecus* is a cercopithecoid after all. *Am. J. Phys. Anthropol.* **50**: 431–2. (Abstract.)

Delson, E. (1986). An anthropoid enigma: historical introduction to the study of *Oreopithecus bambolii*. *J. Hum. Evol.* **15**: 523–31.

Delson, E. (1992). Oreopithecidae. In Jones, S., Martin, R. & Pilbeam, D. (ed.) *The Cambridge Encyclopedia of Human Evolution*, pp. 401–4. Cambridge: Cambridge University Press.

Farris, J. S. (1982). Outgroups and parsimony. *Syst. Zool.* **31**: 328–34.

Fleagle, J. G. (1999). *Primate Adaptation and Evolution*. 2nd edn. San Diego: Academic Press.

Futuyma, D. J. (1986). *Evolutionary Biology*. 2nd edn. Sunderland, MA: Sinauer Associates.

Gentili, S., Mottura, A. & Rook, L. (1998). The Italian fossil primate record: recent finds and their geological context. *Geobios* **31**: 675–86.

Godfrey, L. R. & Sutherland, M. R. (1995). What's growth got to do with it? Process and product in the evolution of ontogeny. *J. Hum. Evol.* **29**: 405–31.

Godfrey, L. R. & Sutherland, M. R. (1996). The paradox of peramorphic paedomorphosis: Heterochrony and human evolution. *Am. J. Phys. Anthrop.* **99**: 17–42.

Gosliner, T. M. & Ghiselin, M. T. (1984). Parallel evolution in opistobranch gastropods and its implications for phylogenetic methodology. *Syst. Zool.* **33**: 255–74.

Gould, S. J. (1977). *Ontogeny and Phylogeny.* Cambridge, MA: The Belknap Press of Harvard University Press.

Gould, S. J. (1980). *The Panda's Thumb.* New York: W. W. Norton.

Gould, S. J. (1989). A developmental constraint in *Cerion*, with comments on the definition and interpretation of constraint in evolution. *Evolution* **43**: 516–39.

Gould, S. J. (1997). The exaptive excellence of spandrels as a term and prototype. *Proc. Nat. Acad. Sci. USA* **94**: 10750–5.

Gould, S. J. & Lewontin, R. C. (1979). The spandrels of San Marco and the Panglossian paradigm: a critique of the adaptationist program. *Proc. Roy. Soc. London* **205**: 581–98. Reprinted: Sober, E. (Ed.) 1984. *Concep. Iss. Evol. Biol.* pp. 252–70. Cambridge. MA: The Mit Press.

Gould, S. J. & Lewontin, R. C. (1982). L'adaptation biologique. *Recherche* **139**: 1494–1502.

Harrison, T. (1985). African oreopithecids and the origin of the family. *Am. J. Phys. Anthrop.* **66**: 180. (Abstract.)

Harrison, T. (1986a). The phylogenetic relationships of the Oreopithecidae. *Am. J. Phys. Anthrop.* **69**: 212. (Abstract.)

Harrison, T. (1986b). A reassessment of the phylogenetic relationships of *Oreopithecus bambolii* Gervais. *J. Hum. Evol.* **15**: 541–83.

Harrison, T. (1986c). New fossil anthropoids from the middle Miocene of East Africa and their bearing on the origin of the Oreopithecidae. *Am. J. Phys. Anthrop.* **71**: 265–84.

Harrison, T. (1989). New estimates of cranial capacity, body size and encephalization in *Oreopithecus bambolii. Am. J. Phys. Anthrop.* **78**: 237. (Abstract.)

Harrison, T. (1991). The implications of *Oreopithecus bambolii* for the origins of bipedalism. In Senut, B. & Coppens, Y. (ed.), *Origine(s) de la Bipédie Chez les Hominidés*, pp. 235–44. Paris: CNRS.

Harrison, T. S. & Harrison, T. (1989). Palynology of the late Miocene *Oreopithecus*-bearing lignite from Baccinello, Italy. *Palaeogeog. Palaeoclimatol. Palaeoecol.* **76**: 45–65.

Harrison, T. & Rook, L. (1997). Enigmatic anthropoid or misunderstood ape? The phylogenetic status of *Oreopithecus bambolii* reconsidered. In Begun, D. R., Ward, C. V. & Rose, M. D. (ed.) *Function, Phylogeny and Fossils: Miocene Hominoid Evolution and Adaptation*, pp. 327–62. New York: Plenum Press.

Harvey, P. H. & Partridge, L. (1998). Different routes to similar ends. *Nature* **392**: 552–3.

Hennig, W. (1966). *Phylogenetic Systematics.* Urbana: University of Illinois Press.

Hetch, M. K. (1989). Development, trends and phylogenetic inference. *Geobios* (mémoire spécial) **12**: 207–16.

Hürzeler, J. (1954). Zur systematischen Stellung von *Oreopithecus. Verhandl. Naturforch. Gesell. Basel* **65**: 88–95.

Hürzeler, J. (1956). *Oreopithecus*, un point de repère pour l'histoire de l'humanité a l'ère Tertiaire. *Centre National de la Recherche Scientifique. Colloque International No. 60 – Avril 1955. 'Problèmes Actuels de Paléontologie'*, pp. 115–21.

Hürzeler, J. (1958). *Oreopithecus bambolii* Gervais. A preliminary report. *Verhandl.*

Naturforch. Gesell. Basel **69**: 1–48.

Hürzeler, J. (1960). Importancia del oreopiteco para la genealogía humana. *Triangulo* **4**: 164–74.

Hürzeler, J. (1968). Questions et reflexions sur l'histoire des Anthropomorphes. *Ann. Paléontol. (Vert.)*, **54**: 195–233.

Jernvall, J. (1995). Mammalian molar cusp patterns: Developmental mechanisms of diversity. *Acta Zool. Fenn.* **198**: 1–61.

Jungers, W. L. (1987). Body size and morphometric affinities of the appendicular skeleton in *Oreopithecus bambolii* (IGF 11778). *J. Hum. Evol.* **16**: 445–56.

Jungers, W. L. (1990). Problems and methods in reconstructing body size in fossil primates. In Damuth, J. & MacFadden, B. J. (ed.), *Body Size in Mammalian Paleobiology: Estimation and Biological Implications*, pp. 103–18. Cambridge: Cambridge University Press.

Kay, R. F. & Ungar, P. S. (1997). Dental evidence for diet in some Miocene catarrhines with comments on the effects of phylogeny on the interpretation of adaptation. In Begun, D. R., Ward, C. V. & Rose, M. D. (ed.), *Function, Phylogeny and Fossils: Miocene Hominoid Evolution and Adaptation*, pp. 131–51. New York: Plenum Press.

Klingenberg, C. P. (1998). Heterochrony and allometry: the analysis of evolutionary change in ontogeny. *Biol. Rev.* **73**: 79–123.

Köhler, M. & Moyà-Solà, S. (1997a). Fossil muzzles and other puzzles. *Nature* **388**: 327–8.

Köhler, M. & Moyà-Solà, S. (1997b). Ape-like or hominid-like? The positional behavior of *Oreopithecus bambolii* reconsidered. *Proc. Nat. Acad. Sci. USA* **94**: 11747–50.

Larson, S. G. (1998). Parallel evolution in the hominoid trunk and forelimb. *Evol. Anthropol.* **6**: 87–99.

Lieberman, D. E., Wood, B. A. & Pilbeam, D. R. (1996). Homoplasy and early *Homo*: an analysis of the evolutionary relationships of *H. habilis sensu stricto* and *H. rudolfensis*. *J. Hum. Evol.* **30**: 97–120.

Losos, J. B., Jackman, T. R., Larson, A., de Queiroz, K. & Rodríguez-Schettino, L. (1998). Contingency and determinism in replicated adaptive radiations of island lizards. *Science*, 279: 2115–18.

Martin, R. D. (1996). Recursos energéticos y la evolución del volumen cerebral de los primates hominoideos. *Sobre la Evolución de los Homínidos. Barcelona, 29 y 30 de octubre de 1996. Resúmenes*, pp. 10–11.

McHenry, H. M. & Corruccini, R. S. (1981). *Pan paniscus* and human evolution. *Am. J. Phys. Anthrop.* **54**: 355–67.

McKinney, M. L. & MacNamara, K. J. (1991). *Heterochrony: The Evolution of Ontogeny*. New York: Plenum Press.

McNamara, K. J. (1997). *Shapes of Time. The Evolution of Growth and Development*. Baltimore: The Johns Hopkins University Press.

Moore, J. & Willmer, P. (1997). Convergent evolution in invertebrates. *Biol. Rev.* **72**: 1–60.

Moyà, S. & Köhler, M. (1993). *Dryopithecus* y el origen de los grandes monos actuales. *Invest. Cien.* **207**: 30–1.

Moyà-Solà, S. & Köhler, M. (1993). Recent discoveries of *Dryopithecus* shed new light on evolution of great apes. *Nature* **365**: 543–5.

Moyà-Solà, S. & Köhler, M. (1995). New partial cranium of *Dryopithecus* Lartet, 1863 (Hominoidea, Primates) from the upper Miocene of Can Llobateres, Barcelona,

Spain. *J. Hum. Evol.* **29**: 101–39.

Moyà-Solà, S. & Köhler, M. (1997). The phylogenetic relationships of *Oreopithecus bambolii* Gervais, 1872. *Com. Rend. Acad. Sci. Paris* **324**: 141–8.

Moyà-Solà, S., Köhler, M. & Rook, L. (1999a). Evidence of hominid-like precision grip capability in the hand of the Miocene ape *Oreopithecus. Proc. Nat. Acad. Sci. USA* **96**: 313–17.

Moyà-Solà, S., Quintana, J., Alcover, J. A. & Köhler, M. (1999b). Endemic island faunas of the Mediterranean Miocene. In G.E. Rössner & K. Heissig (ed.) *The Miocene Land Mammals of Europe*. München: Verlag Fritz Pfeil.

Neff, N. A. (1986). A rational basis for a priori character weighting. *Syst. Zool.* **35**: 110–23.

Pickford, M., Moyà-Solà, S. & Köhler, M. (1997). Phylogenetic implications of the first African Middle Miocene hominoid frontal bone from Otavi, Namibia. *Com. Rend. Acad. Sci. Paris,* **325**: 459–66.

Pilbeam, D. (1996). Genetic and morphological record of the Hominoidea and hominid origins: A synthesis. *Mol. Phylogen. Evol.* **5**: 155–68.

Pilbeam, D. (1997). Research on Miocene hominoids and hominid origins. The last three dedaces. In Begun, D. R., Ward, C. V. & Rose, M. D. (ed.), *Function, Phylogeny and Fossils: Miocene Hominoid Evolution and Adaptation*, pp. 13–28. New York: Plenum Press.

Raff, R. A. (1996). *The Shape of Life. Genes, Development, and the Evolution of Animal Form*. Chicago: The University of Chicago Press.

Raff, R. A. & Kaufman, T. C. (1983). *Embryos, Genes and Evolution*. New York: Macmillan Publishing.

Rook, L., Bondioli, L., Köhler, M., Moyà-Solà, S. & Macchiarelli, R. (1999). *Oreopithecus* was a bipedal ape after all: evidence from the iliac cancellous architecture. *Proc. Nat. Acad. Sci. USA* **96**: 8795–9.

Rose, M. D. (1997). Functional and phylogenetic features of the forelimb in Miocene hominoids. In Begun, D.R., Ward, C. V. & Rose, M. D. (ed.), *Function, Phylogeny and Fossils: Miocene Hominoid Evolution and Adaptation*, pp. 79–100. New York: Plenum Press.

Rosenberger, A. L. & Delson, E. (1985). The dentition of *Oreopithecus bambolii*: systematic and paleobiological implications. *Am. J. Phys. Anthrop.* **66**: 222–3.

Sarmiento, E. (1983). Some functional specializations in the skeleton of *Oreopithecus bambolii. Am. J. Phys. Anthrop.* **60**: 248–9. (Abstract.)

Sarmiento, E. (1987). The phylogenetic position of *Oreopithecus* and its significance in the origin of the Hominoidea. *Am. Mus. Novit.* **2881**: 1–44.

Schwartz, J. H. (1990). *Lufengpithecus* and its potential relationship to an orang-utan clade. *J. Hum. Evol.* **19**: 591–605.

Shea, B. T. (1983). Paedomorphosis and neoteny in the pygmy chimpanzee. *Science* **222**: 521–2.

Shea, B. T. (1984a). An allometric perspective on the morphological and evolutionary relationships between pygmy (*Pan paniscus*) and common (*Pan troglodytes*) chimpanzees. In Susman, R. L. (ed.), *The Pygmy Chimpanzee: Evolutionary Biology and Behavior*, pp. 89–130. New York: Plenum Press.

Shea, B. T. (1984b). Ontogenetic allometry and scaling: a discussion based on growth and form of the skull in African apes. In Jungers, W. L. (ed.), *Size and Scaling in Primate Biology*, 175–205. New York: Plenum Press.

Shea, B. T. (1988). Heterochrony in Primates. In McKinney, M. L. (ed.), *Heterochrony in Evolution. A Multidisciplinary Approach*, pp. 237–66. New York: Plenum Press.
Shea, B. T. (1989). Heterochrony in human evolution: the case for neoteny reconsidered. *Ybk Phys. Anthropol.* **32**: 69–101.
Shubin, N. (1998). Evolutionary cut and paste. *Nature* **394**: 12–13.
Sidow, A. (1994). Parsimony or statistics? *Nature* **367**: 26.
Siebert, D. J. (1992). Tree statistics; trees and 'confindence'; consensus trees; alternatives to parsimony; character weighting; character conflict and its resolution. In Forey, P. L., Humphries, C. J., Kitching, J. J., Scotland, R. W., Siebert, D. J. & Williams, D. M. (ed.), *Cladistics. A Practical Course in Systematics*, pp. 72–88. New York: Oxford University Press.
Simons, E. L. (1960). *Apidium* and *Oreopithecus. Nature* **186**: 824–6.
Simpson, G. G. (1961). *Principles of Animal Taxonomy.* New York: Columbia University Press.
Sondaar, P. Y. (1977). Insularity and its effect on mammal evolution. In Hetch, M. K., Goody, P. C. & Hecht, B. M. (ed.), *Major patterns in vertebrate evolution*, pp. 671–707. New York: Plenum Press.
Stewart, C.-B. (1993). The powers and pitfalls of parsimony. *Nature* **361**: 603–7.
Straus, W. L., Jr. (1957). *Oreopithecus bambolii. Science* **126**: 345–6.
Straus, W. L., Jr. (1958a). Is *Oreopithecus bambolii* a primitive hominid? *Anat. Rec.* **132**: 511–12.
Straus, W. L., Jr. (1958b). A new *Oreopithecus* skeleton. *Science* **128**: 523.
Straus, W. L., Jr. (1963). The classification of *Oreopithecus.* In Washburn, S. L. (ed.), *Classification and human evolution*, 146–177. Chicago: Aldine.
Susman, R. L. & Jungers, W. L. (1981). Bonobos: generalized hominid prototypes or specialized insular dwarfs? (Comments). *Curr. Anthrop.* **22**: 363–75.
Szalay, F. S. & Delson, E. (1979). *Evolutionary History of the Primates.* San Diego: Academic Press.
Ungar, P. S. (1996). Dental microwear of European Miocene catarrhines: evidence for diets and tooth use. *J. Hum. Evol.* **31**: 335–66.
Vogel, G. (1998). For island lizards, history repeats itself. *Science* **279**: 2043.
Wake, D. B. (1991). Homoplasy: the result of natural selection, or evidence of design limitations? *Am. Nat.* **138**: 543–67.
Ward, S. (1997). Functional anatomy and phyletic implications of the hominoid trunk and hindlimb. In Begun, D. R., Ward, C. V. & Rose, M. D. (ed.), *Function, Phylogeny and Fossils: Miocene Hominoid Evolution and Adaptation*, pp. 101–30. New York: Plenum Press.
Willmer, P. G. (1990). *Invertebrate Relationships. Patterns in Animal Evolution.* Cambridge: Cambridge University Press.
Zanaga, M. (1998). Macaca majori, *primate endemico del Pleistocene della Sardegna.* Unpublished thesis, University of Florence.

14 The late Miocene hominoid from Georgia

Leo Gabunia, Ekaterine Gabashvili, Abesalom Vekua and David Lordkipanidze

Introduction

More than half a century since the first announcement of the late Miocene discovery from Udabno (Gare-Kaxheti, East Georgia; Figure 14.1), remains of a hominoid-like primate were described by N.O. Burtshak-Abramovitz and E. G. Gabashvili (1945,1950) as a new taxon – *Udabnopithecus garedziensis.* In the years that followed, other specialists in the field changed its taxonomy. Piveteau (1957) noted its close resemblance to the genus *Dryopithecus,* and later publications claimed that the hominoid ape from Udabno is actually synonymous with the species *Dryopithecus fontani* (Simons & Pilbeam, 1965; Szalay & Delson 1979; Andrews *et al.*, 1996), although without any substantial discussion. Yet other opinions also were expressed (Reshetov, 1966, Nesturkh, 1968). In all the publications, the stratigraphic position of *Udabnopithecus* is rather approximate Middle/Late Miocene or Sarmatian.

A short description of the actual remains

A detailed examination of the hominoid remains from Udabno has convinced us that even though it shows significant similarities to *Sivapithecus,* a similarity which will be touched upon later, it is also very close in its attributes to *Dryopithecus..*

Udabnopithecus is represented by the right P^4 and M^1 imbedded in a maxilla fragment (Figure 14.2). During the extraction of the maxilla fragment, the P^4 has been detached. It seems that the same happened earlier to M^2 as evidenced by the remaining part of its alveolar cavity in the maxilla, and this tooth is now lost. The attrition of the teeth is slight to medium.

The P^4 is oval shaped, rather squeezed in on its medial and distal sides (Figure 14.3). Its labial facet is shorter than the lingual, while the dorsal surface is less convex than the ventral. The paracone is slightly higher and sharper than the protocone. These two cusps are separated by a central groove, which joins the *fovea posterior,* and the groove interrupts the transversal medial crest, coming down from the top of the paracone. The protocone comprises the larger part of the occlusal surface of the tooth and is more eroded than the paracone. The *fovea anterior* is located at the medial

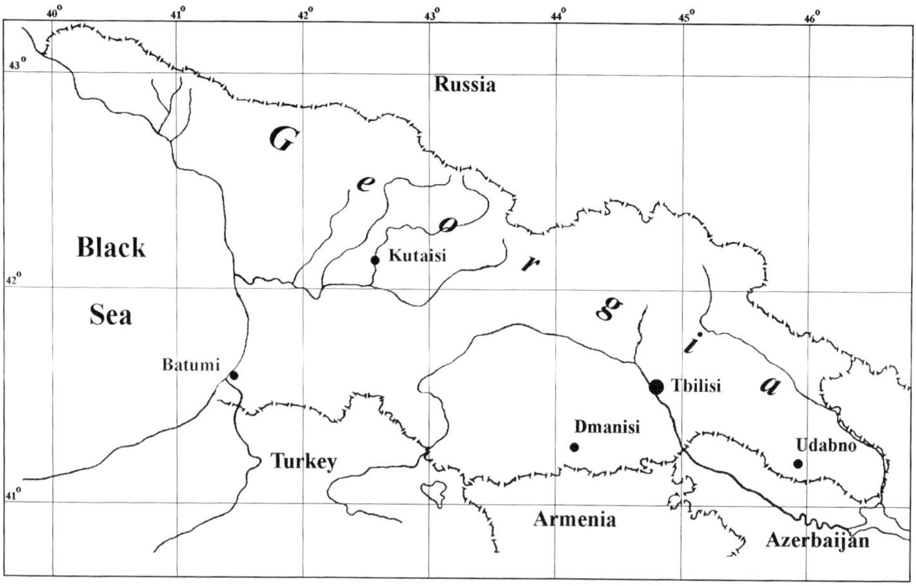

[Figure 14.1]
Location map of Udabno, Georgia.

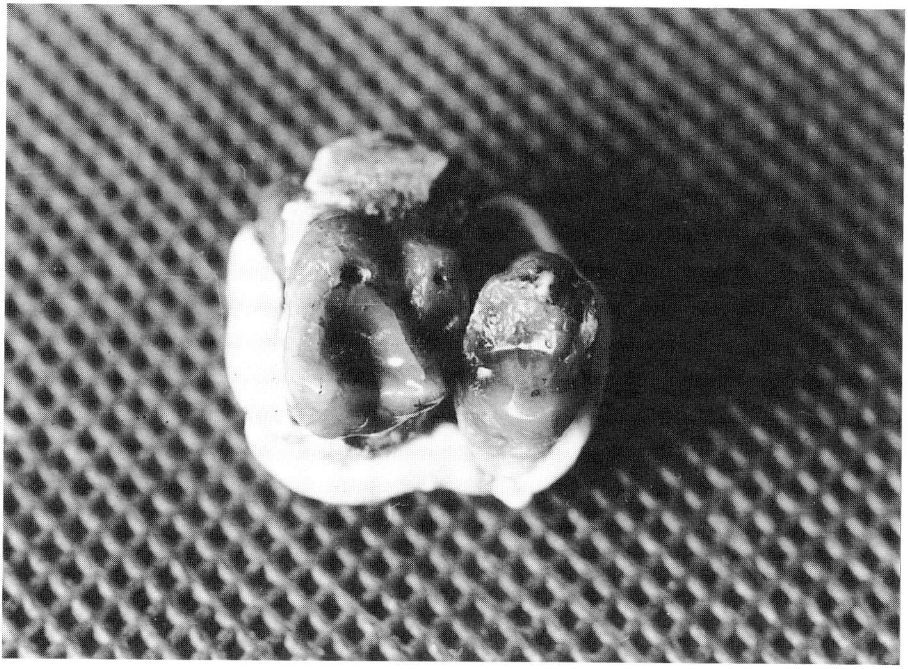

[Figure 14.2]
The maxilla of *Udabnopithecus* from Udabno with crowns of P^4 and M^1.

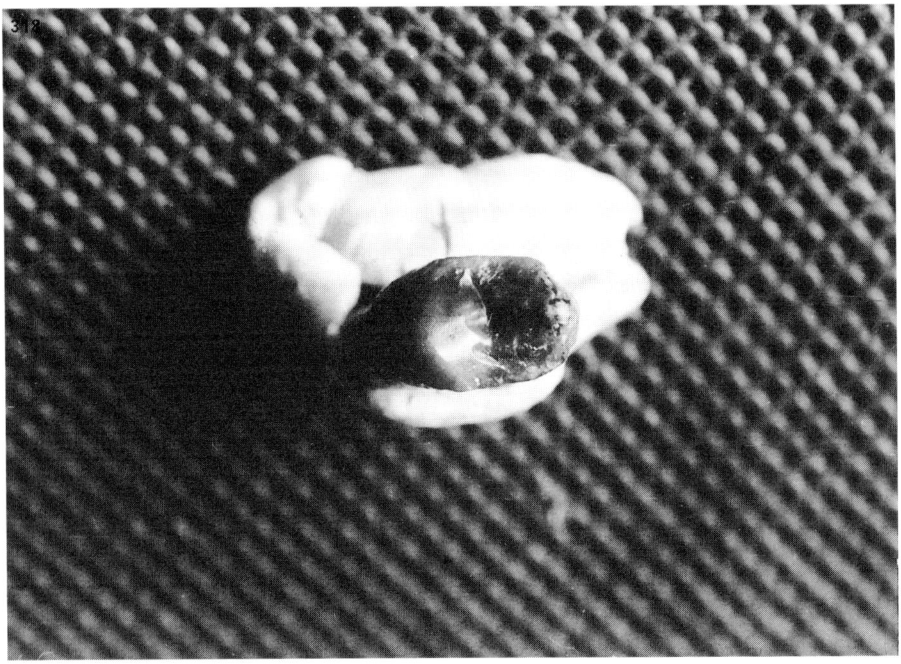

[Figure 14.3]
P^4 of *Udabnopithecus*.

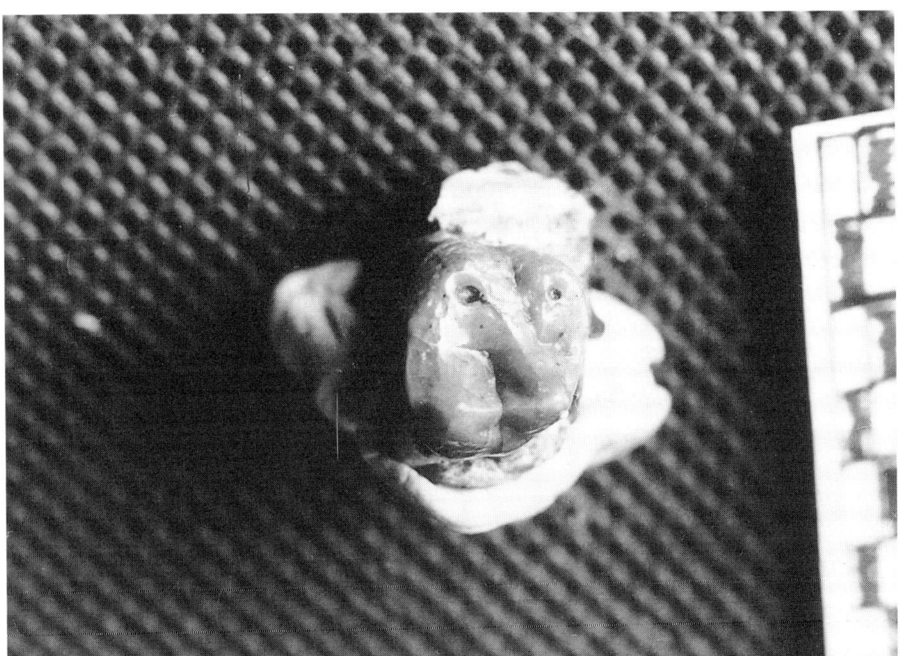

[Figure 14.4]
M^1 of *Udabnopithecus*.

edge of the tooth, bordering with the medial crest coming down from the top of the paracone in the lingual direction. The *fovea posterior* is quite deep, situated at the distal edge of the crown and continues transversely up to the middle of the occlusal surface. There is a small ridge coming down from the top of the paracone, in the distal-lingual direction, which joins the *fovea posterior*. On the labial surface of the crown, though it is quite eroded, a slight convexity of its middle part could be observed, flattening out and widening down towards the base of the crown. Slight grooves can be observed. The distal groove is deeper than medial and both disappear towards the base of crown. The protocone has three facet surfaces – a frontal one inclined towards the medial edge, a central and a dorsal one inclined slightly towards the distal side. The enamel is rather thin as inferred from the clearly observed dentine patch in the central facet of the protocone of this only slightly eroded tooth. At the base of the convex lingual surface of the tooth, there is a barely observed 'neck' or swelling. On the distal side of the tooth there is a large, clearly outlined inter-proximal wear facet from contact with M^1, while on the medial side there is a smaller and less distinct one from contact with P^3. The dimensions (in mm) of the tooth are: maximum length of the crown – 6.5; maximum width – 10.1; and the height of the slightly eroded crown – 7.3.

The M^1 has a squarish appearance and its occlusal surface is of a rhomboid shape (Figure 14.4). The paracone and metacone are located at the very edge of the crown, while the labial crest connecting them is deeply cut in the middle, its deepest point being still high above the central *fossa* of the trigonid. The labial cusps are nearly the same height (the paracone is only slightly larger than the metacone and slightly smaller than the protocone). Distinct crests connect the paracone and metacone with the protocone, which, though only lightly eroded, shows a clear dentine exposure. A similar dentine exposure, though of smaller dimensions, is seen on the surface of the hypocone. The latter, though of the same height as the protocone, is smaller. It is almost isolated from any occlusal contacts though a very thin line of attrition connects it with the distal-lingual edge of the protocone. The trigon *fossa* is divided by a deep transversal line, continuing along the labial wall of the crown, where it touches upon the longitudinal notch which delineates the attrition facets along the external edges. In the central part of the trigonid is a weakly developed longitudinal notch, separating the labial convexities from the protocone.

The *fossa anterior*, which can be traced in the labial part of the tooth, is narrow and is situated between the medial part of the crown and the transversal crest which connects the paracone with the protocone. The *fovea posterior* is quite deep, delineated in the front by the lingual crest of the metacone and at the back by the distal edge of the crown. An obliquely located notch in the *fovea posterior* constitutes the lingual border of the hypocone, approached from the medial–labial side, by a slightly developed transversal swelling (apparently a remnant of the eroded labial crest of the hypocone). Relatively narrow transversal facets of attrition are clearly noticeable on the distal and medial edges of the occlusal surface, as well as along the crest connecting the metacone with the protocone. The attrition surfaces of the protocone and the hypocone are nearly horizontal. The labial surface of the crown bears two weak vertical convexities, matching the paracone and metacone. At the base of the lingual side of the crown a very weak cingulum could be observed. The dimensions of the tooth (in mm) are: maximum length of the crown – 9; maximum width – 10.7; and the height of the slightly eroded crown – 6.

The stratigraphic position of the Udabno hominoid remains

The Hipparion dominated fauna of Udabno was found *c.* 70 km north-east of Tbilisi (Figure 14.1) in a dry valley between the mountains of Dodo (to the north-east) and Udabno (to the south-east). Here, not far from the old sixteenth-century monastery of David Garedji, solid fossilised sand formations are gradually being replaced by clayey-sandy sediments containing a middle Sarmatian mollusc fauna, followed by thick (more than 400 m) sediments of continental origins, interspersed by colourful clay layers with inclusions of sand dunes and conglomerates. This thick unit is covered by layers of the 'Shiraqi' faunal complex (600 m thick), consisting of bright-grey sand dunes with exposures of grey clay and sandy clay, which are covered transgressively by sandy-loamy layers containing the akchagilyan (middle Pliocene) mollusc fauna. All along the continental layers containing the 'Shiraqi' fauna in various stratigraphic locations one can find vertebrate bones but the richest deposits are to be found mainly in two horizons (Wangenheim *et al.*, 1989) – *c.* 110 m below the basis of the 'Shiraqi' complex (Udabno I) and *c.* 40–50 m beneath the contact of this complex with the continental layers below (Udabno II).

The Udabno I faunal assemblage comprises: *Deinotherium giganteum; Tetralophodon* cf. *longirrostris; Miohyaenotherium bessarabicum (Ictitherium hipparionum* var. *garedziensis); Plessiogulo* cf. *brachygnathus; Adcrocuta eximia; Hipparion garedzicum; Aceratherium* cf. *incisivum; Chalicotherium* sp.; *Microstonyx erymanthius; Giraffidae* gen. cf. *Paleotragus; Cervidae* gen.; *Tragocerus* sp.; *Gazella schlosseri; Udabnopithecus garedzien-*

sis (Burtshak-Abramovitz & Gabashvili, 1945,1950; Gabunia, 1959; Meladze, 1985; Tsiskarishvili, 1987).

The maxillary fragment of the *Udabnopithecus* garedziensis (Burtshak-Abramovitz & Gabashvili 1945) was recovered *c.* 40 m above this first bone-bearing horizon. Close to the maxillary fragment a skull fragment of *Deinotherium giganteum* was recovered.

The fauna of Udabno II differs little in its context from its preceeding complex of Udabno I. Besides the species mentioned earlier such as *Deinotherium giganteum* and *Adcrocuta eximia* this complex also contains *Hyaenotherium magnum, Udabnocerus georgicus, Paleotragus* sp. and a very small variety of a bovine. In addition, the *Hipparion* cf. *garedzicum* from this upper level of Udabno seems to be more advanced than the typical species, and judging by its metapodial proportions it is very close to the group of *H. moldavicum-mediterraneum.*

It seems that the faunal complexes of Udabno I and Udabno II, even though quite similar, reflect two consecutive stages of the evolutionary sequence of the Hipparion fauna in Georgia. In general they portray a more or less analogous natural surrounding, characterised by moderately lush forest biotypes interplaced with relatively more open settings.

According to the faunal composition of the two assemblages – Udabno I and Udabno II – both could be assigned, especially the latter, to the Lower Turolian or MN 11, which is also supported by their differences from the more archaic faunal complex of Eldari (Gabunia, 1959, 1980). This faunal complex is found in the Upper Sarmatian marine sediments of eastern Georgia correlated with the Upper Valesian (MN 10). This observation is not contradictory to results from the palaeomagnetism studies (Wangenheim *et al.*, 1989), according to which Udabno I dates to an intermediate period which binds a normal to reversed polarity chrons. Udabno II is attributed to a reversed chron, identified with stage 8 in the palaeomagnetic stratigraphic sequence of the Eastern Paratethys, or the lower part of stage C4 of the European palaeomagnetic stratigraphic sequence (Sen, 1997). Even though the lower part of the reversed magnetic stage 8 is correlated with the Upper Sarmatian (Wangenheim *et al.*, 1989), it is impossible at present to pinpoint the exact palaeomagnetic boundary between the Sarmatian and the Meotic era. Thus we are obliged to rely on the Hipparion fauna which relates the Udabno I fauna to the most Upper Sarmatian or Lower Meotis, while the fauna of Udabno II is identified as Lower Meotis.

Thus the fossil ape *Udabnopithecus gardziensis* should be dated to the Upper Sarmatian or Lower Meotic, which correlates with the Lower Turolian (MN 11) of the continental Neogene sequence.

General discussion

While accepting the opinion of Burtshak-Abramovitz & Gabashvili (1950) as to the unique taxonomic position of the *Udabnopithecus*, Gremiatski (1957) enlarged upon the differences existing between this hominoid and the genera of *Proconsul* and *Dryopithecus*. *Udabnopithecus* differs from the former in the cingulum and its position on the P^4 and M^1, the less complex topography of the M^1, occlusal surface and the overall general rather than specialised character of these teeth. The simplicity of the M^1 surface is illustrated by its weakly developed crest connecting the paracone with the metacone, the slight transversal medial torus and the absence of the distal crest connecting the metacone with hypocone. All these characteristics, in particular the simplicity of the M^1 occlusal surface, the height of the tooth crowns and the different development morphology of the cingulum, separate *Udabnopithecus* from *Dryopithecus* as well as bearing evidence to the isolated position of the *Udabnopithecus*. We do not think these unique attributes are enough to be used as markers of a seperate genus.

Some authors thought *Udabnopithecus* should be identified with *Sivapithecus* (*Ramapithecus)* (Gremiatski, 1957). It should be pointed out that Simons & Pilbeam (1965; Pilbeam, 1969) had claimed in the past that *Dryopithecus* and *Ramapithecus* which are of similar dimensions, differ very little in their molar teeth; and indeed this is true. The comparison of the *Udabnopithecus* teeth with those of URM 13799 (*Ramapithecus punjabensis*), which are very similar dimensions-wise and in the degree of attrition, reveals a great resemblace of the P^4 and even a greater one of the M^1, the differences being higher crowns and more pronounced coni and ridges on the *Udabnopithecus* teeth as well as the presence on both teeth of a lingual cingulum, and a thinner layer of the enamel. The P^4 of *Udabnopithecus* differs from that of the *Ramapithecus* also in the shape of the crown (its labial length is much shorter than the lingual) and its more pronounced convexity of the labial side. There is also a difference in the attrition pattern of the tooth (the *Udabnopithecus* P^4 nearly lacks the lingual facet which is clearly present in the *Ramapithecus* tooth); the M^1 labial promontories of *Udabnopithecus* are more conus-like and they are inflated on both sides – the inner and outside alike. Its hypocone is more distinct, the paracone is of greater dimension and there is a difference in the attrition pattern of the crown (a distal–labial inclination of the hypocone attrition surface, while it is nearly horizontal in the *Ramapithecus* tooth and it has weakly developed facets of the labial edge). In general, the relief of the occlusal surfaces of both teeth is more pronounced than that of the *Ramapithecus* teeth, with a deeper groove separating the labial half of the teeth from the lingual. All of these taken together distance *Udabnopithecus* from *Sivapithecinae* and can be used as evidence for the former being part of the *Dryopithecinae* (Andrews *et al.*, 1996).

But first we have to exclude the possibility of associating *Udabnopithecus* with the earlier subfamily of *Kenyapithecini* (Andrews, 1992) from which it differs, especially if compared with their middle Miocene representatives – *Griphopithecus* from Paşalar (Alpagut *et al.*, 1990) – in its smaller dimensions, the small cingulum and the thinner enamel. It is quite obvious that *Udabnopithecus* is closer to *Dryopithecinae,* namely to *D. fontani, D. laietanus, D. carinthiacus* and *D. crusafonti* (Andrews *et al.*, 1996). Unfortunately, since there are only the two teeth to compare, it is difficult to determine which of the above is closest to *Udabnopithecus.* As already stated, some authors indeed identify the latter with *D. fontani* (Simons & Pilbeam, 1965; Szalay & Delson, 1979) but there is not enough evidence to bear this out. *Udabnopithecus* is as similar to *D. fontani* as to *D. laiteanus* or *D. carinthiacus* while at the same time showing some particularities which are not present in any of them.

Udabnopithecus differs from *D. fontani* (Simons & Pilbeam, 1965; Szalay & Delson, 1979; Andrews *et al.*, 1996) in the significant reduction of the cingulum, a somewhat different shape of the teeth crowns and some minor differences in the location and development of the P^4 and M^1 coni. It differs from *D. carianthicus* from Rudabanya (Kretzoi, 1975; Begun & Kordos, 1993; Andrews *et al.*, 1996) in the relatively reduced length of the lingual part of the P^4, the romboid shape of the M^1 crown, the more labial position of the para- and metacone of M^1, the traces of a 'neck', and the somewhat smaller dimensions and possibly higher crowns of the cheek teeth. The difference between the *Udabnopithecus* and *D. laiteanus* from Can Llobateres (Begun *et al.*, 1990; Moyà-Solà & Köhler, 1997) is that in the latter, the hypocone is nearly the largest of the molar cusps, the protocone is well defined and there is no cingulum to speak of. Besides, the teeth of this Spanish *Dryopithecus* are larger and most probably relatively longer than those of the Ubadno *Dryopithecus* M^1. It is difficult to compare *Udabnopithecus* and the Valesian *D. crusafonti* from Can Ponsic (Begun, 1992) since the latter is represented only by mandibular molars. Yet judging by the relatively long and narrow P^4 (Begun, 1992; Andrews *et al.*, 1996) and the relatively large dimensions of the mandibular molars, it seems that *Udabnopithecus* differs also from this hominoid.

The differences between the *Udabnopithecus* and each of the hominoids detailed above do not suffice to make a convincing case for taxonomic segregation, yet bearing in mind the meager remains at hand, they do prevent its identification with any of the aforementioned *Dryopithecinae.*

It is not just one of the latest *Dryopithecinae* but also one of the smallest variants of this family, characterised by a remarkable reduction of the cingulum in the cheek teeth (slight traces of a neck on the lingual surface of P^4 and M^1) as well as the smaller differences which were detailed above (as for example the relatively prolonged lingual part of the P^4, absence of the protocone in M^1 and its rhomboid crown).

Conclusion

The Late Miocene (MN 11) locality of Udabno, Georgia, has the potential to yield precious information concerning hominoid evolution. If our estimate of age (8–8.5 million years) is correct, *Dryopithecus garedziensis* appears to be youngest known Dryopithecine in Eurasia. The geographical position of Georgia at the crossroads between Africa, Europe and Asia, permits us to consider this region as corridor for faunal movement during the Late Miocene.

The particular morphological traits of *Udabnopithecus* combined with its relatively late stratigraphic context and its isolated geographic position in the Trans-Caucasus region may indeed afford it a separate taxonomic classification, and until more data accumulates, we classify this fossil as *Dryopithecus garedziensis* Burtshak and Gabashvili.

Acknowledgements

We would like to thank Prof. Peter Andrews and Anna Belfer-Cohen for their help in editing the manuscript. D. L. thanks Prof. Louis de Bonis and George Koufos for inviting him on Nikiti meeting.

References

Alpagut, B., Andrews, P. & Martin, L. (1990). New hominoid specimens from the middle Miocene site at Paşalar, Turkey. *J. Hum. Evol.* **19**: 397–422.

Andrews, P. (1992). Evolution and environment in the Hominoidea. *Nature* **360**: 641–6.

Andrews, P., Harrison, T., Delson, E., Bernor, R. L. & Martin, L. (1996). Distribution and biochronology of European and Southwest Asian Miocene cattarhines. In (Bernor, R. L., Fahlbusch, V., Mittman H-V. (ed.) *The Evolution of Western Eurasian Neogene Mammal Faunas*, pp. 169–207.

Begun, D., Moyà-Solà, S. & Köhler, M. (1990). New Miocene hominoid specimens from Can Llobastres (Valles Penedes, Spain) and their geological and paleoecological context. *J. Hum. Evol.* **19**: 255–68.

Begun, D. (1992). *Dryopithecus crusafonti sp.*nov a new Miocene hominid species from Can-Poncic (North-eastern Spain). *Amer. J. Phys. Anthrop.* **87**: 291–309.

Begun, D. & Kordos, L. (1993). Revision of *Dryopithecus* brancoi Schlosser, 1901, based on the fossil Hominoid material from Rudabanya. *J. Hum. Evol.* **25**: 271–85.

Burtschak-Abramovitsch, N. O. & Gabachvili, E. G. (1945). Vishaia cheloveko-obraznaia obeziana iz verkhnetretichnikh otlojenii Vostochnoi Gruzii. *Soobshenia AN.Gruzii*, **6**: 458–64.

Burtschak-Abramovitsch, N. O. & Gabachvili, E. G. (1950). Otkritie isskopaemogo antropoida v Gruzii. *Priroda* **9**: 70–2.

Gabunia, L. K. (1959). *Kistorii Gipparionov.* Moskva: Nauka.

Gabunia, L. K. (1980). Traits essentiels de l'evolution des faunes de Mammiferes de la region mer Noire-Caspienne. *XXVI Congress Geol. Paris*, pp. 195–204.

Gremiatski, M. A. (1957). Isskopaemaia obeziana na territori SSSR. *Sovetskaia Antrop.* **1**: 35–45.

Kretzoi, M. (1975). New ramapithecines and Pliopithecus from the lower Pliocene of Rudabanya in north-Eastern Hungary. *Nature* **257**: 578–81.

Meladze, G. (1985). *Obzor gipparionovoi fauni Kavkaza.* Tbilisi: Metsniereba.

Moyà-Solà, S. & Köhler, M. (1995). A new partial cranium of *Dryopithecus* Lartet,1863 (Hominoidea,Primates) from the upper Miocene of Can Llobastres(Valles Penedes, Spain). *J. Hum. Evol.*, **29**: 101–39.

Nesturkh, M. F. (1968). Razvitie primatologii v SSSR. *Voprosi Antropologii.* **28**: 3–20.

Pilbeam, D. R. (1969). The Tertiary Pongidae of East Africa: evolutionary relationships and taxonomy. *Bull. Peabody Mus. Nat. Hist. New Haven* **31**: 1–185.

Piveteau, J. (1957). *Traite de Paleontologie Humaine.* Paris: Masson.

Reshetov, I. G. (1966). *Priroda Zemli i proischojdenia cheloveka.* Moskva: Misl.

Sen, S. (1997). Magnetostratigraphic calibration of the European Neogene mammal chronology. *Paleogeog. Paleoeclimatol. Paleoecol.* **133**: 181–204.

Simons, E. L. & Pilbeam, D. R. (1965). Preliminary revision of the *Dryopithecinae (Pongidae, Anthropoidea). Folia Primat.* **3**: 81–152.

Szalay, F. & Delson, E. (1979). *Evolutionary History of the Primates.* London: Academic Press.

Tsiskarishvili, G. (1987). *Pozdnetretichnie nosorogi Kavkaza.* Tbilisi: Metsniereba.

Wangenheim, E., Gabunia, L., Pevzner, M. & Tsiskarishvili, G. (1989). Stratigraficheskoe polojenie mestonakhojdenii gipparionovoi fauni Zakavkazia v svete dannikh magnetostrarigrafii. *Izv. AN SSSR, Ser. Geolog.* **8**: 70–7.

15 Forelimb function, bone curvature and phylogeny of *Sivapithecus*

Brian G. Richmond and Michael Whalen

Introduction

Rarely have functional inferences been as significant to phylogenetic reconstructions as in the case of *Sivapithecus*. Although fossils now considered to be *Sivapithecus* were once thought to be hominid ancestors, the discovery of a facial skeleton (GSP 15000) attributed to *Sivapithecus* provided strong evidence of an exclusive relationship with *Pongo* because of a suite of shared, derived traits (Pilbeam, 1982; Ward & Kimbel, 1983; Ward, 1997b). However, the discovery of *Sivapithecus* postcranial remains, notably two humeral shafts, cast doubt on these relationships (Pilbeam *et al.*, 1990). The anteriorly-convex and medially-concave humeral shaft morphology of these fossils revealed that the orang-utan-like face of *Sivapithecus* is combined with a pronograde quadrupedal skeletal design unlike that of extant apes. Pilbeam *et al.* (1990) posited that the fossil evidence is consistent with two phylogenetic scenarios, each requiring a considerable amount of homoplasy. First, *Pongo* and *Sivapithecus* may be sister taxa, in which case many postcranial features shared among extant hominoids must have evolved in parallel (Figure 15.1A). Secondly, *Pongo* and *Sivapithecus* may not form a clade, in which case the similarities in the face and palate are convergently derived or primitive for great apes (Figure 15.1B). A scenario in which the palatal and facial similarities are primitive for great apes requires subsequent reversals in extant African apes and humans. A third possibility, that *Pongo* and *Sivapithecus* are related and that the supposed primitive postcranial features are in fact reversals from a more modern ape-like condition, was considered by Pilbeam *et al.* (1990) to be unlikely in the light of postcranial similarities between *Sivapithecus* and other fossil hominoids such as *Proconsul* and *Equatorius* (formerly *Kenyapithecus africanus*; see Ward *et al.*, 1999). Thus, interpretations of hominoid phylogeny depend on the relative strength of the cranial versus postcranial evidence (Andrews & Martin, 1987; Pilbeam *et al.*, 1990; Rose, 1986, 1989, 1994, 1997; McCrossin & Benefit, 1994; Ward, 1997a; Begun *et al.*, 1997; Larson, 1998).

Moyà-Solà & Köhler (1996) have offered a novel solution to this problem. They re-analysed the *Sivapithecus* humeri and proposed that the genus *Sivapithecus* is comprised of species practicing different locomotor behaviours. Specifically, they suggest that, based on their reconstruction, the smaller humerus (GSP30730) attributed to *S. indicus* is unlike the highly curved large humerus (GSP 30754) of *S. parvada*. Instead, they argue that

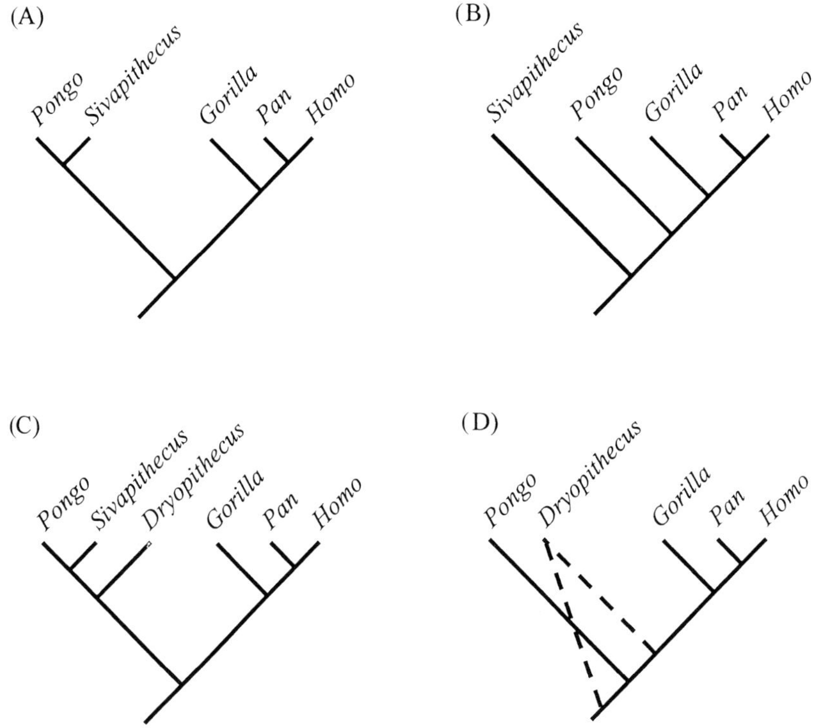

[Figure 15.1]
Cladograms of potential relationships of *Sivapithecus* and *Dryopithecus* to extant great apes and humans. Craniofacial fossils of *Sivapithecus* support (A) a *Sivapithecus-Pongo* clade, while the postcranial evidence suggests that (B) *Sivapithecus* is not a sister taxon of *Pongo*. The reverse is true for *Dryopithecus*, which lacks the derived craniofacial features of *Pongo*, but resembles *Pongo* in many postcranial features. Moyà-Solà & Köhler (1996) suggest that the *Sivapithecus* humerus GSP 30730 provides a postcranial link between *Dryopithecus* and *Pongo*, thus favoring a cladogram in which (C) *Dryopithecus* is the sister taxon of a *Sivapithecus–Pongo* clade. Other phylogenies (D) consider *Dryopithecus* to be a either a stem great ape, or the sister taxon of African apes and humans.

the *S. indicus* humerus is straighter, like the humeral shafts of *Pan*, *Pongo* and the humerus from St. Gaudens attributed to *Dryopithecus*. From this, they infer that the smaller species was adapted primarily for climbing and suspension (like *Pongo* and *Dryopithecus*), while the larger species was substantially more quadrupedal. If true, this would help clarify the phylogenetic uncertainties surrounding *Sivapithecus* because *S. indicus* would share many cranial and postcranial derived traits with *Pongo*, and *Dryopithecus* would share a postcranial body form with the *Sivapithecus–Pongo* clade but lack the derived facial characteristics (Figure 15.1C). This phylogeny would therefore require that *S. parvada* be secondarily derived for pronograde quadrupedalism. It should be noted that other phylogenies (Begun, 1992a; Andrews, 1992; Begun *et al.*, 1997; Harrison & Rook, 1997; Fleagle, 1998) differ from that proposed by Moyà-Solà and Köhler in that they consider *Dryopithecus* to be a sister taxon of African apes and humans, or a stem great ape (Figure 15.1D).

Despite the central role played by humeral shaft morphology in reconstructions of the locomotor adaptations and phylogenetic relationships of these Miocene hominoids, there has been no attempt to quantify this morphology. This study develops novel morphometric methods to quantify the humerus curvature shape in order to examine the variation in extant anthropoids, and test three hypotheses. Moyà-Solà & Köhler's (1996) proposal includes two separate, but related hypotheses. The first hypothesis is that the locomotor diversity within the genus *Sivapithecus* is such that different species practice different modes of locomotion, specifically that the humerus of *S. indicus* is designed for an orthograde positional behaviour like that of *Pongo*, while the humerus of *S. parvada* resembles the humeri of pronograde quadrupeds.

The second hypothesis states that there are locomotor similarities between *S. indicus* and *Dryopithecus*. The partial skeleton attributed to *D. laietanus* shows clear adaptations for orthograde, suspensory positional behaviour, but this specimen lacks a well-preserved humeral shaft, thereby precluding direct comparisons with the *Sivapithecus* humeri. Instead, Moyà-Solà & Köhler (1996) compared the *Sivapithecus* specimens to the humerus from St. Gaudens, the type site of *D. fontani* (Lartet, 1856; Pilbeam & Simons, 1971; Begun, 1992b). Although the St. Gaudens humerus itself has been the subject of disagreement over its functional and phylogenetic affinities (see review in Begun, 1992b), the second hypothesis states that the St. Gaudens and GSP 30730 humeri resemble one another and those of modern great apes.

The first two hypotheses bear on the third hypothesis that *Sivapithecus* is the sister taxon of *Pongo*. Postcranial similarities between *S. indicus*, *Dryopithecus* and extant great apes would add postcranial support to the cranial evidence of a *Sivapithecus–Pongo* relationship. On the other hand, if the postcranium of *S. indicus* differs from those of the other taxa, then one is left with a choice between whether the cranial or postcranial evidence is more reliable for inferring the systematic relationships of *Sivapithecus*.

Other anatomical regions provide evidence regarding the locomotor behaviour of *Sivapithecus* and *Dryopithecus*. While *S. parvada* has been interpreted as an arboreal pronograde quadruped, the considerable body mass of about 69 kg reconstructed for this species suggests a preference for terres-

Table 15.1. *Humerus comparative sample*

Locomotor behavior	Taxa*	Sample size
Terrestrial quadruped	*Erythrocebus patas*	10
Terrestrial quadruped	*Papio sp.*	10
more terrestrial ↑	*Macaca mulatta*	13
Arboreal quadruped	*Presbytis sp.*	9
more	*Colobus sp.*	14
arboreal/climb ↓	*Alouatta seniculus*	10
Climb/knuckle-walk	*Pan troglodytes*	10
Suspension	*Pongo pygmaeus*	11
Suspension	*Hylobates sp.*	10

* Taxa with mixed subspecies including the following: *Papio hamadryas, P. anubis* and *P. cynocephalus; Presbytis melalophos, P. obscura* and *P. rubicunda; Colobus badius* and *C. polykomos; Hylobates hoolok* and *H. agilis*.

trial supports (Rose, 1993). The phalangeal evidence is reassessed here to determine whether the morphology, especially phalangeal curvature, is more consistent with use of the hands in an arboreal or terrestrial environment. The phalangeal and humeral evidence are then compared to other anatomical regions.

Methods

The small *Sivapithecus* humerus GSP 30730 is crushed anteroposteriorly, but appears undistorted in the transverse plane. Because of the damage to the anterior surface of the proximal shaft of GSP 30730, comparisons are restricted to mediolateral curvature. Casts of the fossil hominoid humeri GSP 30730, GSP 30754, and St. Gaudens were compared to a sample of extant catarrhines that range from specialised terrestrial quadrupeds to arboreal quadrupeds and suspensory apes (Table 15.1; Figure 15.2).

Traditional measures of curvature in long bones reflect the magnitude of curvature, but do not record information about curvature shape. Morphometric open contour methods were developed here to capture the shape of the medial shaft outline (Rohlf, 1990). Each humerus was video-recorded with a Sony Hi-8 camera oriented directly perpendicular to the transverse plane of the humerus, using established video-based morphometric methods (Spencer & Spencer, 1995; Richmond, 1998b). MacMorph Image Analysis software (Spencer & Spencer, 1994) was used to video-capture and calibrate each humerus image, and to superimpose a grid of 15 evenly-spaced vertical lines over the humeral shaft with the first and last lines

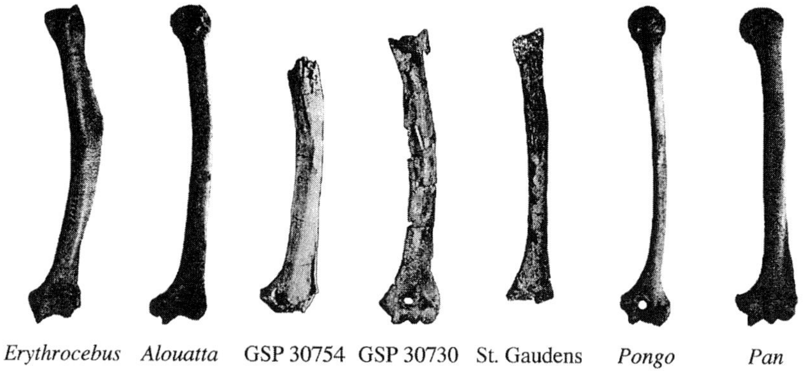

[Figure 15.2]
Humeri of *Erythrocebus patas*, *Alouatta seniculus*, GSP 30754, GSP 30730, St. Gaudens, *Pongo pygmaeus*, and *Pan troglodytes*, shown at approximately the same length. Note the magnitude and shape of the curvature in the *Sivapithecus* specimens compared to the relatively straight shaft in the specimen attributed to *Dryopithecus*.

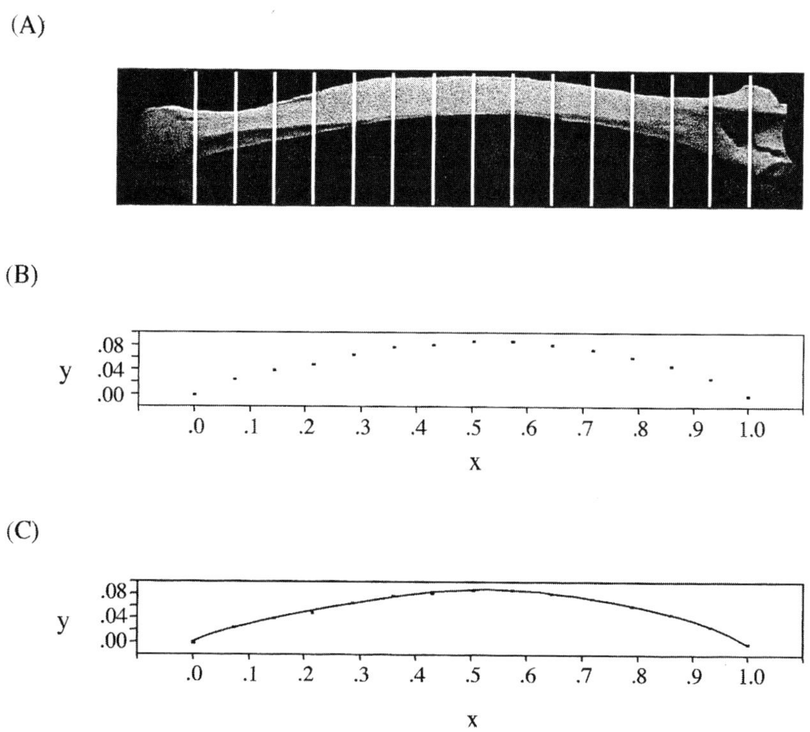

[Figure 15.3]
Humerus curvature methods. (A) The video-captured image of each humerus is superimposed with an evenly-spaced grid consisting of 15 vertical lines. The medial border (lower border as shown) is digitized at these points, resulting in (B) coordinates at comparable positions along the shaft between specimens. (C) A polynomial curve is used to illustrate shape.

adjacent to the proximal and distal articular margins, respectively (Figure 15.3). The 15 points of intersection between the vertical lines and medial border of the humerus were then digitised as '*x*' and '*y*' coordinates (Figure 15.3). For each humerus, the digitised points were translated, rotated and scaled such that the first (most superior) and last (most inferior) points were transformed into '0,0' and '1,1', respectively, using standard transformation equations (Bookstein, 1991). In this way, each humerus was standardised to the same length, or 'size'. After this transformation, the new '*x*' coordinates represent the same relative distance along the shafts of each humerus; that is, the transformed '*x*' coordinates are comparable between humeri. Thus, the '*y*' coordinates describe the shape of the curve. In this way, the '*y*' coordinates are treated as variables at comparable positions along the humeral shaft. Averaging the '*y*' coordinates at each '*x*' position for each taxon provides an average curvature shape for each species, along with statistics of variation, such as standard deviations. Thus, it is possible to test statistically whether one species or specimen is significantly different from another species relative to its variation. Mean curvature shapes were displayed by fitting a sixth-order polynomial line to the average '*y*' coordinates (see Figures 15.4, 15.5).

Unfortunately, the proximal articular surface of GSP 30754 is not preserved. Rose (pers. comm.) estimates a total length of approximately 36 cm, meaning that approximately nine-tenths of the humeral shaft is preserved. This estimate agrees with humeral morphology in the comparative sample. Extrapolating the missing portion could have an undue effect on the curvature shape. Therefore, to allow comparison of this critical specimen to others, the comparative data set (including GSP 30730 and St. Gaudens) was limited to the distal majority of each humeral shaft by removing the first, most superior point from each humerus outline. Direct comparisons of the incomplete GSP 30754 with complete extant and fossil humeri yielded results that were almost identical to the analysis above, indicating that possible errors in estimating length have negligible effects on the results.

A second analysis was performed on the central portion of the shaft to examine curvature shape apart from the influential projection of the medial epicondyle (see below). In this analysis of the central shaft, both the first coordinate (due to the incompleteness of GSP 30754) and last two coordinates (to omit the medial epicondyle) were removed from each humeral outline, and the remaining 12 coordinates were translated, rotated and scaled as described above. Thus, the second analysis was performed on the central 12 coordinates at comparable positions on the humeral shafts in each taxon.

To better assess curvature shape, the '*y*' (shape) coordinates were input into a canonical variates analysis (CVA), a commonly-used multivariate

Table 15.2. *Proximal phalanx comparative sample*

Taxa*	Sample size
Macaca mulatta	29
Cercopithecus sp.	7
Colobus sp.	3
Homo sapiens	43
Gorilla gorilla gorilla	31
Pan troglodytes	13
Pongo pygmaeus pygmaeus	23
Pongo pygmaeus abelii	13
Hylobates lar	25

* Taxa with mixed species include the following: *Cercopithecus campbelli, C. diana* and *C. petaurista*; *Colobus polykomos*, (*Piliocolobus*) *badius* and (*Procolobus*) *verus*.

method that maximises between-group differences (Klecka, 1980). The CVA illustrates how taxa discriminate in the shape of the humerus shaft. Because the first and last coordinate of each outline are '0,0' and '1,0', respectively, these have no variance, provide no independent shape information and are not appropriate as data in a multivariate analysis. Therefore, the central 10 '*y*' coordinates were used in the CVA.

Other anatomical regions have received less attention, but none the less provide important data for reconstructing positional behaviour in these fossil taxa. One example is the functional morphology, including shaft curvature, of manual proximal phalanges. Phalangeal curvature in two fossil manual proximal phalanges, one (GSP 19700) attributed to *S. parvada* and the other (RUD 78) attributed to *Dryopithecus* (probably *D. brancoi*), were compared to a large sample of adult proximal phalanges of extant taxa (Table 15.2). Curvature was measured as the included angle of the arc of a circle encompassed by the phalanx (Stern *et al.*, 1995). Included angle was computed in MacMorph from the digitised endpoints and midshaft of video images of each phalanx (Richmond, 1998b). The comparative sample consists of third manual proximal phalanges. RUD 78 is probably also a median digit (Begun, 1993). Although GSP 19700 appears to be a fourth proximal phalanx (Rose, 1986), there are little to no differences in curvature among the proximal phalanges within a hand (Susman *et al.*, 1984). Because RUD 78 is missing its base, curvature was measured in two ways. First, curvature was taken directly from the preserved fragment, but this is an underestimate because the measurement of included angle is sensitive to length (Stern *et al.*, 1995). The second measure is an estimate that reconstructs the missing portion.

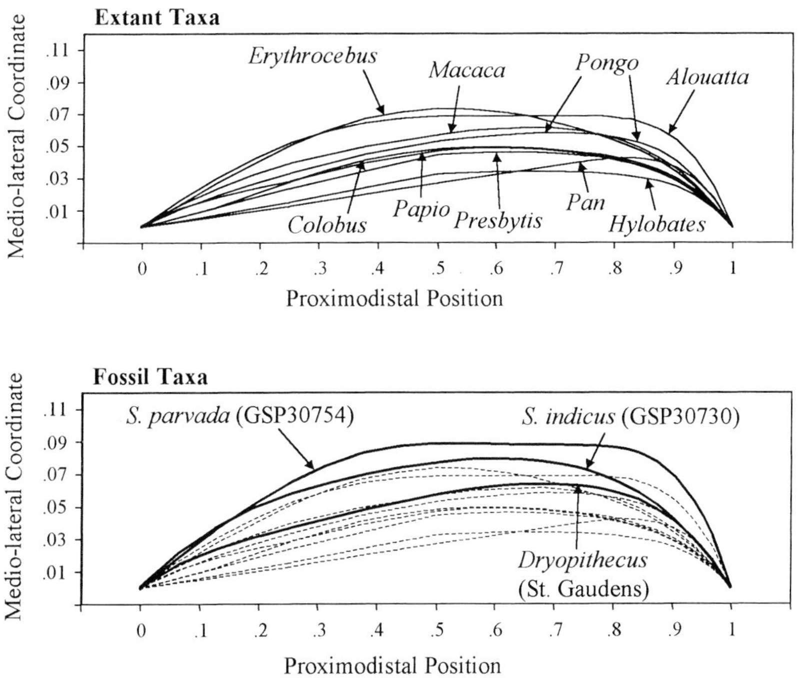

[Figure 15.4]
Mean mediolateral curvature of humeral shaft in extant taxa (top) and fossil and extant taxa (bottom). Note that *Erythrocebus* and *Alouatta* have the most curved humeri among the extant taxa, but differ in shape. The sharp curve at the distal end (right in graph) of the humerus in *Alouatta* reflects a large, projecting medial epicondyle. *Pan* and *Hylobates* have the least curved humeri. Both *Sivapithecus* specimens are more curved than the mean for any extant taxon, as well as *Dryopithecus*, which resembles both *Pongo* and *Macaca*. However, GSP 30754 has a more massive medial epicondyle than does GSP 30730. Bottom plot: fossil (solid line) and extant (dashed line) taxa.

Results

In general, the mean curves of extant taxa (Figure 15.4) support existing descriptions of pronograde quadrupeds having more medially-concave humeri and suspensory primates having straighter humerus shafts. For example, *Erythrocebus patas* humeri are on average the most curved, and *Hylobates* and *Pan* humeri are least curved. However, the mediolateral curvature of the humerus is not completely successful in discriminating among locomotor categories. For example, baboon humeri are less curved than the arboreally quadrupedal macaques and howler monkeys and the suspensory orang-utan (Figure 15.4). Close inspection shows that many taxa differ not only in magnitude, but also in shape. Although *Alouatta* is second only to *Erythrocebus* in degree of curvature, *Alouatta* humeri curve most along the distal-most shaft (right Figure 15.4) whereas *Erythrocebus* humeri are more evenly curved. *Pongo* and *Pan* have fairly straight shafts that curve sharply at the distal end as in *Alouatta*. This sharp distal curve reflects a strongly-projecting, medially-oriented medial epicondyle in these taxa, a feature associated with pronounced forearm flexor musculature for powerful grasping throughout a range of pronation and supination (Fleagle, 1998).

Both *Sivapithecus* humeri exhibit more marked curvature than the mean for any extant taxon (Figure 15.4). In contrast, the St. Gaudens humerus is less curved, but perhaps more so than would be expected. It departs from the shape of *Hylobates* and *Pan*, but closely follows the mean *Pongo* form. The remarkable curvature of GSP 30754 is strongly influenced by the large size of the medial epicondyle, and in its shape most closely resembles the much smaller *Alouatta* humeri. Although well-developed, the medial epicondyle of GSP 30730 is not as extreme as that in the large *Sivapithecus* humerus (Figure 15.4). However, both GSP 30730 and GSP 30754 are most like pronograde monkeys and differ from extant hominoids, whereas the St. Gaudens humerus resembles both orang-utans and the quadrupedal macaques.

To examine curvature shape apart from the strong influence of the medial epicondyle, the second analysis involves only the central portion of the shaft (see above). This analysis provides a clearer distinction of the extreme taxa, namely the highly curved, medially-concave patas humeral shafts and the straight-shafted humeri of chimpanzees (Figure 15.5). Averages of all other taxa lie between these extremes.

In central shaft morphology, the two *Sivapithecus* humeri closely resemble one another and are as curved as the terrestrially-adapted patas monkey (Figure 15.5). The *Dryopithecus* humerus, on the other hand, is much less curved and resembles orang-utans, macaques and baboons.

The discrimination of taxa based on central shaft curvature is better illustrated with the CVA of humeral coordinates (Figure 15.6). The first canonical axis discriminates roughly according to the magnitude of curvature, while the second axis reflects more subtle aspects of shape. Taxa that fall near the top of the plot (i.e. those with high canonical scores on Axis 2) have humeri in which curvature is generally greatest in the distal shaft. Taxa with low scores on the Axis 2, such as *Alouatta* and *Macaca* are characterised by humeral shafts in which the greatest curvature occurs more proximally. This contrast is visible in the mean humeral contours in Figure 15.5.

The CV plot makes clear the morphological similarity between the two *Sivapithecus* humeri. In degree of curvature, they most closely resemble *Erythrocebus*, but in shape they are more similar to the humeri of macaques and howler monkeys in that curvature is greatest in the proximal portion of

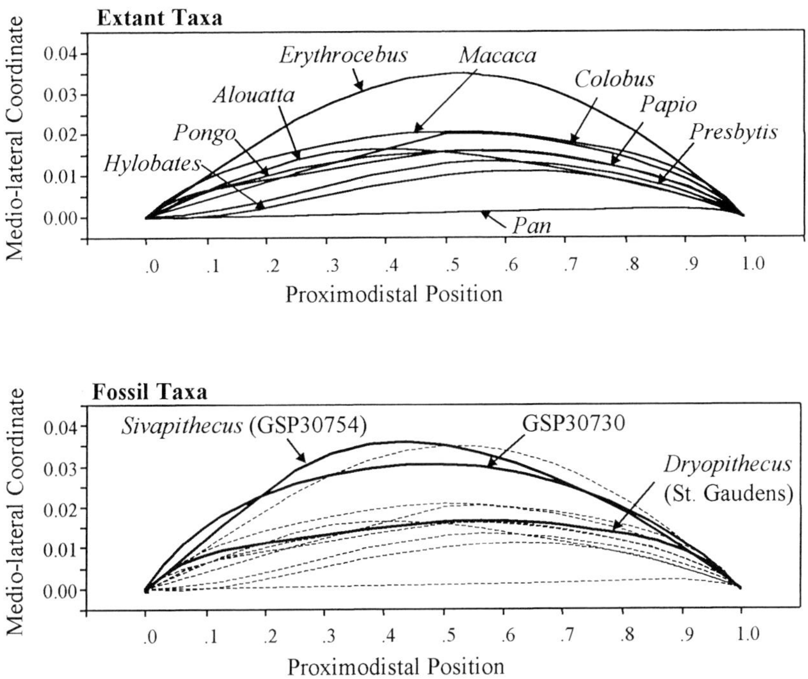

[Figure 15.5]
Mean mediolateral curvature of central humeral shaft in extant (top) and fossil (bottom) taxa. *Erythrocebus* stands from other taxa in its extreme curvature, and *Pan* humeri are very straight. Averages for all other taxa are intermediate. The *Sivapithecus* humeri resemble *Erythrocebus* in their magnitude of curvature, and differ from *Dryopithecus*. Note that the close similarities among *Pongo*, *Macaca*, and *Papio* indicate that mediolateral humeral shaft shape does not consistently discriminate between primates of different locomotor adaptations. CVA of these coordinates shows that the *Sivapithecus* humeri resemble *Macaca* and *Alouatta* in that the greatest curvature occurs proximally along the humeral shaft. Bottom plot: fossil (solid line) and extant (dashed line) taxa.

the shaft. In contrast, the St. Gaudens humerus is markedly different from the Siwalik humeri. Its curvature is more modest (see its central position along canonical axis 1, Figure 15.6), and its shape more typical of orang-utans, baboons and some macaques, in which the position of greatest curvature occurs near midshaft.

Data on phalangeal curvature yield results consistent with the locomotor reconstructions based on humerus shaft shape. As noted by Rose (1986) in his description of the material, the proximal phalanx attributed to *S. indicus* resembles chimpanzees in its size, proportions and degree of longitudinal curvature. However, the included angle value of 34.4° reported for GSP 19700 suggests a fairly straight phalanx, straighter than the mean for *Gorilla*. Remeasurement here resulted in an included angle of 51°, suggesting that the original measurement (provided by Stern) may have been 54.4° (Figure

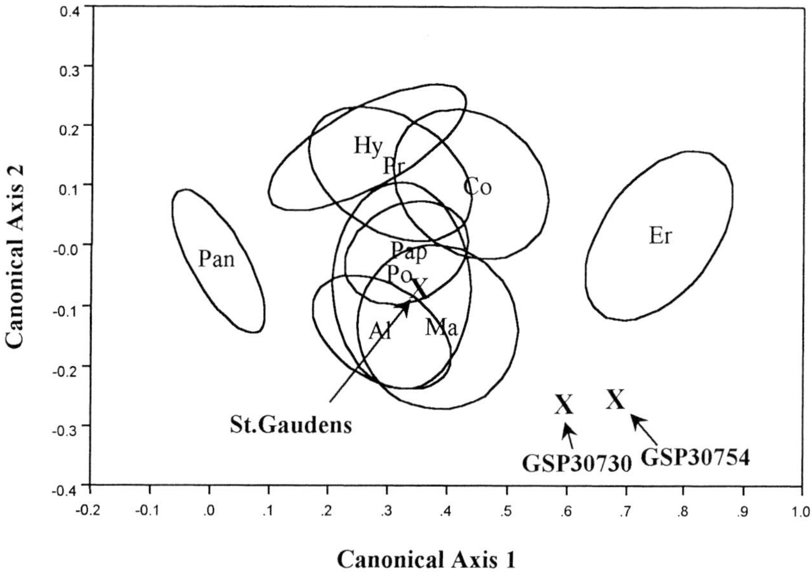

[Figure 15.6]
Canonical variates plot of mediolateral curvature of central humeral shaft. Taxon means (positions of abbreviated names) and 50% confidence ellipses are shown for extant taxa, along with the positions of the fossil hominoid humeri (Xs). Taxa with the greatest medial humeral curvature lie in the lower right of the plot, and those with the straightest humeri lie at the upper left. The *Sivapithecus* humeri GSP 30730 and GSP 30754 are among the most curved in the sample, but the curvature shape is not identical to that in *Erythrocebus*. Both humeri differ from the St. Gaudens humerus. Pan: *Pan*; Hy: *Hylobates*; Pr: *Presbytis*; Co: *Colobus*; Pap: *Papio*; Po: *Pongo*; Al: *Alouatta*; Ma: *Macaca*; Er: *Erythrocebus*.

15.7). However, the values obtained here for comparative taxa are also somewhat different, although the taxa sort in the same fashion (i.e. *Pongo*→*Pan*→*Gorilla*). In either case, the *Sivapithecus* phalanx clearly possesses an intermediate level of curvature, resembling chimpanzees and large colobines (especially the red colobus monkey).

Discussion

Functional morphology

The inability of mediolateral humerus shaft shape alone to completely distinguish locomotor modes from one another must be emphasised. If a taxon has extremely straight or curved humeri, the locomotor implications are reasonably clear. For example, the pronounced medial curvature seen in the Siwalik specimens is unlike any extant hominoid, and only matched in magnitude and shape by pronograde quadrupeds. However, some extant taxa, such as *Pongo*, *Papio*, *Alouatta* and *Macaca*, have diverse locomotor adaptations and skeletal designs, and yet exhibit similar humeral outlines. In fossils that resemble these taxa, such as the St. Gaudens humerus, locomotor behaviour cannot be reliably inferred from medial humerus shaft shape alone. Other *Dryopithecus* species have a wealth of fossil remains from which to infer an orthograde, suspensory skeletal design. However, using these data to infer adaptations in *D. fontani* is problematic because the paucity of evidence for this species makes its systematic position relatively insecure. Also, some have suggested the Klein Hadersdorf humerus belongs with the St. Gaudens material in *D. fontani* (Pilbeam & Simons, 1971), while others argue these specimens are too different to belong to the same species (Aiello, 1981; Begun, 1992b). Because of these uncertainties, locomotor reconstructions for the St. Gaudens fossils should be made strictly from the postcranial material from this site, namely the humerus. Reconstructions of the proximal end indicate minimal torsion of the humeral head, and thus a scapulohumeral complex unlike that of modern apes (Rose, 1994; Larson, 1996). Combined with more suspensory features, including its anteriorly concave shaft and gracile deltopectoral crest (Aiello, 1981; Begun, 1992b; B. Richmond, unpublished data), the St. Gaudens humerus apparently belonged to an animal that was adapted for orthograde positional behaviour but lacked some suspensory specialisations seen in extant apes.

Despite the limited functional discrimination provided by mediolateral humerus shaft curvature, several conclusions can be made regarding the functional anatomy of *Sivapithecus* and *Dryopithecus* species. The first hypothesis, that there is locomotor diversity within the genus *Sivapithecus*, is not supported by evidence from the humerus shaft. Both Siwalik specimens resemble one another more closely than either resembles other taxa in medial humerus shaft morphology (Figures 15.5, 15.6).

Moyà-Solà & Köhler (1996) based their hypothesis on their reconstruction of the crushed and somewhat distorted GSP 30730. From this reconstruction, they argue that GSP 30730 is slender, not retroflexed, lacks a flat deltoid plane and has a lateral side that is almost flat. While it is clear that this specimen is crushed anteriorly, there is no indication of any significant postmortem distortion in the transverse plane aside from isolated indentations of bone. Indeed, the close resemblance in shaft morphology to the larger GSP 30754 (see Figures 15.5, 15.6) suggests that GSP 30730 is not significantly different from other *Sivapithecus* fossils and is not in need of reconstruction. Although Moyà-Solà and Köhler specifically called attention to the lateral side of the shaft, the curvature on one side of the shaft mirrors the opposite side in humeri examined here. Visual comparison of casts of the two specimens reinforces the similarity in curvature on both sides. The deltoid plane of GSP 30730 is not quite as pronounced as in GSP 30754, but the difference is not greater than would be expected of extant congeneric individuals of a comparable difference in size. Larson (1998, p. 88) also considers GSP 30730 to be 'similar to GSP 30754 in its curvature and development of crests'. GSP 30730 is more slender than its larger cousin, but again, not more so than might be expected based on size variation in congeneric quadrupedal taxa. The two humeri also differ in the development of the medial epicondyle, but the relatively smaller epicondyle in GSP 30730 does not support an interpretation that it is better-adapted for suspension. The medial projection of the medial epicondyle in both specimens is consistent with the view that both are adapted for arboreal rather than terrestrial quadrupedalism (Fleagle, 1998). The well-known allometric relationship between body mass (a volume) and the cross-sectional strengths (areas) of muscles and long bones may help explain the differences observed in robusticity and development of muscle attachments (Fleagle, 1985).

In keeping with the morphology of GSP 30730, there is little evidence of orang-utan-like suspensory adaptations in other postcranial remains of *S. indicus*. For example, the proximal humerus GSP 28062 resembles cebids and lacks the derived morphology of extant apes associated with enhanced mobility (Rose, 1989). Lastly, the suggestion that the smaller, chimpanzee-sized species was adapted for climbing and suspension, while the female gorilla-sized species was adapted for more quadrupedal locomotion, conflicts with allometric patterns observed in extant catarrhines, in which larger taxa tend to distribute their weight from multiple overhead supports while smaller taxa tend to walk quadrupedally above branches (Fleagle, 1985). It is very unlikely that GSP 30730 represents an individual adapted for suspensory locomotion.

Concluding that neither *Sivapithecus* species was adapted for orthograde positional behaviour bears on the second hypothesis that *Sivapithecus* and *Dryopithecus* share locomotor adaptations. As discussed above, the mediolateral shape of the St. Gaudens humerus shaft is consistent with several locomotor modes (see Figures 15.4 to 15.6). Other characteristics, including the curvature in the anteroposterior plane (Begun, 1992b; B. Richmond, unpublished data), indicate that it was designed for orthograde suspensory behaviour, but perhaps with less shoulder mobility than that in extant apes (Rose, 1994; Larson, 1996). Locomotor reconstructions that include suspensory behaviours are consistent with skeletal evidence from other *Dryopithecus* remains, including the pronounced curvature in RUD 78

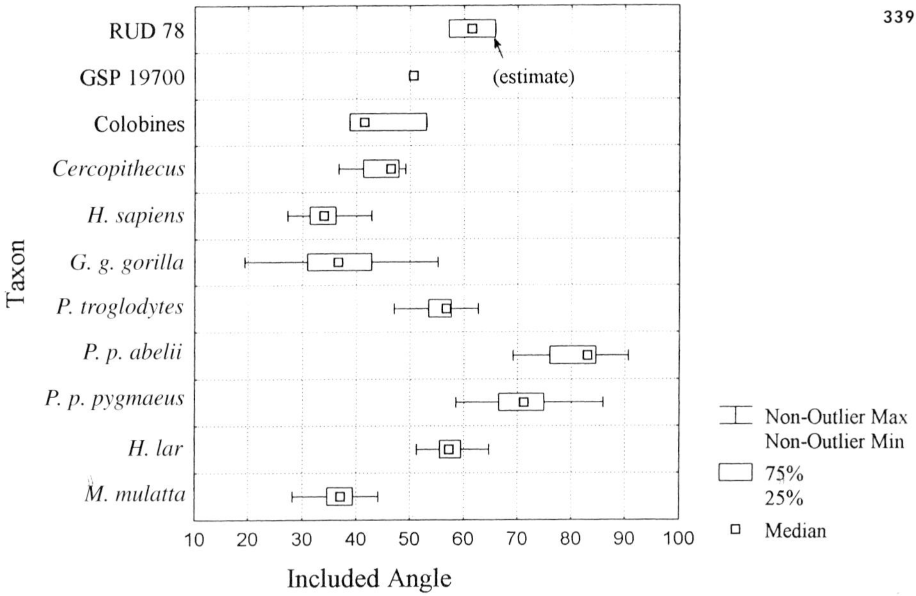

[Figure 15.7]
Curvature of manual proximal phalanges of *Sivapithecus*, *Dryopithecus*, and a broad sample of extant catarrhines. RUD 78 preserves most of the shaft and distal end of a manual proximal phalanx attributed to *Dryopithecus* (Begun, 1993). A range of curvature values is shown for RUD 78. The low value (57.2°) is the included angle measure taken directly from the preserved fragment. However, included angle is sensitive to length, and therefore underestimates the true curvature of the complete phalanx. Included angle is estimated at 66° (the high value shown). Unless this estimate is grossly incorrect, the *Dryopithecus* proximal phalanx is more highly curved than any extant hominoid save *Pongo*, supporting interpretations that orthograde suspension was an important component of the locomotor repertoire of *Dryopithecus*. In contrast, the complete fourth manual proximal phalanx (GSP 19700) attributed to *S. parvada* (Rose, 1986) resembles some colobines (especially *Colobus badius*) and chimpanzees in curvature. Combined with evidence from other anatomical regions, *Sivapithecus* was a large, pronograde arboreal quadruped. (Species list: *Homo sapiens*; *Gorilla gorilla gorilla*; *Pan troglodytes*; *Pongo pygmaeus abelii*; *Pongo pygmaeus pygmaeus*; *Hylobates lar*; *Macaca mulatta*.)

of *Dryopithecus brancoi* (Figure 15.7; Begun, 1993) and the proportions and morphology of the partial skeleton of *D. laietanus* from Can Llobateres (Moyà-Solà & Köhler, 1996). The Siwalik and St. Gaudens humeri are distinct enough from one another to be confident about their locomotor differences.

Sivapithecus locomotor behaviour

The mediolateral curvature of the *Sivapithecus* humeri suggests a pronograde quadrupedal manner of locomotion. The magnitude of curvature is most comparable to the terrestrial patas monkey, but in shape the resemblance is closer to arboreal quadrupeds such as howler monkeys and macaques (Figure 15.6). Other anatomical features provide more detailed functional information. For example, curvature in the proximal manual phalanx suggests that *Sivapithecus* habitually used arboreal supports for quadrupedalism and climbing, but was not specialised for suspensory postures like those used in extant gibbons and orang-utans (Figure 15.7). Furthermore, the proximal joint morphology is dorsoproximally oriented, indicating that *Sivapithecus* hand posture typically involved extension at the metacarpophalangeal joint, a posture (and morphology) characteristic of pronograde quadrupeds (Rose, 1986). Therefore, despite the large size of *S. parvada,* it appears that this species shows signs of locomotion in an arboreal setting. Of course, this joint and shaft morphology does not preclude locomotion on the ground, but shows that they were not adapted to a terrestrial form of locomotion. Moreover, the sensitivity of phalangeal shaft curvature to mechanical loading suggests that *Sivapithecus* actively used arboreal supports (Richmond, 1998b, 1999).

Although no comparable phalangeal fossils are known for *S. indicus*, a reconstructed positional behaviour of arboreal quadrupedalism is consistent with humeral shaft and joint morphology (Rose, 1989; Pilbeam *et al.*, 1990; Figures 15.4 to 15.6). It would be surprising if there were no differences in locomotor behaviour and substrate use given the body size differences between *S. indicus* and *S. parvada*. In extant primates, significant differences in locomotor behaviour between congeneric species are well-documented (e.g. Fleagle, 1977; Ward & Sussman, 1979). Female and male chimpanzees in the same population even differ in their positional behaviour, probably as a result of size differences (Doran, 1993). However, while the postcranial evidence for *S. indicus* is unfortunately limited, at present there is no evidence that species of *Sivapithecus* practiced fundamentally different modes of locomotion.

In contrast to the pronograde quadrupedalism of *Sivapithecus*, *Dryopithecus* exhibits clear suspensory adaptations. Phalangeal curvature of RUD 78 probably exceeds that in gibbons, and the estimated value of 66° compares best to extant orang-utans, indicating that *Dryopithecus* was well-adapted to resist the stresses involved in suspensory grasping (Preuschoft, 1973; Richmond, 1998a,b).

Sivapithecus phylogeny

Conclusions reached above lead to the third, outstanding hypothesis regarding whether or not *Sivapithecus* is the sister taxon of *Pongo*. In the light of the lack of evidence for suspensory skeletal design in *Sivapithecus*, we are left with the debate as it has stood for the past decade, namely a conflict between the cranial or postcranial evidence. Pilbeam has pointed out that most hominoid systematists are 'craniophilic', that is, they base their phylogenetic conclusions almost entirely on craniodental evidence. In fairness, this is primarily due to the differential preservation of craniodental remains relative to other anatomical regions. However, Pilbeam's point that we must not ignore the postcranial evidence is well-taken. Given our current knowledge, Pilbeam (1996) believes the phylogenetic reconstruction from postcranial evidence to be more likely. Others, notably Ward (1997a,b) and Larson (1998), favour the craniodental evidence for various reasons.

How, then, to decide between the cranial and postcranial evidence? Three lines of evidence suggest that some of the postcranial features shared among extant great apes are homoplasious and the craniofacial architecture shared between *Sivapithecus* and *Pongo* is homologous. First, biomechanical data provide evidence of the plastic nature of certain aspects of the postcranial skeleton, such as long bone curvature. Lanyon (1980) demonstrated the importance of normal activity levels to the development of normal long bone curvature, cortical thickness and mineral density. In rats with the sciatic nerve cut in one leg, the tibia in the inactive leg grew to the same length, but was virtually straight compared to the pronounced sigmoidal curvature of the contralateral, weight-bearing tibia (Lanyon, 1980). Curvature in manual phalanges has been shown to be influenced by more subtle differences in activity. Specifically, taxa such as African apes and rhesus macaques that undergo changes in the amount of arboreal hand use during growth, experience significant changes in levels of phalangeal curvature through modeling and remodeling, while species like gibbons that remain arboreal throughout life show no signs of postnatal changes in phalangeal curvature (Paciulli, 1995; Richmond, 1998b, 1999). However, phalangeal curvature differs among taxa well before birth, suggesting that there is some predetermined degree of curvature that is subsequently modified by activity (Richmond 1998b). These data indicate that, while bone features such as length are under tighter genetic control and not influenced by normal variation in activity, features such as curvature and cross-sectional shaft geometry have a genetic component but are influenced by both the level and nature of activity. Even aspects of joint shape are not immune to modeling and remodeling activity in response to the biomechanical environment. Japanese macaques that regularly walk bipedally experience a variety of osteological changes, including a lumbar lordosis, vertebral wedging and expansion of the femoral head articular surface (Nakatsukasa *et al.*, 1995). The sensitivity of shaft morphology to activity makes unlikely the recent suggestion that GSP 30754 may have functioned in a novel manner despite retaining a primitive shaft morphology (Madar, in Ward, 1997b). Morphological features that model and remodel in response to *in vivo* activity provide the most reliable information about functional behaviour. Thus differences in the curvature of the humerus between *Sivapithecus* and *Pongo* are best explained by differences in biomechanical, locomotor use.

Secondly, there are examples among modern taxa of convergence in postcranial morphology analogous to the hypothesised convergence between orang-utans and African apes. Larson (1998) compiled a list of skeletal traits reported to be diagnostic of hominoids, and showed that many are convergently derived in other anthropoid taxa, especially in the most suspensory New World monkey, *Ateles*. Straight humeral shafts (here, referring to anterior curvature) occur in numerous taxa. Thus, Larson provides a good example in which postcranial characteristics such as humeral shaft morphology evolve convergently in distantly-related taxa that independently evolved similar locomotor behaviours. Larson favours the scenario in which *Sivapithecus* is the sister taxon of *Pongo*, *Sivapithecus* retains a primitive postcranial morphology, and many of the postcranial similarities between orang-utans and African apes and humans are the result of parallel evolution rather than being homologous. This scenario is also supported by what has been argued to be close similarity between the postcranial morphology of *Sivapithecus* and other Miocene hominoids, such as *Proconsul* and *Equatorius* (Pilbeam *et al.*, 1990). However, the apparent ease with which some aspects of functional anatomy change with the evolution of new positional repertoires is also consistent with the scenario in which the common ancestor of great apes and humans possessed a generalised modern great ape-like skeletal design, and *Sivapithecus* convergently evolved a more quadrupedal body form.

Lastly, Ward's description of the detailed nature of the derived palatal, facial and deciduous dental morphology shared between *Sivapithecus* and extant orang-utans suggests that these anatomical structures are homologous. These features include interorbital constriction, tall and narrow orbits, the structure of the premaxillary alveolar clivus and subnasal/palatal region, frontal invasion of the maxillary sinus, supraorbital rim/temporal line morphology, the relationship between the postglenoid process and tympanic tube, dp_4 occlusal geometry, tall and vertical mandibular rami, heteromorphic upper incisors, I^1 lingual crenulation morphology and extreme airorhynchy (Pilbeam, 1982; Ward & Kimbel, 1983; Ward, 1997b). However, it is not only the number of similarities, but also the nature of the similarities that so convincingly suggests that the shared characteristics are homologous. Unlike the postcranial evidence, there is currently no other known primate that shares the unique palate and facial morphology seen in *Sivapithecus* and *Pongo* (Ward & Pilbeam, 1983). If *Pongo* and *Sivapithecus* are distantly related (Pilbeam *et al.*, 1990; Pilbeam, 1996; McCrossin & Benefit, 1997), then either this morphology evolved in parallel or was primitive for the great ape/human clade. Relatively simple changes between more orthograde and more pronograde positional repertoires can explain the postcranial homoplasies in extant great apes or *Sivapithecus*. In contrast, there are no clear biomechanical factors to explain the independent evolution of the unique craniofacial morphology of *Pongo* and *Sivapithecus*. However, we must note that there has been very little research on hominoid craniofacial biomechanics that might reveal functional and developmental complexes underlying the highly derived morphology in orang-utans. At present, it is not clear what environmental or mechanical stimuli would lead to the convergent evolution of the palatofacial morphology in *Sivapithecus* and *Pongo*.

Homoplasy is no less problematic if the craniodental anatomy of *Sivapithecus* represents the primitive condition for great apes and humans (Figure 15.1B). In this case, all the features (e.g. broad orbits, wide interorbital distance, subnasal morphology) shared between primitive fossil hominoids and the African ape/human clade would be reversals in the latter. Therefore, for the reasons outlined above, it appears most likely that the shared craniodental anatomy is homologous in *Sivapithecus* and *Pongo*, and that it faithfully represents their shared ancestry to the exclusion of other hominoids (Figure 15.8).

If *Sivapithecus* is the sister taxon of *Pongo*, then two scenarios are possible: either (1) *Sivapithecus* retains quadrupedal postcranial morphology from the last common ancestor of great apes, and orang-utans and the African ape/human clade evolved their similarities convergently; or (2) the common ancestor of great apes and humans had an essentially ape-like postcranial skeleton, and *Sivapithecus* is secondarily derived (i.e. underwent a reversal) towards a more pronograde arboreal repertoire. Considering that some hominoids (e.g. *Morotopithecus* and perhaps *Dryopithecus*) exhibit orthograde adaptations, but may be more primitive than the common ancestor of extant great apes and humans, the second scenario appears most likely. In this scenario, the close similarities in postcranial form between orang-utans and African apes (and, in some features, ancestral hominids) are explained by common ancestry. Moreover, a recent analysis using published phylogenies to reconstruct the locomotor anatomy of common ancestors concludes on the basis of parsimony that the last common ancestor of great apes and humans was adapted for generalised orthograde quadrumanous locomotion, regardless of the phylogenetic position of

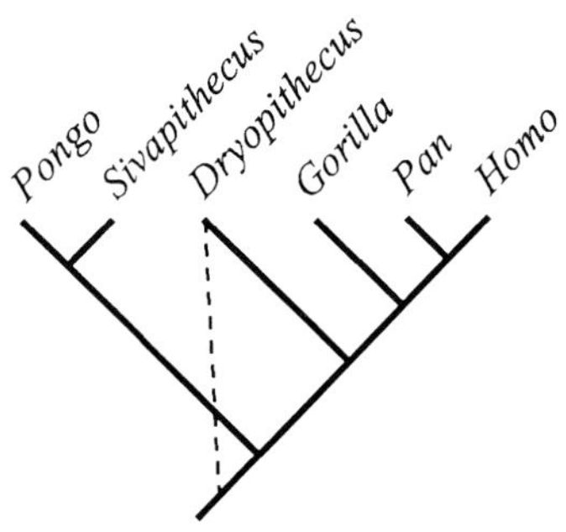

[Figure 15.8]
Cladogram preferred here for *Sivapithecus* and extant hominoids based on the likelihood that the primitive-like skeletal features reflecting pronograde quadrupedalism are reversals from an orthograde common ancestor of extant great apes and humans.

Sivapithecus (Strait *et al.*, 2000). This common ancestor was probably adapted for cautious climbing, suspension and orthograde postures (Sarmiento, 1995), but lacked the extreme suspensory specialisations of *Pongo* and the terrestrial adaptations of extant African apes. Although it has been argued that the postcranial morphologies of *Sivapithecus* and other early and middle Miocene hominoids are too similar for a scenario involving a reversal in *Sivapithecus* (Pilbeam *et al.*, 1990), Rose (1993) characterizes *Sivapithecus* as resembling *Proconsul* in some respects, but being more similar in the bulk of features to extant hominoids, particularly the African apes. This is consistent with the hypothesis that *Sivapithecus* is the sister taxon of *Pongo* and that its pronograde quadrupedal locomotor anatomy represents a reversal from a more orthograde condition. The likelihood that this scenario is correct depends upon postcranial reconstructions of the last common ancestor of great apes and humans. These reconstructions, in turn, depend upon the phylogenetic relationships and functional locomotor anatomy of fossil hominoids, including *Ankarapithecus, Equatorius, Dryopithecus* and *Ouranopithecus* (Alpagut *et al.*, 1996; Pilbeam, 1996; Begun *et al.*, 1997; Bonis & Koufos, 1997; McCrossin & Benefit, 1997; Ward *et al.*, 1999).

Regardless of which phylogeny is correct, there are many interesting questions concerning the ecological conditions leading to homoplasy. If *Sivapithecus* is not related to modern orang-utans, then what dietary and biomechanical conditions or developmental pathways may have led to such a remarkable convergence in craniofacial architecture? If *Sivapithecus* represents the primitive postcranial condition, then why did orang-utans evolve such unique suspensory adaptations when *Sivapithecus* lived successfully as a large-bodied pronograde quadruped in an arboreal habitat

(Kappelman, 1988)? Finally, if *Sivapithecus* is secondarily derived, then why did this clade of moderately-sized to large-bodied apes undergo a reversal from an orthograde suspensory kind of locomotion to pronograde quadrupedalism when they existed in a forested habitat? The answer to these questions lie in future fossil discoveries, novel analyses of existing fossil evidence and research on the biomechanics and development of the postcranial and craniodental skeleton.

Acknowledgements

We would like to thank Louis de Bonis and George Koufos for organizing a productive and enjoyable conference, and for the invitation to attend it. We are grateful to Susan Larson, Mike Rose, David Strait, and Carol Ward for reading and improving the manuscript. We also thank John Fleagle for his encouragement, David Pilbeam, Laura MacLatchy, Eric Delson, and Jay Kelley for access to casts of humeri and phalanges, and Bill Jungers for advice and use of computer equipment. This project was supported by a Student Research Grant (to BGR) from the Doctoral Program in Anthropological Sciences of the State University of New York at Stony Brook, NSF SBR-9624726 (to W. L. Jungers and BGR), and the Henry Luce Foundation.

References

Aiello, L. C. (1981). Locomotion of Miocene Hominoidea. In Stringer, C. B. (ed.), *Aspects of Human Evolution*, pp. 63-97. London: Taylor & Francis.

Alpagut, B., Andrews, P., Fortelius, M., Kappelman, J., Temizsoy, I., Çelebi, H. & Lindsay, W. (1996). A new specimen of *Ankarapithecus meteai* from the Sinap Formation of central Anatolia. *Nature* **382**: 349–51.

Andrews, P. J. (1992). Evolution and environment in the Hominoidea. *Nature* **360**: 641–6.

Andrews, P. J. & Martin, L. B. (1987). Cladistic relationships of extant and fossil hominoids. *J. Hum. Evol.* **16**: 101–18.

Begun, D. R. (1992a). Miocene fossil hominids and the chimp-human clade. *Science* **257**: 1929–33.

Begun, D. R. (1992b). Phyletic diversity and locomotion in primitive European hominids. *Am. J. Phys. Anthrop.* **87**: 311–40.

Begun, D. R. (1993). New catarrhine phalanges from Rudabanya (Northeastern Hungary) and the problem of parallelism and convergence in the hominoid postcranial morphology. *J. Hum. Evol.* **24**: 373–402.

Begun, D. R., Ward, C. V. & Rose, M. D. (1997). Events in hominoid evolution. In Begun, D. R., Ward, C. V. & Rose, M. D. (eds.), *Function, Phylogeny, and Fossils:*

Miocene Hominoid Evolution and Adaptations, pp. 389–415. New York: Plenum Press.

Bonis, L. de & Koufos, G. D. (1997). The phylogenetic and functional implications of *Ouranopithecus macedoniensis*. In Begun, D. R., Ward, C. V. & Rose, M. D. (eds.), *Function, Phylogeny, and Fossils: Miocene Hominoid Evolution and Adaptations*, pp. 317–26. New York: Plenum Press.

Bookstein, F. L. (1991). *Morphometric Tools for Landmark Data: Geometry and Biology*. Cambridge: Cambridge University Press.

Doran, D. M. (1993). Sex differences in adult chimpanzee positional behavior: the influence of body size on locomotion and posture. *Am. J. Phys. Anthrop.* **91**: 99–115.

Fleagle, J. G. (1977). Locomotor behavior and skeletal anatomy of sympatric Malaysian leaf-monkeys (Presbytis obscura and Presbytis melalophos). *Yrbk Phys. Anthrop.* **20**: 440–53.

Fleagle, J. G. (1985). Size and adaptation in primates. In Jungers, W. L. (ed.), *Size and Scaling in Primate Biology*, pp. 1–19. New York: Plenum Press.

Fleagle, J. G. (1998). *Primate Adaptation and Evolution*. San Diego: Academic Press.

Harrison, T. & Rook, L. (1997). Enigmatic anthropoid or misunderstood ape? The phylogenetic status of *Oreopithecus bambolii* reconsidered. In Begun, D. R., Ward, C. V. & Rose, M. D. (eds.), *Function, Phylogeny, and Fossils: Miocene Hominoid Evolution and Adaptations*, pp. 327–62. New York: Plenum Press.

Kappelman, J. (1988). Morphology and locomotor adaptations of the bovid femur in relation to habitat. *J. Morph.* **198**: 119–30.

Klecka, W. R. (1980). *Discriminant Analysis*. Beverley Hills: Sage Publications.

Lanyon, L. E. (1980). The influence of function on the development of bone curvature. An experimental study on the rat tibia. *J. Zool., London* **192**: 457–66.

Larson, S. G. (1996). Estimating humeral torsion on incomplete fossil anthropoid humeri. *J. Hum. Evol.* **31**: 239–57.

Larson, S. G. (1998). Parallel evolution in the hominoid trunk and forelimb. *Evol. Anthrop.* **6**: 87–99.

Lartet, E. (1856). Note sur un grand singe fissile qui se rattache au groupe des singes superieurs. *C. R. Acad. Sci.* **43**: 219–23.

McCrossin, M. L. & Benefit, B. R. (1994). Maboko Island and the evolutionary history of Old World monkeys and apes. In Corruccini, R. S. & Ciochon, R. L. (eds.), *Integrative Pathways to the Past: Paleoanthropological Advances in Honor of F. Clark Howell*, pp. 95–122. Englewood Cliffs, NJ: Prentice-Hall.

McCrossin, M. L. & Benefit, B. R. (1997). On the relationships and adaptations of *Kenyapithecus*, a large-bodied hominoid from the Middle Miocene of eastern Africa. In Begun, D. R., Ward, C. V. & Rose. M. D. (eds.), *Function, Phylogeny, and Fossils: Miocene Hominoid Evolution and Adaptations*, pp. 241–67. New York: Plenum Press.

Moyà-Solà, S. & Köhler, M. (1996). A Dryopithecus skeleton and the origins of great-ape locomotion. *Nature* **379**: 156–9.

Nakatsukasa, M., Hayama, S. & Preuschoft, H. (1995). Postcranial skeleton of a macaque trained for bipedal standing and walking and implications for functional adaptation. *Folia Primatol.* **64**: 1–29.

Paciulli, L. M. (1995). Ontogeny of phalangeal curvature and positional behavior in chimpanzees. *Am. J. Phys. Anthrop.* **20** (Suppl.): 165.

Pilbeam, D. R. (1982). New hominoid skull material from the Miocene of Pakistan. *Nature* **295**: 232–4.

Pilbeam, D. R. (1996). Genetic and morphological records of the Hominoidea and hominid origins: A synthesis. *Mol. Phyl. Evol.* **5**: 155–68.

Pilbeam, D. R., Rose, M. D., Barry, J. C. & Shah, S. M. I. (1990). New Sivapithecus humeri from Pakistan and the relationship of Sivapithecus and Pongo. *Nature* **348**: 237–9.

Pilbeam, D. R. & Simons, E. L. (1971). Humerus of Dryopithecus from Saint Gaudens, France. *Nature* **229**: 406–7.

Preuschoft, H. (1973). Functional anatomy of the upper extremity. In Bourne (ed.), *The Chimpanzee*, vol.6, pp. 34–120. Basel: Karger.

Richmond, B. G. (1998a). Finite element models of manual proximal phalanges in three modes of locomotion. *Am. J. Phys. Anthrop.* **26** (Suppl.): 188.

Richmond, B. G. (1998b). *Ontogeny and Biomechanics of Phalangeal Form in Primates.* Ph.D. Dissertation, State University of New York at Stony Brook.

Richmond, B. G. (1999). Reconstructing locomotor behavior in early hominids: evidence from primate development. *J. Hum. Evol.* **36**: A20.

Rohlf, F. J. (1990). Fitting curves to outlines. In Rohlf, F. J. & Bookstein, F. L. (eds.), *Proceedings of the Michigan Morphometrics Workshop*, pp. 167–77. Ann Arbor, Michigan: University of Michigan Museum of Zoology.

Rose, M. D. (1986). Further hominoid postcranial specimens from the Late Miocene Nagri Formation of Pakistan. *J. Hum. Evol.* **15**: 333–67.

Rose, M. D. (1989). New postcranial specimens of catarrhines from the Middle Miocene Chinji Formation, Pakistan: descriptions and a discussion of proximal humeral morphology in anthropoids. *J. Hum. Evol.* **18**: 131–62.

Rose, M. D. (1993). Locomotor anatomy of Miocene hominoids. In Gebo, D. L. (ed.). *Postcranial Adaptation in Nonhuman Primates*, pp. 252–272. DeKalb: Northern Illinois University Press.

Rose, M. D. (1994). Quadrupedalism in some Miocene catarrhines. *J. Hum. Evol.* **26**: 387–411.

Rose, M. D. (1997). Functional and phylogenetic features of the forelimb in Miocene hominoids. In Begun, D. R., Ward, C. V. & Rose, M. D. (eds.), *Function, Phylogeny, and Fossils: Miocene Hominoid Evolution and Adaptations*, pp. 79–100. New York: Plenum Press.

Sarmiento, E. E. (1995). Cautious climbing and folivory: a model of hominoid differentiation. *Hum. Evol.* **10**: 289–321.

Spencer, M. A. & Spencer, G. S. (1994). MacMorph Image Analysis software. Stony Brook, NY.

Spencer, M. A. & Spencer, G. S. (1995). Technical note: video-based three-dimensional morphometrics. *Am. J. Phys. Anthrop.* **96**: 443–53.

Stern, J. T., Jungers, W. L. & Susman, R. L. (1995). Quantifying phalangeal curvature: an empirical comparison of alternative methods. *Am. J. Phys. Anthrop.* **97**: 1–10.

Strait, D. S., Richmond, B. G. & Polk, J. D. (2000). The locomotor anatomy of hominoid and hominid ancestors. *Am. J. Phys. Anthrop.* (Suppl.) (In press.)

Susman, R. L., Stern, J. T. & Jungers, W. L. (1984). Arboreality and bipedality in the Hadar hominids. *Folia Primat.* **43**: 283–306.

Ward, C. V. (1997a). Functional anatomy and phyletic implications of the hominoid trunk and hindlimb. In Begun, D. R., Ward, C. V. & Rose, M. D. (eds.), *Function,*

Phylogeny, and Fossils: Miocene Hominoid Evolution and Adaptations, pp. 101–130. New York: Plenum.

Ward, S. C. (1997b). The taxonomy and phylogenetic relationships of *Sivapithecus* revisited. In Begun, D. R., Ward, C. V. & Rose, M. D. (eds.), *Function, Phylogeny, and Fossils: Miocene Hominoid Evolution and Adaptations*, pp. 269–90. New York: Plenum Press.

Ward, S. C., Brown, B., Hill, A., Kelley, J. & Downs, W. (1999). *Equatorius*: A new hominoid genus from the Middle Miocene of Kenya. *Science* **285**: 1382–6.

Ward, S. C. & Kimbel, W. H. (1983). Subnasal alveolar morphology and the systematic position of *Sivapithecus*. *Am. J. Phys. Anthrop.* **61**: 157–71.

Ward, S. C. & Pilbeam, D. R. (1983). Maxillofacial morphology of Miocene hominoids from Africa and Indo-Pakistan. In Ciochon, R. L. & Corruccini, R. S. (eds.), *New Interpretations of Ape and Human Ancestry*, pp. 211–38. New York: Plenum.

Ward, S. C. & Sussman, R. W. (1979). Correlates between locomotor anatomy and behavior in two sympatric species of *Lemur*. *Am. J. Phys. Anthrop.* **50**: 575–90.

16 *Sivapithecus* and hominoid evolution: some brief comments

David R. Pilbeam and Nathan M. Young

Introduction

'I cannot but recall that the Miocene hominoid fossil record seemed a lot less confusing before the discovery of two Sivapithecus *humeri...'*

(Larson, 1998: p. 97)

We address here, briefly, a particular problem, the relationship of Sivapithecus and *Pongo*, but we also review some more general issues involving Miocene apes and the phylogenetic analyses of fossils. As is now well known, certain features suggest a specific *Sivapithecus*–orang-utan link (Pilbeam, 1982; Ward & Kimbel, 1983; Ward & Brown, 1986; Moyà-Solà & Köhler, 1995; Begun & Kordos, 1997), for example: frontal bone shape and internal architecture; orbital shape and interorbital distance; facial profile; and (especially) palatal–premaxillary morphology. But as is also now well known, other features – postcranial, facial, mandibular and dental – either do not show such similarities, or fail to show similarities to any living apes large or small (Corruccini, 1975; Corruccini & McHenry, 1980; Ward & Brown, 1986; Pilbeam *et al.*, 1990; Rose, 1997; Uchida, 1998).

Clearly there are very interesting patterns of homoplasy involving either facial, dental or postcranial regions, or all of the above. One immediate question concerns the region in which homoplasy is most likely to occur – postcranial, cranial, or dental? We detect something of a tendency to point the finger at the postcranium (Begun, 1993; Moyà-Solà & Köhler, 1996; Ward, 1997; Richmond, 1999). The '*Sivapithecus* dilemma' has served as a trigger for a series of recent analyses of living hominoid postcranial morphology addressing the extent to which crown hominoid postcranial similarities are homoplasies. Thus Begun (1993) and Larson (1998) have argued that some of the postcranial characters traditionally used to define the living hominoids are likely to be homoplasies (either convergences or parallelisms). Support for this would come from the well recognised postcranial convergence between *Ateles* and *Brachyteles* and the living hominoids (Erikson, 1963), and also from the consequences of the recent recognition that, based on phylogenetic analyses of a growing number of independent genes, *Brachyteles* is the sister to *Lagothrix* and not *Ateles*; namely, that the postcranial similarities of *Ateles* and *Brachyteles* are themselves likely to be largely homoplasies.

Yet an examination of consistency indices calculated from large data sets of hominoid characters (phenotypic unless otherwise stated, and mainly characters of the hard tissue phenotype) suggests that no particular area is more prone to homoplasy than any other (our analysis). The demonstration (Disotell 1992, 1994, 1996; Harris & Disotell, 1998) that the characteristic facial resemblances of baboons and mandrills, often considered members of the same genus, are almost certainly homoplasies raises interesting questions because, based on abundant genetic data, they are only distantly related papionins. More recently McCollum (1999) has explored ways in which quite substantial facial homoplasies might arise in early hominids. If the facial resemblances between *Paranthropus boisei* and *P. robustus* have indeed arisen independently (as originally suggested as a possibility by Wood, 1988) as the consequences of convergence in a very small number of underlying factors (e.g. McCollum, 1999), then we face potentially as great a problem with parts of the cranium as with the postcranium in extracting phylogenetically informative information.

The problem

Sivapithecus is not the only Miocene hominoid that presents a phylogenetic dilemma resulting in disagreements over relationships. What is the crux of the problem? One is the fossil record itself, and this is perhaps the most frequently cited cause. Perhaps the problem lies with the incompleteness of the fossil record – more fossils will solve the problem. Except for *Oreopithecus* and *Proconsul*, Miocene apes are not well represented. More and better fossils would allow an increase in the number of possible characters through expanding sampling of body parts. This would be preferable to increasing the number of characters for the same anatomical area, which is a sometimes favoured solution but which simply exaggerates the problem of unequal representation.

Another possible reason for disagreement is that it has to do with lack of analytical rigor (Stewart & Disotell, 1998), in the sense that analyses are only worthwhile if based on large numbers of morphological characters analysed cladistically. We believe that the 'problem', such as it is, has little if anything to do with such 'rigor'. Indeed, analytical rigor – the use of 'scientific' approaches (packaged computer programs) – has in our view obscured a more fundamental problem.

Rather, problems reflect the fact that different experts use different characters for the same specimens, and they also reflect our inability to generate objective and biologically plausible reasons for selecting one phylogeny over another when faced with character state incongruence (as with *Sivapithecus*). Picking the tree with the fewest steps is reasonable *if* there is an adequate skeletal sampling and *if* there is agreement on how to parse complex morphology. At present, neither of these preconditions have been met. We believe that disagreements are related most of all to the selection of phenotypic (mostly hard tissue) characters – the different phylogenies reflect different characters, and/or different character states, or polarity assignments. Basically, they reflect different formal descriptions and hypotheses about homology and polarity. We are by no means the first to raise this as a problem (see, for example, Cartmill 1982, 1994; Atchley & Hall, 1991; Sarich, 1993). Indeed, the recognition that this is the principle problem, and a largely unstated one, should play a much larger role in future discussions (and seems to be doing so; e.g. see Köhler *et al.* Chapter 8).

Of course we need to expand the ape fossil record, particularly by adding poorly known or unknown body parts, or sampling geographical areas currently unsampled (for example, those occupied by living apes!). But at least as urgently, we need to spend more time thinking formally about how to describe morphology and especially complex three-dimensional hard tissue phenotypes, and we need to be equally careful about character states. Too often we describe spaces (for example, canine fossas) and shapes (angulated malars) that are the architectural reflections of more important or fundamental biological organising units of the cranium (e.g. McCollum, 1999).

Phylogenetic analysis of extant groups

We might ask ourselves the following question – to what extent should we believe phylogenies based on hard tissue phenotypes? We can begin to answer this question by looking at living groups (hominoids, papionins, atelines) for which there are robust phylogenies based on genetic data, and note the extent to which analyses using the kinds of characters available for fossils are congruent with robust genetic-supported trees.

In several interesting cases, the most probable phylogeny based on genetic data is either not recovered by phenotypic characters, or if it is there are reasons to be sceptical of the results. We will not focus here on the hominoid genetic data base, except to say that the best results are based on analyses of nucleotide sequences of large numbers of independent genes (inherited without recombination because they are either on different chromosomes or separated enough on the same chromosome to be unlinked), and on DNA–DNA hybridisation data (Caccone & Powell, 1989; Pilbeam, 1996; Ruvolo, 1997a). Chimpanzees and humans are sister taxa, with gorillas, orang-utans and gibbons successively more distant. Recent critiques (Deinard *et al.*, 1998) either have raised no new biologically valid points or repeat earlier misunderstandings; they will be discussed fully elsewhere. We continue to be surprised and disappointed at the refusal of some colleagues to accept the inevitable – that *Pan* and *Gorilla* are clearly not sister taxa.

Analyses of the living hominoids using phenotypic characters have generated almost all combinations of relationships. Of more than historical interest are the various 1980s options. Kluge (1983), following what had become the mid-century consensus (e.g. Schultz, 1968; Tuttle, 1975) found characters that supported a great ape clade. However, by the 1980s this was a minority view because early genetic comparisons (Goodman, 1962; Sarich & Wilson, 1967) had shown that humans were clearly linked with African apes to the exclusion of the orang-utan. Others, for example Andrews & Martin (1987), favoured what became the majority view that chimpanzees and gorillas were monophyletic, with humans next closest. Schwartz (1984) argued the distinctly minority position that, in addition to a chimpanzee–gorilla clade, humans and orang-utans were monophyletic. Groves (1986) was the first person in a very long time to propose on the basis of phenotypic characters that humans and chimpanzees were closest relatives.

Clearly, not all of these phylogenies could have been correct. All involved just a listing of characters supporting a phylogeny rather than exhaustive comparisons of all alternatives, and there is a suspicion that, in some cases, the phylogenetic conclusions unconsciously preceded the selection of characters which supported the phylogeny (actually, we suspect that this is more widespread than we collectively care to recognise – see Sarich, 1993). None of these analyses involved what had become by the early 1990s relatively simple (because of PC-compatible computer packages) analyses in which large numbers of characters could be used to evaluate a full range of phylogenetic possibilities. Andrews (1988) argued that this was a cause of disagreement, and in response Groves & Paterson (1991) put together a list of characters combining those favoured by Groves Schwartz, and Andrews and Martin. The results (Groves & Paterson, 1991) of several different analyses varied, depending on assumptions about character state transformation and on the characters chosen (thus the existence or not of a chimpanzee–gorilla link rather than a chimpanzee–human link depended on the inclusion or exclusion of knuckle-walking characters). Broader analyses of similar data bases (Shoshani *et al.*, 1996) yielded a similar pattern – results varied depending upon the characters used, taxa included and body parts sampled. A final data set for living hominoids based exclusively on morphological characters (Begun *et al.*, 1997) yields a tree concordant with the genetic tree (that is, with chimpanzees and *Australopithecus* as sisters), but the alternative chimpanzee–gorilla clade is barely less likely – five putative unambiguous synapomorphies support the former, four the latter. We believe that several of the characters in these lists are questionable because some of the chimpanzee–gorilla differences could be size dependent, and in other instances ('premolar flare') the allometric relationships of these taxa strongly (not reflected in the character lists) suggest that their common ancestor would have been chimp-like, which raises the possibility that the gorilla state is derived relative to that in *Pan*. This analysis also excludes knuckle-walking features. But what is perhaps most surprising here, beyond the ambiguity of the results, is that taxa as morphologically derived as *Homo* or *Australopithecus* should be placed as close as they are to *Pan*, relative to *Gorilla*.

Collard (1997) and Collard & Wood (1998) used the morphological characters that have been widely utilised in phylogenetic analyses of fossil hominids to examine the relationships of living hominoids and papionins, yet were unable to recover the robust genetically-generated trees.

Earlier work on the dentition by Corruccini used a morphometric approach to capture details of cheek tooth occlusal shape (Corruccini, 1975; Corruccini & McHenry, 1980) and yielded a living great ape tree concordant with the current genetic consensus. More recent studies by Hartman (1988, 1989) and Uchida (1998) failed, however, to capture the genetic tree, although one of Hartman's analyses came close. A study by Braga (1995) of non-metric traits in extant hominoids produced a tree which linked bonobos with humans. However, the network from which the tree was derived closely matched the genetic network generated for living hominoids (Ruvolo, 1997b), including the relative positions and distances of gorilla and orang-utan subspecies. Most recently, Gibbs (1999) used soft tissue phenotypic characters of hominoids in a phylogenetic analysis and recovered a tree fully and significantly concordant with the genetic tree.

This suggests that at least some phenotypic features come close to reflecting known genetic relationships, although these tend not to be data in the form of hard tissue phenotypic characters.

The previously mentioned study of extant hominoid postcranial features by Larson (1998) raised an interesting set of issues. Using discrete characters widely used to describe ape postcrania, Larson showed that the hominoids differed among themselves to a surprising degree, noting, as had many others, the convergent similarities of the long-armed atelines and the apes, and he also observed overlap with non-ateline monkeys. While the second conclusion was not surprising, the first and third were, particularly in light of a number of other studies, for example those that tried to capture complex shapes through multivariate rather than discrete character approaches. These show that the living hominoids form a quite homogeneous group, regardless of their similarity to the atelines (Ashton *et al.*, 1965, 1976; Oxnard, 1967, 1977; Corruccini & Ciochon, 1976, 1978).

So, in contrast to the substantial concordance among independent genetic data sets, with phenotypic characters there is considerable disagreement. The results differ not because of analytical differences (unless 'discrete character' versus 'morphometric shape' differences are so counted) but rather because different workers favour different characters.

There are no objective rules governing the choice of phenotypic characters as there are with genetic characters, nor is there any biologically valid and objective measure of independence. It is difficult to take complex shapes (particularly in the case of hard tissue phenotypes) and atomise them – break them unambiguously into objective units (Cartmill, 1982, 1994; Fitch & Atchley, 1987; Sarich, 1993; McCollum, 1999). We see this particularly clearly with fossils. It appears either that osteological and odontological features are not as good as soft tissue characters in recovering the genetic tree, or that the kinds of such hard tissue morphological features with a reasonably unambiguous signal of genetic relationships – non-metric features – are not accessible in fossils because large samples are needed for these frequency-based characters.

A final critical problem remains – the determination of polarity. For example, note the debate (Benefit & McCrossin, 1997 versus Moyà-Solà & Köhler, 1997) on the polarity of catarrhine upper facial morphology. Perhaps the best example in hominoids involves chimpanzee–gorilla–human relationships. Many authors have noted the strong allometric similarity of the two ape genera (Shea, 1984, 1985). A particular case is that of Hartwig-Scherer (1993) who documented the many allometric similarities of chimpanzees and gorillas, and their marked difference from humans. She concluded from this that a chimpanzee–human clade was improbable. Hartwig-Scherer is, we believe, correctly identifying homologous similarities among the African apes. Indeed, allometric analyses, particularly those involving ontogenetic allometry, strike us as providing very suitable characters, in that they can be identified as probable homologies. (An excellent example of this is the use by Davis (1964) of growth trajectories to identify *Ailuropoda* as an ursid, a result much later supported by considerable genetic data – O'Brien *et al.*, 1985.) However, the genetic results clearly show that chimpanzees and humans, not chimpanzees and gorillas, are sister taxa; although the many (allometric) similarities of chimpanzees and gorillas are surely largely homologous, they are symplesiomorphic and not synapomorphic (this implies that the chimpanzee–human ancestor could have been basically chimpanzee-like – that is, assignable to *Pan*).

Fossil hominoids

As a consequence of a less than adequate fossil record, and of failure to agree on the characters to be used in phylogenetic analyses, there is disagreement over the relationships of Miocene apes to each other and to crown hominoids. For example, three different formal (analytically rigorous) phylogenetic analyses have addressed the issue of the relationships among (principally) Eurasian Miocene apes, those of Begun & Kordos (1997), Moyà-Solà & Köhler (1995) and Cameron (1997). One of the few areas of agreement is the position of *Sivapithecus* as sister to *Pongo*, but there is disagreement over the placement of other taxa. Begun & Kordos (1997) have *Dryopithecus* and *Ouranopithecus* as sister taxa, linked next to the *Pan–Gorilla–Australopithecus* clade with *Pongo* and *Sivapithecus* more distant. Moyà-Solà & Köhler (1995) have both *Dryopithecus* and *Ouranopithecus* (*Graecopithecus*) linked instead to the orang-utan clade. Cameron (1997) splits the *Dryopithecus* of Begun & Kordos and Moyà-Solà and Köhler into *Dryopithecus* (Hungary), which falls outside the great ape plus human clade, and *Hispanopithecus* (Spain) which he links with *Pongo* (agreeing partly with Moyà-Solà & Köhler, 1995), and links *Ouranopithecus* with *Gorilla* (agreeing partly, though not completely, with Begun & Kordos, 1997).

These differences reflect (in addition to some taxonomic disagreements), not analytical differences but the fact that the same taxa are being formally described, as discrete characters, differently. A particular expression of this problem is that frequently the exact same body part is described differently. As an example, Begun *et al.* (1997) describe the zygomatic with eight characters, whereas Moyà-Solà & Köhler (1995) use only one character and Cameron (1997) just two characters for the morphology of the same specimens. Although there is some degree of overlap in these characters (e.g. zygomatics are basically described as either 'laterally' or 'anteriorly' oriented in all three), the overall result is that anatomical differences within a region are given different 'weights' by each describer and thus different results emerge. The zygomatic characters of Begun tend to link *Sivapithecus* and *Pongo* to the exclusion of other Miocene hominoids, Moyà-Solà and Köhler's character is a synapomorphy of *Dryopithecus, Ouranopithecus, Sivapithecus* and *Pongo*, and Cameron's two characters link *Hispanopithecus* (Spanish

 Dryopithecus) to *Pongo* and *Sivapithecus* while excluding *Rudapithecus* (Hungarian *Dryopithecus*) from this group. There are other examples – for example, the description of nasoalveolar morphology.

Within characters, some character states are formulated in ways that favour a preferred hypothesis. For example, frontal morphology and its internal architecture differs in *Dryopithecus* from both *Pan* and *Pongo*. Yet Begun & Kordos (1997) define character states in such a way as to make *Dryopithecus* like African apes, while Moyà-Solà & Köhler (1995) make it like *Pongo*. We would score it as equally unlike both. In a second example, Moyà-Solà & Köhler (1995, p. 129, figures 22 & 23) assign character states in ways that do not reflect the distribution of the trait across ape genera; an obvious gap is ignored while state boundaries are drawn between taxa that differ minimally.

We have no space here to review as thoroughly as they deserve the various data sets used to analyse Miocene ape relationships, although we have problems with many of the characters and character states. In the case of the largest set (Begun *et al.*, 1997) we note that, with the possible exception of *Proconsul*, the fossil character sampling of body parts is very incomplete, focusing mainly on facial and dental features rather than postcranial characters. The fact that phylogenetic relationships can vary depending on which body parts or which taxa are used does not give us confidence that a robust phylogeny has yet been achieved. The differentially greater support for *Pan–Australopithecus* over *Pan–Gorilla* relative to the analysis using just living species depends upon the placement of fossils on the tree. Many of these we question and will review in greater depth elsewhere.

Sivapithecus

We return now to our brief discussion of *Sivapithecus*. In terms of degree of preservation of body parts, *Sivapithecus* falls in the 'intermediate' category (Pilbeam, 1996), along with *Afropithecus*, *Morotopithecus*, *Kenyapithecus* and *Dryopithecus*, in that facial (but not cranial), dental and mandibular, and some postcranial remains are sampled (only the sampling of *Proconsul* and *Oreopithecus* can be described as 'adequate').

Similarities of *Sivapithecus* to *Pongo* are best shown in the orbital and subnasal regions, and it is these features which are interpreted as the clearest synapomorphies linking the two taxa. Could these be homoplasies? If the totality of the evidence suggested that this was a plausible option (for example, if more characters were added from known body parts in which *Sivapithecus* and *Pongo* differed, or if new body parts were sampled which also showed patterns incongruent with the aforementioned facial characters), biologically (developmentally, functionally) plausible explanations could be produced. For example, McCollum (1999) has recently shown how, based on a few 'parameters' of nasomaxillary modelling, southern and eastern African robust australopithecine facial features (described in standard cladistic analyses by many discrete characters) could be derived convergently.

What about features in which *Sivapithecus* and *Pongo* differ, or in which *Sivapithecus* differs from all large hominoids or all hominoids? Some features of the face and mandible (Ward & Brown, 1986) would fall under these categories, as would dental (Corruccini & McHenry, 1980; Uchida, 1998) and many postcranial features (Pilbeam *et al.*, 1990). We know of no postcranial features that are clear candidate synapomorphies for a *Sivapithecus–Pongo* relationship. In some features, for example the articular surfaces of the distal metacarpals and proximal phalanges, *Sivapithecus* differs from all living hominoids. We note that, despite attempts to explain away the primitive features of the *Sivapithecus indicus* humerus from the Potwar Chinji Formation (Köhler *et al.*, Chapter 8; Moyà-Solà & Köhler, 1996), the likeliest reconstruction of this admittedly crushed and somewhat distorted specimen yields a proximal and diaphyseal morphology that is non-modern (Rose, unpublished response to Moyà-Solà & Köhler, 1996, refused for publication by *Nature*). Also, let us not forget that for more than two decades it has also been clear that in basic occlusal design, *Sivapithecus* teeth differed from those of *Pongo*, and indeed from all living apes, and resemble some other Miocene hominoids (Corruccini, 1975). These characters are not included in the Begun *et al.* (1997) data set.

These and other features either support a tree in which *Sivapithecus* falls outside the crown hominoid radiation, or indicate that a non-trivial number of extant hominoid postcranial *and* dental, mandibular and facial features are homoplasies. How likely is this? Or, rather, can a biologically plausible case for homoplasy be made? As far as the postcranium is concerned the answer is surely yes (indeed the postcranium has been favoured over the cranium as the likeliest 'source' of homoplasy although as previously mentioned there is less objective evidence to support this view).

As we noted, Larson's analysis published in 1998, of postcranial features makes a couple of points – that hominoid postcranial similarities have been overstated and that hominoids frequently overlap with various monkeys; and that the obviously convergent similarities of hominoids and *Ateles-Brachyteles* make convergence a plausible option for within-hominoid similarities (a point, as noted by Larson, made by many others). However, as we noted earlier, in contrast to Larson's analyses the multivariate analyses of

Ashton *et al.* (1965, 1976), Oxnard (1967, 1977) and of Corruccini & Ciochon (1976, 1978) show that hominoids, including humans, clearly resemble each other. Different characters are sometimes used in the various analyses and analytical techniques differ so it is not easy to compare these results, but it may be that multivariate techniques capture overall shape and its contrasts better than do approaches in which complex shapes are atomised into characters. The multivariate approaches also show that the two atelins are similar to hominoids, although in the Corruccini & Ciochon analysis (1978) they are much less similar to hominoids than in the analyses of Oxnard (1977). The former note that 'the pattern shared among hominoids might also [compared to atelins] have partially arisen in parallel, especially in the case of the gibbon' (Corruccini & Ciochon, 1978, p. 542), although the overall conclusion from the multivariate analyses would be that much of the hominoid similarity, particularly that of the great apes, is likely to be homologous.

However, a further fly in the ointment follows from the recent genetic studies of atelines which show that *Ateles* and *Brachyteles* are unlikely to be sister taxa, *Brachyteles* being closer to *Lagothrix* (Harada *et al.*, 1995; Schneider *et al.*, 1996; von Dornum, 1997; Canavez *et al.*, 1999; Meireles *et al.*, 1999); rather, therefore, their shoulder similarities, relative to *Lagothrix*, are also likely to be homoplasies! So, postcranial similarities among hominoids, and even between *Pan* and *Pongo* (*Pan* being likely to represent the 'ur-African hominoid') could well be, at least in part, homoplasies (Begun, 1993). This would, of course, help solve the '*Sivapithecus* dilemma', but it raises another dilemma for *Dryopithecus* and *Oreopithecus* – their postcranial similarities to extant apes would be equally suspect as characters *linking* them to crown hominoids.

The problem for us is how might we tell whether these cranial or postcranial or dental characters are more or less likely to be homoplasies? One solution is to 'add more characters', either by focusing on parts already well sampled, such as the face and palate (which we do not recommend), or by finding fossils which preserve previously un- or undersampled body parts. But until there is a more equi-dense sampling across more body parts, caution is warranted concerning phylogenetic conclusions based mainly on facial features.

Geographically targeted searches should also be a priority, particularly in areas (geographical and ecological) in which extant apes live (or would have lived during interglacial or preglacial Pliocene and later Miocene times). Would a late Miocene Indonesian ape have looked like *Sivapithecus* postcranially and dentally, or would it have resembled more the orang-utan? Either possibility would likely contribute more to our understanding of

Miocene apes and their relationships than continuing arguments over the pieces we currently have.

Clearly, an essential approach is to take a closer a priori look (Lauder, 1994) at characters and try to make sure that they reflect underlying architectural, functional and developmental processes or modules (Atchley & Hall, 1991; Lieberman, 1999; McCollum, 1999), rather than being selected because a particular body part is represented in the fossil record. Ultimately, though, because there are no objective rules for selecting phenotypic characters, we are likely to remain in the frustrating situation where students can continue talking past each other, because there is no way to settle the argument as to whether 'my characters' are better than 'your characters', except with extant groups where sequencing enough independent genes is the way to proceed. We shall continue to live in a situation where it is easy for a sceptical outsider to believe that we select characters subconsciously in ways that reflect our preferred phylogeny.

Concluding, we have little faith in *any* of the current phylogenetic analyses for Miocene hominoids, 'analytically rigorous' or otherwise, nor do we believe that things will improve without some significant new material (especially from the tropical regions inhabited by extant apes). Accordingly, we have little faith in palaeogeographical reconstructions locating the area of origin for the ancestors of living hominoids in subtropical or warm temperate regions such as Turkey (Begun & Güleç, 1998; Stewart & Disotell, 1998). The living apes are all species addicted to ripe fruit (mountain gorillas are an exception of course; they live in a non-fruit salad bowl, and are able to do so because of large body size), and require habitats which can – because of sufficient rainfall and appropriate temperature – provide enough of such food for a significant fraction of the year. This is likely to be a 'synapomorphy'. Miocene apes living outside the tropics (which is where essentially all the known fossils are located) would have inhabited forests which could *not* have provided such sustenance. The excellent studies of the Turkish Miocene hominoid locality at Paşalar (Andrews, 1990; Quade *et al.*, 1995) show that whatever these Miocene apes were eating it could not have been abundant ripe fruit (similar conclusions are likely to be obtained for other known taxa). We believe it unlikely that the feeding adaptations of living apes would have evolved in parallel, and hence we are sceptical of models of ape palaeobiogeography which would locate the ancestors of extant hominoids outside the geographical tropics, and outside the kinds of tropical forests found today which can provide enough ripe fruits throughout the year. Phylogenetic analyses, as well as the biogeographical models derived from them, need to include such physiological–behavioural–ecological characters as well as morphology.

In summary, given the current state of the Miocene ape record – because sampling (of body parts, and geographical and ecological areas) is less than adequate and because the apes are to a marked extent different from living hominoids – we have little confidence in any phylogenetic analyses of these interesting creatures. For the moment, pursuit of functional, behavioural, and life history reconstructions is likely to be more fruitful.

References

Andrews, P. (1988). Advances in primatology. *Cladistics* **4**: 297–304.

Andrews, P. (1990). Paleoecology of the Miocene fauna from Pasalar, Turkey. *J. Hum. Evol.* **19**: 569–82.

Andrews, P. & Martin, L. (1987). Cladistic relationships of extant and fossil hominoids. *J. Hum. Evol.* **16**: 101–18.

Ashton, E. H., Flinn, R. M., Oxnard, C. E. & Spence, T. F. (1976). The adaptive and classificatory significance of certain quantitative features of the forelimb in primates. *J. Zool., London* **179**: 515–56.

Ashton, E. H., Healy, M. J. R., Oxnard, C. E. & Spence, T. F. (1965). The combination of locomotor features of the primate shoulder girdle by canonical analysis. *J. Zool.* **147**: 406–29.

Atchley, W. R. & Hall, B. K. (1991). A model for development and evolution of complex morphological structures. *Biol. Rev.* **66**: 101–57.

Begun, D. R. (1993). New catarrhine phalanges from Rudabánya (Northeastern Hungary) and the problem of parallelism and convergence in hominoid postcranial morphology. *J. Hum. Evol.* **24**: 373–402.

Begun, D. R. & Güleç, E. (1998). Restoration of the Type and Palate of *Ankarapithecus meteai*: Taxonomic and Phylogenetic Implications. *Am. J. Phys. Anthrop.* **105**: 279–314.

Begun, D. R. & Kordos, L. (1997). Phyletic Affinities and Functional Convergence in *Dryopithecus* and Other Miocene and Living Hominids. In Begun, D. R., Ward, C. V. & Rose, M. D. (eds.), *Function, Phylogeny, and Fossils: Miocene Hominoid Evolution and Adaptations*, pp. 291–316. Plenum Press, New York.

Begun, D. R., Ward, C. V. & Rose, M. D. (1997). Events in Hominoid Evolution. In Begun, D. R., Ward, C. V. & Rose, M. D. (eds.), *Function, Phylogeny, and Fossils: Miocene Hominoid Evolution and Adaptations*, pp. 389–416. Plenum Press, New York.

Benefit, B. R. & McCrossin, M. L. (1997). Earliest known Old World monkey skull. *Nature* **388**: 368–71.

Braga, J. (1995). *Définition de certains caractères discrets crâniens chez Pongo, Gorilla et Pan: Perspectives taxonomiques et phylogénétiques*. Ph.D. Dissertation, University Bordeaux I.

Caccone, A. & Powell, J. R. (1989). DNA divergence among hominoids. *Evolution* **43**: 925–42.

Cameron, D. W. (1997). A revised systematic scheme for the Eurasian Miocene fossil Hominidae. *J. Hum. Evol.* **33**: 449–77.

Canavez, F. C., Moreira, M. A. M., Ladasky, J. J., Pissinatti, A., Parham, P. & Seuánez, H. (1999). Molecular Phylogeny of New World Primates (Platyrrhini) based on β2-Microglobulin DNA Sequences. *Mol. Phylogenet. Evol.* **12**: 74–82.

Cartmill, M. (1982). Assessing tarsier affinities: is anatomical description phylogenetically neutral? *Geobios (mém. spéc.)* **6**: 279–87.

Cartmill, M. (1994). A critique of homology as a morphological concept. *Am. J. Phys. Anthrop.* **94**: 115–23.

Collard, M. (1997). *Morphological Evolution of the Extant Hominoids and Papionins: Implications for Paleoanthropological Cladistics.* Ph.D. Dissertation, University of Liverpool.

Collard, M. & Wood, B. (1998). Cladistics and the estimation of hominid phylogeny. *Am. J. Phys. Anthrop.* **26** (Suppl.): 13.

Corruccini, R. S. (1975). *A Metrical Study of Crown component Variation in the Hominoid Dentition.* Ph.D. Dissertation, University of California – Berkeley.

Corruccini, R. S. & Ciochon, R. L. (1976). Morphometric affinities of the human shoulder. *Am. J. Phys. Anthrop.* **45**: 19–38.

Corruccini, R. S. & Ciochon, R. L. (1978). Morphoclinal variation in the anthropoid shoulder. *Am. J. Phys. Anthrop.* **48**: 539–42.

Corruccini, R. S. & McHenry, H. J. (1980). Cladometric analysis of Pliocene hominids. *J. Hum. Evol.* **9**: 209–21.

Davis, D. (1964). *The Giant Panda: A Morphological Study of Evolutionary Mechanisms.* Fieldiana: Zoology Memoirs Volume 3. Chicago: Chicago Natural History Museum.

Deinard, A. S., Sirugo, G. & Kidd, K. K. (1998). Hominoid phylogeny: inferences from a sub-terminal minisatellite analyzed by repeat expansion detection (RED). *J. Hum. Evol.* **35**: 313–18.

Disotell, T. R. (1992). *Molecular Evolution of the Papionini (Primates: Cercopithecinae).* Ph.D. Dissertation, Harvard University.

Disotell, T. (1994). Generic level relationships of the Papionini (Cercopithecoidea). *Am. J. Phys. Anthrop.* **94**: 47–57.

Disotell, T. (1996). Phylogeny of Old World monkeys. *Evol. Anthrop.* **5**: 18–24.

Erikson, G. E. (1963). Brachiation in New World monkeys and in anthropoid apes. *Symp. Zool. Soc. London* **10**: 135–64.

Fitch, W. M. & Atchley, W. R. (1987). Divergence in inbred strains of mice: a comparison of three different types of data. In Patterson, C. (ed.), *Molecules and morphology in evolution: conflict or compromise?* pp. 203–16. Cambridge: Cambridge University Press.

Gibbs, S. (1999). *Comparative Soft Tissue Morphology of the Extant Hominoidea, Including Man.* Ph.D. Dissertation, University of Liverpool.

Goodman, M. (1962). Immunochemistry of the primates and primate evolution. *Ann. N.Y. Acad. Sci.* **102**: 219–34.

Groves, C. P. (1986). Systematics of the Great Apes. In Swindler, D. R. & Erwin, J. (eds.), *Comparative Primate Biology*, vol. I: Systematics, Evolution, and Anatomy pp. 187–217. New York: Alan R. Liss.

Groves, C. P. & Paterson, J. D. (1991). Testing hominoid phylogeny with the PHYLIP programs. *J. Hum. Evol.* **20**: 167–83.

Harada, M. L., Schneider, H., Schneider, M. P. C., Sampaio, I. M., Czelusniak, J. & Goodman, M. (1995). DNA evidence on the phylogenetic systematics of the

New World monkeys: support for the sister-grouping of *Cebus* and *Saimiri* from two unlinked nuclear genes. *Mol. Phylogenet. Evol.* **4**: 331–49.

Harris, E. E. & Disotell, T. R. (1998). Nuclear gene trees and the phylogenetic relationships of the mangabeys (Primates: Papionini). *Mol. Biol. Evol.* **15**: 892–900.

Hartman, S. E. (1988). A cladistic analysis of hominoid molars. *J. Hum. Evol.* 17: 489–501.

Hartman, S. E. (1989). Stereophotogrammetric analysis of occlusal morphology of extant hominoid molars: phenetics and function. *Am. J. Phys. Anthrop.* **80**: 145–66.

Hartwig-Scherer, S. (1993). *Allometry in Hominoids: A Comparative Study of Skeletal Growth Trends.* Ph.D. Dissertation, University Zürich.

Kluge, A. G. (1983). Cladistics and the classification of the great apes. In Ciochon, R. & Corruccini, R. (eds.), *New Interpretations of Ape and Human Ancestry,* pp. 151–77. New York: Plenum.

Larson, S. (1998). Parallel evolution in the hominoid trunk and forelimb. *Evol. Anthrop.* 6: 87–99.

Lauder, G. V. (1994). Homology, form, and function. In Hall, B. K. (ed.), *Homology: The Hierarchical Basis of Comparative Biology,* pp. 151–96. San Diego: Academic Press.

Lieberman, D. E. (1999). Homology and hominid phylogeny: problems and potential solutions. *Evol. Anthrop.* **7**: 142–51.

Meireles, C. M., Czelusniak, J., Schneider, M. P. C., Muniz, J. A. P. C., Brigido, M. C., Ferreira, H. S. & Goodman, M. (1999). Molecular phylogeny of ateline New World monkeys (Platyrrhini, Atelinae) based on g-globin gene sequences: evidence that *Brachyteles* is the sister group of *Lagothrix. Mol. Phylogenet. Evol.* **12**: 10–30.

McCollum, M. A. (1999). The robust australopithecine face: a morphogenetic perspective. *Science* 284: 301–5.

Moyà-Solà, S. & Köhler, M. (1995). New partial cranium of *Dryopithecus lartet,* 1863 (Hominoidea, Primates) from the upper Miocene of Can Llobateres, Barcelona, Spain. *J. Hum. Evol.* **29**: 101–39.

Moyà-Solà, S. & Köhler, M. (1996). The first *Dryopithecus* skeleton: origins of great ape locomotion. *Nature* **379**: 156–9.

Moyà-Solà, S. & Köhler, M. (1997). Fossil muzzles and other puzzles. *Nature* **388**: 327–8.

O'Brien, S. J., Nash, W. G., Wildt, D. E., Bush, M. E. & Benveniste, R. E. (1985). A molecular solution to the riddle of the giant panda's phylogeny. *Nature* 317: 140–4.

Oxnard, C. E. (1967). The functional morphology of the primate shoulder as revealed by comparative anatomical, osteometric and discriminant function techniques. *Am. J. Phys. Anthrop.* **26**: 219–40.

Oxnard, C. E. (1977). Morphometric affinities of the human shoulder. *Am. J. Phys. Anthrop.* **46**: 367–74.

Pilbeam, D. R. (1982). New hominoid skull material from the Miocene of Pakistan. *Nature* **295**: 232–4.

Pilbeam, D. R. (1996). Genetic and morphological records of the Hominoidea and hominid origins: a synthesis. *Mol. Phylogenet. Evol.* **5**: 155–68.

Pilbeam, D. R., Rose, M. D., Barry, J. C. & Shah, S. M. I. (1990). New *Sivapithecus* humeri from Pakistan and the relationship of *Sivapithecus* and *Pongo*. *Nature* **348**: 237–9.

Quade, J., Cerling, T. E., Andrews, P. & Alpagut, B. (1995). Paleodietary reconstruction of Miocene faunas from Pasalar, Turkey using stable carbon and oxygen isotopes of fossil tooth enamel. *J. Hum. Evol.* **28**: 373–84.

Richmond, B. (1999). Eurasian hominoid evolution. *Evol. Anthrop.* **7**: 194–6.

Rose, M. D. (1997). Functional and Phylogenetic Features of the Forelimb in Miocene Hominoids. In Begun, R. D., Ward, C. V. & Rose, M. D. (eds.) *Function, Phylogeny, and Fossils: Miocene Hominoid Evolution and Adaptations*, pp. 79–100. New York: Plenum Press.

Ruvolo, M. (1997a). Molecular phylogeny of the hominoids: inference from multiple independent DNA sequence data sets. *Mol. Biol. Evol.* **14**: 248–65.

Ruvolo, M. (1997b). Genetic diversity in hominoid primates. *Ann. Rev. Anth.* **26**: 515–40.

Sarich, V. M. (1993). Mammalian systematics: twenty-five years among their albumins and transferrins. In Szalay, F., Novacek, M. J. & McKenna, M. (eds.), *Mammal Phylogeny: Placentals*, pp. 103–14. New York: Springer-Verlag.

Sarich, V. M. & Wilson, A. C. (1967). Immunological time scale for hominoid evolution. *Science* **158**: 1200–3.

Schneider, H., Sampaio, I, Harada, M. L., Barroso, C. M. L., Schneider, M. P. C., Czelusniak, J. & Goodman, M. (1996). Molecular phylogeny of the New World monkeys (Platyrrhini, Primates) based on two unlinked nuclear genes: IRBP intron I and e-globin sequences. *Am. J. Phys. Anthrop.* **100**: 153–80.

Schultz A (1968). The recent hominoid primates. In Washburn, S. L. & Jay, P. C. (eds.), *Perspectives on Human Evolution*, pp. 122–95. New York: Holt, Rinehart, & Winston.

Schwartz, J. H. (1984). The evolutionary relationships of man and orang-utans. *Nature* **308**: 501–5.

Shea, B. T. (1984). An allometric perspective on the morphological and evolutionary relationships between pygmy (*Pan paniscus*) and common (*Pan troglodytes*) chimpanzees. In Susman, R. (ed.), *The Pygmy Chimpanzee*, pp. 89–130. New York: Plenum Press.

Shea, B. (1985). On aspects of skull form in African apes and orangutans, with implications for hominoid evolution. *Am. J. Phys. Anthrop.* **68**: 329–42.

Shosani, J., Groves, C. P., Simons, E. L. & Gunnell, G. F. (1996). Primate phylogeny: morphological vs. molecular results. *Mol. Phylogenet. Evol.* **5**: 101–53.

Stewart, C. & Disotell, T. R. (1998). Primate evolution – in and out of Africa. *Curr. Biol.* **8**: 582–8.

Tuttle, R. H. (1975). Parallelism, brachiation, and hominoid phylogeny. In Luckett, W. P. & Szalay, F. (eds.) *Phylogeny of the Primates: A multidisciplinary approach*, pp. 447–80. New York: Plenum Press.

Uchida, A. (1998). Design of the mandibular molar in the extant great apes and Miocene fossil hominoids. *Anthrop. Sci.* **106** (Suppl.): 119–26.

von Dornum, M. (1997). *DNA Sequence Data from Mitochondrial COII and Nuclear G6PD Loci and a Molecular Phylogeny of the New World Monkeys (Primates, Platyrrhini)*. Ph.D. Dissertation, Harvard University.

Ward, S. C. (1997). The taxonomy and phylogenetic relationships of *Sivapithecus* revisited. In Begun, R. D., Ward, C. V. & Rose, M. D. (eds.), *Function, Phylogeny, and Fossils: Miocene Hominoid Evolution and Adaptations*, pp. 269–90. New York: Plenum Press.

Ward, S. C. & Brown, B. (1986). The facial skeleton of *Sivapithecus indicus*. In Swindler, & Erwin, (eds.), *Comparative Primate Biology*, vol. I: *Systematics, Evolution, and Anatomy*, pp. 413–52. New York: Alan R. Liss.

Ward, S. C. & Kimbel, W. (1983). Subnasal alveolar morphology and the systematic position of *Sivapithecus*. *Am. J. Phys. Anthrop.* **61**: 157–71.

Wood, B. (1988). Are 'Robust' australopithecines a monophyletic group? In Grine, F. (ed.), *Evolutionary History of the 'Robust' Australopithecines*, pp. 269–84. New York: Aldine de Gruyter.

Index

Page numbers in bold indicate figures and tables.